FUNDAMENTALS OF PHYSICS (9th), CONDENSED

A study guide to accompany

FUNDAMENTALS OF PHYSICS

Ninth Edition

FUNDAMENTALS OF PHYSICS (9th), CONDENSED

Thomas E. Barrett
Ohio State University

A study guide to accompany

FUNDAMENTALS OF PHYSICS

Ninth Edition

David Halliday
University of Pittsburgh

Robert Resnick
Rensselaer Polytechnic Institute

Jearl Walker
Cleveland State University

JOHN WILEY & SONS, INC.

ISBN 978- 0-470-55182-0

Printed in the United States of America

10 9 8 7 6 5 4 3

Printed and bound by Bind-Rite Graphics, Inc.

Introduction

Physics books are big. Sometimes they can be intimidating.

Like any condensed work, the goal of this book is to get the plot, while sacrificing the depth and artistry of the original. To do this, I walk through each chapter with the following approach:

- Identify the most important concepts in the chapter.
- Describe how they translate into problem-solving techniques.
- Show how these techniques work through the use of a few detailed examples.

The examples here are organized as a series of questions and responses. First learn to answer the questions, as posed by the "tutor." Then look for patterns in the questions, and learn to ask yourself the questions. As you tackle other problems, try to ask yourself similar questions as you work toward an answer.

After a while you'll find that the questions themselves are easy to answer, almost insulting to ask. Many students of physics struggle because they try to learn physics by learning how to answer the questions. The successful students learn to ask themselves the questions.

Styling used in this book:

Key points are in bold.

Important terms are underlined.

When the student or tutor makes a mistake in an equation, it is marked with a question mark at the right side of the equation, like

$$2 + 2 = 5 \quad ?$$

Remember that you should not take equations out of examples, because these equations apply to the particular problem. Instead use the examples to understand the equations and techniques.

Contents

Five Things to Remember When Doing Physics

- The goal is not possession of the answer, but mastery of the technique.
- Expect problems to take multiple steps. Don't skip the steps.
- Be able to say what your variable means in words. You can make up a variables for things you want or need and don't have.
- Use the drawing, not the formula, for the direction.
- Minus signs and units are important.

Chapter 1

Measurement

The important skills to get from this chapter are dealing with prefixes and units. We deal with both of these the same way, by **multiplying by one**.

When we multiply something by one, we get the same thing we started with. This means that we can multiply anything by one, all we want.

The key is choosing the correct "one." Anything divided by the same thing is one (unless the two things are zero). For example, if I have three dozen eggs and I want to know how many eggs I have, I choose 12 and 1 dozen as my two things that are equal to each other.

$$3\,\cancel{\text{dozen}} \times \frac{12}{1\,\cancel{\text{dozen}}} = 36$$

To determine the correct "one," see what you have that you want to get rid of. Put that in the numerator or denominator so that it cancels. Then find something that's equal to put in the other half of the fraction.

Handle prefixes the same way. For example, there are 1000 milli in one, no matter what they are. So to turn 0.126 seconds into milliseconds:

$$0.126 \text{ seconds} \times \frac{1000 \text{ milli}}{1} = 126 \text{ milliseconds} = 126 \text{ ms}$$

or

$$0.126 \text{ seconds} \times \frac{1000 \text{ milliseconds}}{1 \text{ second}} = 126 \text{ milliseconds} = 126 \text{ ms}$$

Three things to watch out for:

- If a unit is squared or to some power other than one, then you'll need to convert all of them (there are three feet in a yard but nine square feet in a square yard).
- Don't skip steps; experienced people may do the conversions in their heads, but that's a good way to make mistakes.
- Check your work: If you end with a bigger unit then you should have a smaller number, and vice versa.

EXAMPLE

How many inches are there in 5 km?

Tutor: How are you going to attack this problem?
Student: I'm going to divide 5 km by one inch.
Tutor: Correct, but does it make sense that the answer is 5?
Student: No, a kilometer is much longer than an inch.
Tutor: We need to have the two lengths in the same units.
Student: Should I use kilometers or inches?
Tutor: You could use either, or some other length like meters.
Student: I'll use meters.
Tutor: How many meters is 5 km equal to?
Student: One kilo is one thousand.

$$5 \text{ km} \times \frac{10^3}{1 \text{ k}} = 5000 \text{ m}$$

Tutor: How many meters is one inch?
Student: One inch is 2.5 centimeters, and there are 100 centi in one.

$$1 \text{ in.} \times \frac{2.5 \text{ cm}}{1 \text{ in.}} = 2.5 \text{ cm} \times \frac{1}{100 \text{ c}} = 0.025 \text{ m}$$

Tutor: What is 5 km divided by an inch?

$$\frac{5000 \cancel{\text{m}}}{0.025 \cancel{\text{m}}} = 2 \times 10^5$$

Student: There are 200,000 inches in 5 kilometers.
Tutor: What are the units of your answer?
Student: The meters cancelled the meters and it doesn't have any.
Tutor: What units should you get when you divide a length by a length?
Student: The ratio of two lengths should be a number, without any units.

EXAMPLE

How many square feet (ft^2) are there in an acre?

Tutor: Do square feet and acres measure the same thing?
Student: A square foot is an area, and an acre is an area, so yes.
Tutor: Then we should be able to convert between them.
Student: How big is an acre?
Tutor: There are 640 acres in a square mile, and a mile is 5280 feet.
Student: So an acre is

$$1 \text{ acre} = \frac{1 \text{ square mile}}{640} \times \frac{5280 \text{ feet}}{\text{miles}} \quad ?$$

Tutor: A square mile is 1 mile $\times$ 1 mile, so what you've done is

$$1 \text{ acre} = \frac{1 \text{ mile}^2}{640} \times \frac{5280 \text{ feet}}{\text{miles}} = 8.25 \text{ mile feet}$$

Student: I need to cancel both miles.

$$1 \text{ acre} = \frac{\cancel{1 \text{ mile}^2}}{640} \times \left(\frac{5280 \text{ feet}}{\cancel{\text{miles}}}\right)^2 = 43,560 \text{ feet}^2$$

EXAMPLE

Gravity is 9.8 m/s^2. What is this in mi/h^2?

Tutor: What do you have to do?
Student: First I need to turn meters into miles.
Tutor: There are 1610 meters in a mile.
Student: So (1 mi/1610 m) and (1610 m/1 mi) are equal to one. I need to get rid of meters in the top, so meters has to go on the bottom.

$$9.8\ \mathrm{m/s^2} \times \left(\frac{1\ \mathrm{mi}}{1610\ \mathrm{m}}\right)$$

Student: And I need to change seconds into hours. Seconds is on the bottom so seconds has to go on the top.
Tutor: How many seconds are there in an hour?
Student: There are 60 seconds in a minute and 60 minutes in an hour.

$$9.8\ \mathrm{m/s^2} \times \left(\frac{1\ \mathrm{mi}}{1610\ \mathrm{m}}\right) \times \left(\frac{60\ \mathrm{s}}{1\ \mathrm{minute}}\ \frac{60\ \mathrm{minutes}}{1\ \mathrm{h}}\right) \quad ?$$

Tutor: But there are two factors of seconds.
Student: So I need to convert both of them.

$$9.8\ \cancel{\mathrm{m}}/\cancel{\mathrm{s}}^2 \times \left(\frac{1\ \mathrm{mi}}{1610\ \cancel{\mathrm{m}}}\right) \times \left(\frac{60\ \cancel{\mathrm{s}}}{\cancel{1\ \mathrm{minute}}}\ \frac{60\ \cancel{\mathrm{minutes}}}{1\ \mathrm{h}}\right)^2 = 7.9 \times 10^4\ \mathrm{mi/h^2}$$

Tutor: That seems like a big number.
Student: It's bigger than 9.8, so mi/h^2 must be a smaller unit than m/s^2.
Tutor: It is.
Student: What is m/s^2, anyway?
Tutor: Meters per second (m/s) is a speed, and means that each second you go so many meters. Meters per second (m/s^2) is an acceleration, and means that each second your speed changes by that many meters per second. So, 9.8 m/s^2 means that each second your speed changes by 9.8 m/s, or 9.8 meters per second each second.

EXAMPLE

A machine draws narrow parallel lines, spaced 1200 per millimeter (mm). What is the distance between lines, measured in nanometers (nm)?

Tutor: What do you need to do?
Student: I need to take mm in the denominator and make it nm in the numerator.
Tutor: To do that you would need to multiply by two lengths; that is, two distances in the numerator and none in the denominator. Would that be equal to one?
Student: No. To end with a length in the numerator I need to start with the length in the numerator. How about

$$d = \frac{1}{1200/\ \mathrm{mm}} = \frac{1\ \mathrm{mm}}{1200}$$

Tutor: If there are 1200 in each millimeter, then the distance between them is 1 mm divided by 1200, yes. Now we need it in nanometers.
Student: How many nanometers in a millimeter?

Tutor: There are 10^3 milli in one, and 10^9 nano in one. Skipping steps is a good way to make mistakes.
Student: So I'll turn mm into m, and m into nm.

$$d = \frac{1\ \cancel{\text{mm}}}{1200} \times \left(\frac{1\ \cancel{\text{m}}}{10^3\ \cancel{\text{mm}}}\right) \times \left(\frac{10^9\ \text{nm}}{1\ \cancel{\text{m}}}\right) = 833\ \text{nm}$$

Chapter 2

Motion Along a Straight Line

The most important thing to learn from this chapter is how to solve constant acceleration problems.

We start with five quantities: the displacement $\Delta x = x - x_0$, the initial velocity v_0, the final velocity v, the acceleration a, and the time t. The first four are vectors, so they have direction. In one dimension, the direction means positive versus negative. They **must all use the same axis**, so any two that are in the same direction will have the same sign.

To solve constant acceleration problems, we use the equations

	includes	omits
$v - v_0 = at$	$v_0,\ v,\ a,\ t$	displacement Δx
$\Delta x = \frac{1}{2}(v_0 + v)\,t$	$\Delta x,\ v_0,\ v,\ t$	acceleration a
$\Delta x = v_0 t + \frac{1}{2}at^2$	$\Delta x,\ v_0,\ a,\ t$	final velocity v
$\Delta x = vt - \frac{1}{2}at^2$	$\Delta x,\ v,\ a,\ t$	initial velocity v_0
$v^2 - v_0^2 = 2a\,\Delta x$	$\Delta x,\ v_0,\ v,\ a$	time t

If we know **any three of the five** quantities, and have a fourth that we want to find, but don't care about the fifth, then pick the equation that has the three you know and the one you want but not the one you don't care about.

EXAMPLE

If you drop a penny from the top of the Empire State Building (381 m tall), how fast will it be going when it hits the ground below?

Tutor: What is happening here?
Student: The penny is undergoing constant acceleration.
Tutor: How do we solve constant acceleration problems?
Student: If we know three of the five variables, we can solve for the other two.
Tutor: Begin by writing down the five variables.
Student: Okay.

$$
\begin{array}{rcl}
\text{displacement} & \Delta x & = \\
\text{initial velocity} & v_0 & = \\
\text{final velocity} & v & = \\
\text{acceleration} & a & = \\
\text{time} & t & =
\end{array}
$$

Tutor: Do we know the displacement?
Student: Yes, it's 381 m.
Tutor: Is it positive or negative 381 m?
Student: Does it matter?
Tutor: Displacement is a vector, so it is important. We haven't chosen either direction as positive yet.
Student: The displacement is downward, so I choose down as the positive direction, and $\Delta x = +381$ m.
Tutor: Do we know the initial velocity?
Student: The penny is dropped, so the initial velocity is zero.
Tutor: Is it positive or negative?
Student: It doesn't matter, because $+0 = -0$.
Tutor: Do we know the final velocity?
Student: It ends on the ground, so the final velocity is zero.
Tutor: To use our equations we need a constant acceleration. As the penny falls it accelerates downward, but when it hits the ground the acceleration is upward.
Student: How is the acceleration upward?
Tutor: Just before it hits the ground, it was going downward. After it hits the ground it isn't moving.
Student: So the change in the velocity is upward. Does this mean that the acceleration isn't constant?
Tutor: It is constant until the instant that the penny hits the ground. So the final velocity is the velocity it has as it hits the ground. Do we know this velocity?
Student: No. Does this mean that we can't use the constant acceleration equations?
Tutor: Not necessarily. We need three of the five, and we already have two. Do we know the acceleration?
Student: The acceleration is g downward, so it's 9.8 m/s^2.
Tutor: Is it $+9.8$ m/s^2 or -9.8 m/s^2?
Student: I chose downward as positive and the acceleration is downward, so it's $+9.8$ m/s^2.
Tutor: Last, do we know the time?
Student: No; we want to find the time.

$$\begin{array}{rcll} \text{displacement} & \Delta x & = & +381 \text{ m} \\ \text{initial velocity} & v_0 & = & 0 \\ \text{final velocity} & v & = & \\ \text{acceleration} & a & = & 9.8 \text{ m/s}^2 \\ \text{time} & t & = & ? \end{array}$$

Tutor: How do we find the time?
Student: I don't care about the final velocity v, so I'll use the equation that doesn't contain v.

$$\Delta x = v_0 t + \frac{1}{2}at^2$$

$$(+381 \text{ m}) = (0)t + \frac{1}{2}(+9.8 \text{ m/s}^2)t^2$$

$$t = \sqrt{\frac{2(+381 \text{ m})}{(+9.8 \text{ m/s}^2)}} = 8.82 \text{ s}$$

EXAMPLE

A volleyball player hits a volleyball 2.1 m above the floor so that it reaches a maximum height of 4.0 m above the floor. How long is the volleyball in the air?

Tutor: What is happening here?
Student: The volleyball is undergoing constant acceleration.

Tutor: How do we solve constant acceleration problems?
Student: If we know three of the five variables, we can solve for the other two.
Tutor: Begin by writing down the five variables.
Student: Okay.

$$\begin{array}{rcc} \text{displacement} & \Delta x & = \\ \text{initial velocity} & v_0 & = \\ \text{final velocity} & v & = \\ \text{acceleration} & a & = \\ \text{time} & t & = \end{array}$$

Tutor: Do we know the displacement?
Student: First it goes up 1.9 m and then it goes down 4.0 m.
Tutor: The displacement is the difference between the final position and the initial position.
Student: It ends 2.1 m below where it started, so that is the displacement.
Tutor: Is it positive or negative 2.1 m?
Student: I choose up as the positive direction, and the displacement is downward, so $\Delta x = -2.1$ m.
Tutor: Do we know the initial velocity?
Student: No.
Tutor: Do we know the final velocity?
Student: No.
Tutor: Last, do we know the time?
Student: No; we want to find the time.

$$\begin{array}{rccl} \text{displacement} & \Delta x & = & -2.1 \text{ m} \\ \text{initial velocity} & v_0 & = & \\ \text{final velocity} & v & = & \\ \text{acceleration} & a & = & 9.8 \text{ m/s}^2 \\ \text{time} & t & = & ? \end{array}$$

Tutor: Can we find the time?
Student: We only know two of the five, so we can't solve for anything.
Tutor: Is there anything else we know?
Student: We know how high it goes, but that doesn't happen at either the start or finish.
Tutor: So we need a new start or finish.
Student: I'll do the way up. I know the displacement (+1.9 m), the final velocity (0), and the acceleration (-9.8 m/s^2).

$$\begin{array}{rccl} \text{displacement} & \Delta x & = & +1.9 \text{ m} \\ \text{initial velocity} & v_0 & = & \\ \text{final velocity} & v & = & 0 \\ \text{acceleration} & a & = & 9.8 \text{ m/s}^2 \\ \text{time} & t & = & \end{array}$$

Tutor: We know three of the five so we can solve, but how does this help us to find the total time?
Student: If I find the initial velocity, then it is the same initial velocity as for the whole trip. Alternatively, I could find the time up, then do the down problem to find the time down and add them.

$$v^2 - v_0^2 = 2a\ \Delta x$$
$$(0)^2 - v_0^2 = 2(-9.8\ \text{m/s}^2)(+1.9\ \text{m})$$
$$v_0^2 = 6.10\ \text{m/s}$$

Tutor: Is it +6.10 m/s or −6.10 m/s?
Student: Does it matter?
Tutor: You just took a square root, which could be positive or negative.
Student: The initial velocity is up, which is my positive direction, so $v_0 = +6.10$ m/s for the whole trip.

displacement	Δx	=	−2.1 m
initial velocity	v_0	=	+6.10 m/s
final velocity	v	=	
acceleration	a	=	9.8 m/s^2
time	t	=	?

$$\Delta x = v_0 t + \frac{1}{2} a t^2$$
$$-2.1\ \text{m} = (+6.10\ \text{m/s})t + \frac{1}{2}(-9.8\ \text{m/s}^2)t^2$$
$$(+4.9\ \text{m/s}^2)t^2 + (-6.10\ \text{m/s})t + (-2.1\ \text{m}) = 0$$
$$t = \frac{-b \pm \sqrt{b^2 - 4ac}}{2a}$$
$$t = \frac{-(-6.10\ \text{m/s}) \pm \sqrt{(-6.10\ \text{m/s})^2 - 4(+4.9\ \text{m/s}^2)(-2.1\ \text{m})}}{2(+4.9\ \text{m/s}^2)}$$
$$t = \frac{6.10\ \text{m/s} \pm 8.85\ \text{m/s}}{(+9.8\ \text{m/s}^2)}$$
$$t = \frac{6.10\ \text{m/s} \pm 8.85\ \text{m/s}}{(+9.8\ \text{m/s}^2)}$$
$$t = -0.28\ \text{s or } 1.53\ \text{s}$$

Tutor: There is a way to avoid the quadratic formula.
Student: Really? How?
Tutor: We know three of the five so we can solve for the other two. Solve first for the final velocity v, then we'll know four of the five.
Student: And I can choose a different equation to solve.

$$v^2 - v_0^2 = 2a\ \Delta x$$
$$v^2 - (+6.10\ \text{m/s})^2 = 2(-9.8\ \text{m/s}^2)(-2.1\ \text{m})$$
$$v = \sqrt{78.37\ \text{m}^2/\text{s}^2} = \pm 8.85\ \text{m/s}$$

Student: That number looks familiar.
Tutor: It should. We're really doing the same math, but in two easy steps instead of one hard step.

Student: And I have to pick the sign for the square root. It's going down, which is the negative direction, so it's negative.

$$v - v_0 = at$$
$$(-8.85 \text{ m/s}) - (+6.10 \text{ m/s}) = (-9.8 \text{ m/s}^2)t$$
$$t = 1.53 \text{ m/s}$$

Tutor: If you had chosen the positive square root, then your time would have been −0.28 s.

EXAMPLE

A speeder at 80 mph (in a 35 mph zone) passes a policeman. The policeman, initially at rest, accelerates at 10 mi/h/s. How long will it take the policeman to catch the speeder?

Tutor: What is happening here?
Student: The policeman is undergoing constant acceleration.
Tutor: What about the speeder?
Student: He has a constant velocity.
Tutor: Is a constant velocity also constant acceleration?
Student: He has zero acceleration, which is constant.
Tutor: So we can use all of the same techniques for him too. Where do we start?
Student: We write down the five symbols for each person.

	Speeder		Policeman	
displacement	Δx_S	=	Δx_P	=
initial velocity	v_{S0}	=	v_{P0}	=
final velocity	v_S	=	v_P	=
acceleration	a_S	=	a_P	=
time	t_S	=	t_P	=

Student: What's Δx_S and where did it come from?
Tutor: It's the displacement of the speeder, as opposed to the displacement of the policeman, and we invented it. We can use all of the constant acceleration formulas on it, like $v_S - v_{S0} = a_S t_S$.
Student: Aren't the times the same for the speeder and the policeman?
Tutor: Yes, so we can use t for both. Do we know the speeder's displacement?
Student: No. We don't know the policeman's displacement either.
Tutor: What event "starts" our constant acceleration?
Student: The speeder passes the policeman.
Tutor: What event "ends" our constant acceleration?
Student: The policeman catches the speeder.
Tutor: Since they start at the same place, and they end at the same place...
Student: the displacements are equal, whatever they are.

	Speeder			Policeman	
displacement	Δx	=	$\longleftrightarrow$	Δx	=
initial velocity	v_{S0}	=		v_{P0}	=
final velocity	v_S	=		v_P	=
acceleration	a_S	=		a_P	=
time	t	=	$\longleftrightarrow$	t	=

Tutor: Do we know the speeder's initial velocity?
Student: It's 80 mph. Do I need to convert this into meters per second?
Tutor: Not necessarily. We can keep the units with the numbers and convert them when we need to. Do we know the speeder's final velocity?
Student: His velocity isn't changing, so it's also 80 mph. His acceleration is zero. Now we know three of the five and we can solve for the time.
Tutor: Unfortunately, this is the exception to the rule. If the acceleration is zero and we know both velocities, that is not enough. We need either the time or the displacement.
Student: And we don't have either.
Tutor: What about the policeman?
Student: His initial velocity is zero, his acceleration is 10 mi/h/s, and we don't know his final velocity. What kind of a unit is mi/h/s?
Tutor: Each second his velocity increases by 10 miles per hour, so 10 miles per hour per second.

	Speeder				Policeman		
displacement	Δx	$=$		$\longleftrightarrow$	Δx	$=$	
initial velocity	v_{S0}	$=$	80 mph		v_{P0}	$=$	0
final velocity	v_{S}	$=$	80 mph		v_{P}	$=$	
acceleration	a_{S}	$=$	0		a_{P}	$=$	10 mi/h/s
time	t	$=$		$\longleftrightarrow$	t	$=$	

Tutor: We could write an equation for the policeman, including his displacement and time as variables.

$$\Delta x_{\mathrm{P}} = v_{\mathrm{P0}} t_{\mathrm{P}} + \frac{1}{2} a_{\mathrm{P}} t_{\mathrm{P}}^2$$

$$\Delta x = (0)t + \frac{1}{2}(10\ \mathrm{mi/h/s})t^2$$

Student: But it has two variables and we can't solve it.
Tutor: We could write an equation for the speeder, but it would also have two variables, Δx and t.
Student: So we'd have two equations and two unknowns. We could solve them.

$$\Delta x_{\mathrm{S}} = v_{\mathrm{S0}} t_{\mathrm{S}} + \frac{1}{2} a_{\mathrm{S}} t_{\mathrm{S}}^2$$

$$\Delta x = (80\ \mathrm{mph})t + \frac{1}{2}(0)t^2$$

Tutor: Since we want the time and don't care about the displacement, let's eliminate that.

$$\frac{1}{2}(10\ \mathrm{mi/h/s})t^2 = (80\ \mathrm{mph})t$$

Student: The time cancels, and we don't get a quadratic after all.

$$\frac{1}{2}(10\ \mathrm{mi/h/s})t^{\not 2} = (80\ \mathrm{mph})\not t$$

$$\left(5\ \frac{\mathrm{mi}}{\mathrm{h}\cdot\mathrm{s}}\right) t = \left(80\ \frac{\mathrm{mi}}{\mathrm{h}}\right)$$

Student: The miles and hours cancel.

$$\left(5\ \frac{\not{\mathrm{mi}}}{\not{\mathrm{h}}\cdot\mathrm{s}}\right) t = \left(80\ \frac{\not{\mathrm{mi}}}{\not{\mathrm{h}}}\right)$$

$$t = 16\ \mathrm{s}$$

EXAMPLE

How must the initial speed of a car change so that the car is able to stop in only half the distance?

Student: How can I solve a problem without any numbers?
Tutor: Use variables. What equation would work here?
Student: I need to identify three of the five. The final velocity is zero, so I know that. I don't know any of the others.
Tutor: Which one don't you care about?
Student: Initial velocity and displacement are mentioned, so it must be the acceleration or the time.
Tutor: The acceleration for the car is the same, no matter what the initial speed.
Student: So I don't want that one?
Tutor: No, you do. Whatever it is, it is the same for both, and that is a piece of useful information. Given that we don't care about the time, what is the equation to use?
Student: That would be

$$v^2 - v_0^2 = 2a\ \Delta x$$

Tutor: Good. This applies to any car with constant acceleration. So let's come up with a set of variables a_1 and Δx_1 and so on for one car, and then another set a_2 and Δx_2 and so on for a second car.

$$\cancelto{0}{v_1^2} - v_{1,0}^2 = 2a_1\ \Delta x_1 \qquad \longrightarrow \qquad -v_{1,0}^2 = 2a_1\ \Delta x_1$$

$$\cancelto{0}{v_2^2} - v_{2,0}^2 = 2a_1\ \Delta x_1 \qquad \longrightarrow \qquad -v_{2,0}^2 = 2a_2\ \Delta x_2$$

Student: But I can't solve either equation.
Tutor: True, but $a_1 = a_2$, and the distance Δx_2 is half of Δx_1.
Student: So I substitute those.

$$-v_{2,0}^2 = 2a_1 \left(\frac{1}{2}\Delta x_1\right)$$

Student: I still can't solve it.
Tutor: No, but look at what you have. Does it look like anything else you have?
Student: It looks similar to the first equation, except for the $\frac{1}{2}$.
Tutor: Then take the $\frac{1}{2}$ out, and substitute.

$$-v_{2,0}^2 = \frac{1}{2}\left(2a_1\ \Delta x_1\right) = \frac{1}{2}\left(-v_{1,0}^2\right)$$

Student: Well, the minus sign cancels.

$$v_{2,0}^2 = \frac{1}{2}v_{1,0}^2$$

Tutor: Your goal is to find $v_{2,0}$, so take the square root and see what you have.

$$v_{2,0} = \sqrt{\frac{1}{2}v_{1,0}^2} = \sqrt{\frac{1}{2}}v_{1,0}$$

Student: The initial speed the second time has to be $\sqrt{\frac{1}{2}}$ times the initial speed the first time?
Tutor: Yes, that's exactly what the equation says. $\sqrt{\frac{1}{2}}$ is about 0.70, so the car has to be going 30% slower to stop in half the distance.
Student: Where did you get 30%?
Tutor: The speed has to be 70% of the original speed, so the change is 30% of the original speed.
Student: And it doesn't matter how fast the car was going, or even what the acceleration was?
Tutor: As long as the acceleration is the same each time. In science we call problems like these "scaling"

problems. You want to know how one thing changes when another changes, without any values. One way to solve these problems is to write down the equation both before and after, then substitute anything that you know, either because it stays the same or you know how it compares to before.

Chapter 3

Vectors

Many things in physics have direction and are expressed as vectors. These include displacement, velocity, acceleration, and force. We need to be able to add and subtract vectors.

Adding two parallel vectors is easy.

$$3\hat{\imath} + 4\hat{\imath} = (3+4)\hat{\imath} = 7\hat{\imath}$$

Adding two vectors that are perpendicular can be done with the Pythagorean theorem. Adding two vectors that are neither parallel nor perpendicular is more difficult. It can be done using the law of cosines, but this becomes unwieldy with more than two vectors.

Instead, we **divide each vector into components**. The x components of each vector are parallel, and are easy to add. The y components of each vector are likewise parallel to each other, and are likewise easy to add. Then the x and y components are perpendicular to each other and can be added using the Pythagorean theorem.

There is more than one way to find the components. I review the two most popular here.

Draw the right triangle with the vector as the hypotenuse and the other two sides parallel to the axes. The side of the triangle that is adjacent to the angle is cosine, and the side of the triangle that is opposite to the angle is sine. If the component is in the opposite direction as the axis, then the component is negative.

Always draw the angle from the x axis toward the y axis. Then the x component is cosine and the y component is sine.

Let's see how each method works.

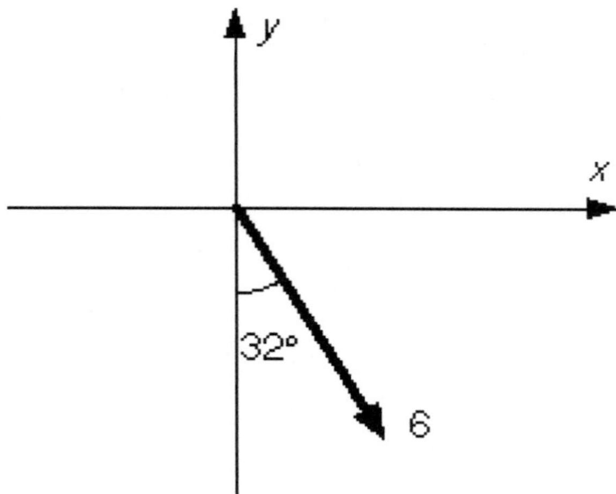

We draw the right triangle as shown. The x component is the side opposite to the angle, so the x component is $x = 6\sin(32°)$. The y component is the side adjacent to the angle, but it is down and the y axis is up, so the y component is $y = -6\cos(32°)$.	or	The angle given is not measured from the x axis. We draw a new angle measured from the x axis initially toward the y axis. This new angle is $270° + 32° = 302°$. Then the x component is $x = 6\cos(302°)$ and the y component is $y = 6\sin(302°)$.

I shall use the left-hand method.

EXAMPLE

What is the sum of the four vectors?

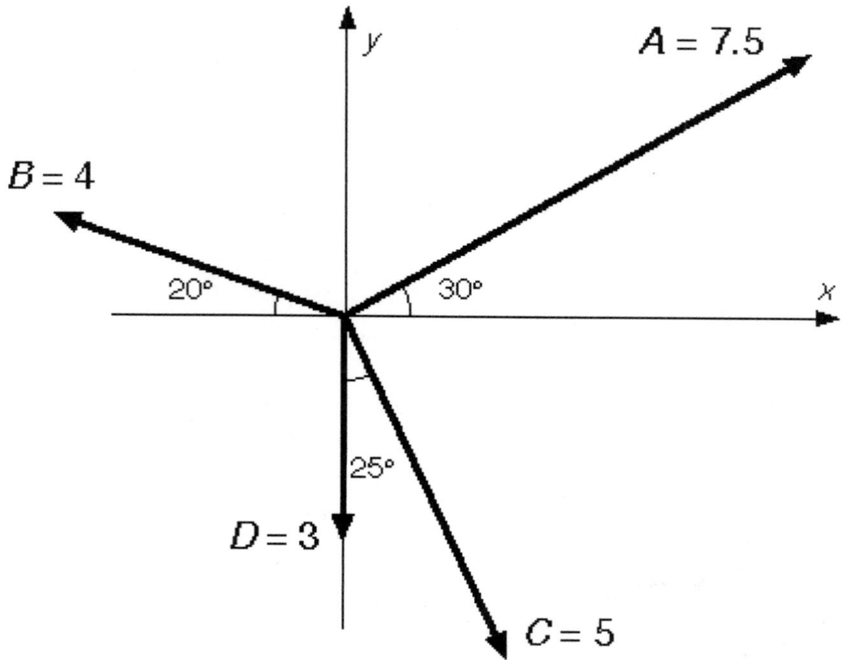

Tutor: Are the vectors all parallel?
Student: No.
Tutor: Are the vectors all perpendicular?
Student: No.
Tutor: So how do we add them?
Student: We break each one into components and add the components.
Tutor: What are you going to use for your axes?
Student: Uh, there's already axes in the problem.
Tutor: Yes, but there won't always be.
Student: I'm going to use the axes provided.
Tutor: Good. What is the x component of vector A?
Student: First I draw the triangle.

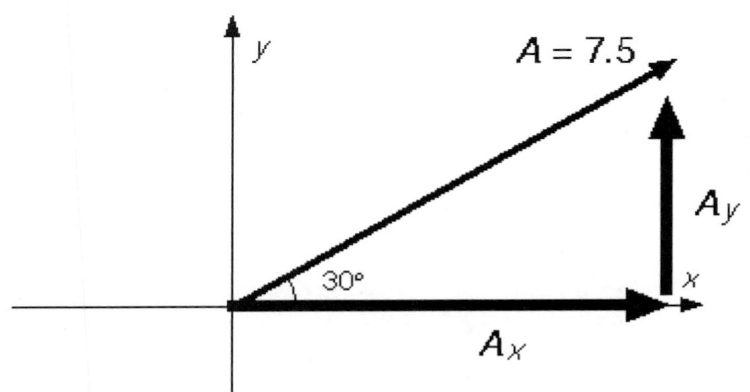

Student: The x component is the adjacent side of the triangle, so it's cosine.

$$A_x = A\cos 30^\circ = 7.5\cos 30^\circ = 6.50$$

Tutor: Good. What is the y component of vector A?
Student: The y component is the opposite side of the triangle, so it's sine.

$$A_x = A\sin 30^\circ = 7.5\sin 30^\circ = 3.75$$

Tutor: What is the x component of vector B?
Student: First I draw the triangle.

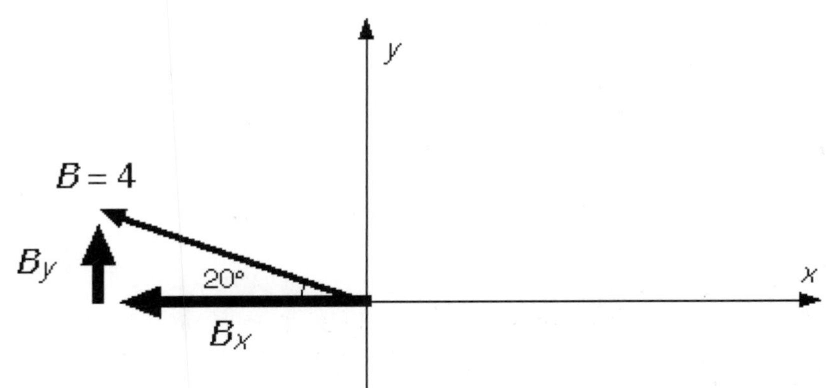

Student: The x component is the adjacent side of the triangle, so it's cosine.

$$B_x = B\cos 20^\circ = 4\cos 20^\circ = 3.76 \quad ?$$

Tutor: But the x component goes to the left, and the x axis goes to the right, so it's negative.

$$B_x = -B\cos 20^\circ = -4\cos 20^\circ = -3.76$$

Student: Doesn't the math take care of that automatically?
Tutor: Yes, if you always measure the angle from the x axis.

$$B_x = B\cos 160^\circ = 4\cos 160^\circ = -3.76$$

Tutor: In many of the things we'll use vectors for, it's more convenient to draw the triangle and add minus signs when determining components.
Student: Okay. The y component is opposite the angle, so it's sine.

$$B_y = B\sin 20^\circ = 4\sin 20^\circ = 1.37$$

Tutor: What is the x component of vector C?
Student: First I draw the triangle.

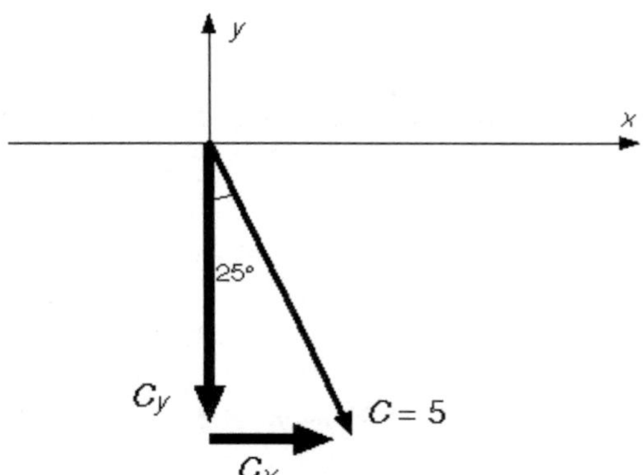

Student: The x component is cosine.

$$C_x = C\cos 25^\circ = 5\cos 25^\circ = 4.53 \quad ?$$

Tutor: The x component is opposite the angle this time, so it's sine.

$$C_x = C\sin 25^\circ = 5\sin 25^\circ = 2.11$$

Student: Isn't the x component always cosine?
Tutor: No, the component adjacent to the angle is always cosine, and opposite is sine. More often than not these will be the x and y components, respectively, but sometimes it's the other way around.
Student: And the y component of C is adjacent, so it's cosine. The y component is down, away from the y axis, so it's negative.

$$C_y = -C\cos 25^\circ = -5\cos 25^\circ = -4.53$$

Tutor: What is the x component of vector D?
Student: How do I draw the triangle for vector D?
Tutor: Because D is already parallel to one of the axes, you don't need to draw the triangle. The x component of D is zero, because it doesn't go to the left or right at all.
Student: And the y component is -3.

Tutor: Good, now we can add them.
Student: The x components add

$$(A+B+C+D)_x = A_x + B_x + C_x + D_x = 6.50 + (-3.76) + 2.11 + (0) = 4.85$$

Student: And the y components add

$$(A+B+C+D)_y = A_y + B_y + C_y + D_y = 3.75 + 1.37 + (-4.53) + (-3) = -2.41$$

Student: Do we need to find the magnitude and direction or are the components enough?
Tutor: Often the components are enough, but let's find the magnitude and direction for practice.
Student: The magnitude or absolute value is

$$\sqrt{(X)^2 + (Y)^2} = \sqrt{(4.85)^2 + (-2.41)^2} = \sqrt{23.52 - 5.81} = \sqrt{17.71} = 4.21 \quad \textbf{?}$$

Tutor: That can't be right, because it's less than one of the components. Remember to square the negative sign too.

$$\sqrt{(X)^2 + (Y)^2} = \sqrt{(4.85)^2 + (-2.41)^2} = \sqrt{23.52 + 5.81} = \sqrt{29.33} = 5.42$$

Tutor: Some people find it easier to make a table:

	x	y
A	6.50	3.75
B	−3.76	1.37
C	2.11	−4.53
D	0	−3
	4.85	−2.41

Tutor: We can also express this vector using "unit vectors."

$$\overrightarrow{(A+B+C+D)} = \vec{A} + \vec{B} + \vec{C} + \vec{D} = \vec{A} + \vec{B} + \vec{C} + \vec{D} = (4.85)\hat{\imath} + (-2.41)\hat{\jmath}$$

Student: What's a unit vector?
Tutor: A unit vector is a vector of length 1 with no units. $\vec{\hat{\imath}}$ points in the x direction and $\vec{\hat{\jmath}}$ points in the y direction.
Student: How do I find the angle?
Tutor: Draw the triangle, but this time starting with the components.

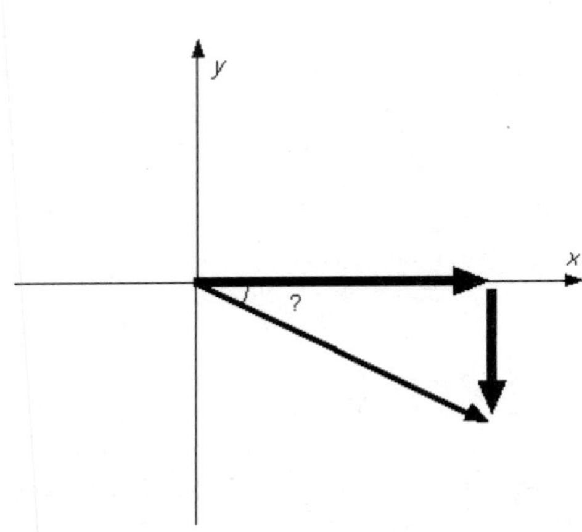

Student: So the angle is below the x axis. Does that mean the angle is negative?
Tutor: Some people are happy with that, but it's best to show the angle that you mean with a drawing.
Student: The tangent of the angle is opposite over adjacent, so the angle is

$$\theta = \arctan \frac{\text{opposite}}{\text{adjacent}} = \arctan \frac{2.41}{4.85} = \arctan 0.497 = 26.4^\circ$$

Student: ... below the x axis. Is adding vectors always so repetitive?
Tutor: Usually, but with practice it will go faster.

EXAMPLE

Find the vectors $A + B$ and $A - B$ graphically.

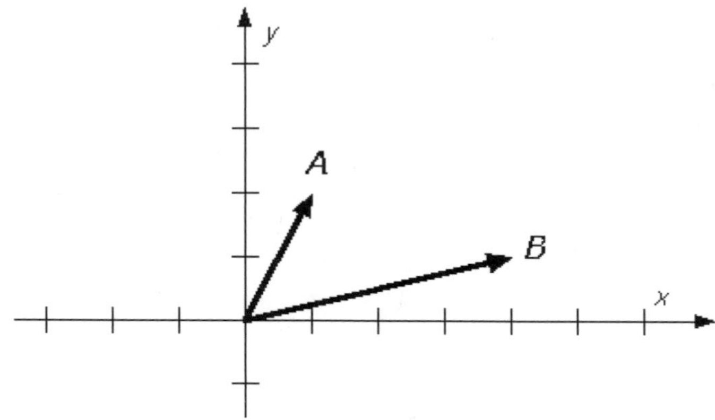

Tutor: How do we add vectors?
Student: We connect the ends, like this:

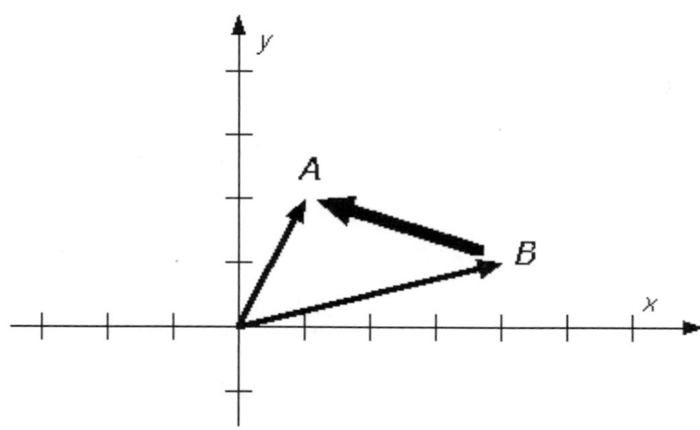

Tutor: Both vectors A and B point to the right, and your sum points to the left. Can that be right?
Student: No, so it must be the other way.
Tutor: Then A and B point up, but the sum points down. When adding vectors, we use "tip to tail," meaning that we move one vector so that it starts where the previous one left off.

Student: Don't vectors have to start at the origin?
Tutor: No. A vector seen this way is a displacement, or a change, and it doesn't matter where it starts.
Student: So I add the vectors like this:

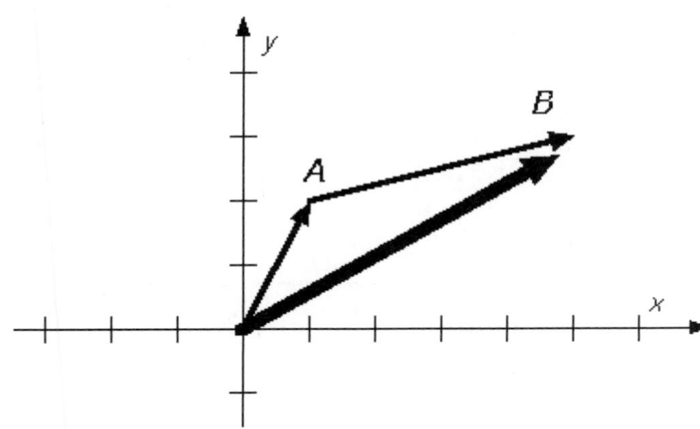

Tutor: Yes. The vector you drew first is the difference of the two vectors.
Student: But is it $A - B$ or $B - A$?
Tutor: Look for the "tip to tail" combination.
Student: So B plus my vector equals A, and my vector is

$$B + \text{mine} = A \quad \rightarrow \quad \text{mine} = A - B$$

Tutor: An easier way to subtract vectors is to add the negative vector.

$$A - B = A + (-B)$$

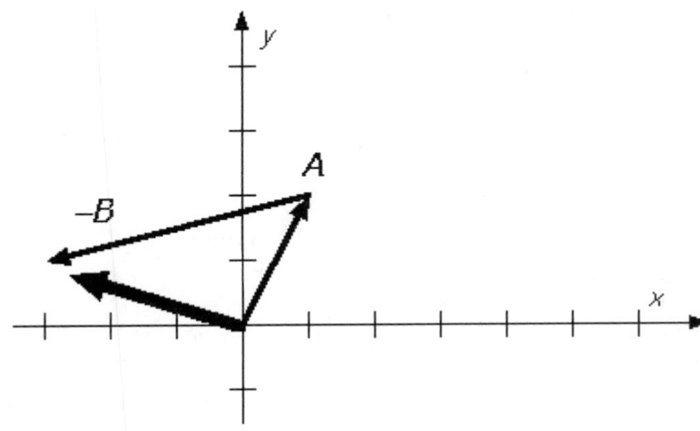

Student: A plus negative B equals my first vector, so it was $A - B$.
Tutor: Drawing it the other way would give $B - A$.

EXAMPLE

Find a vector of length 4 that, when added to the vector shown, sums to a horizontal vector (parallel to the x axis).

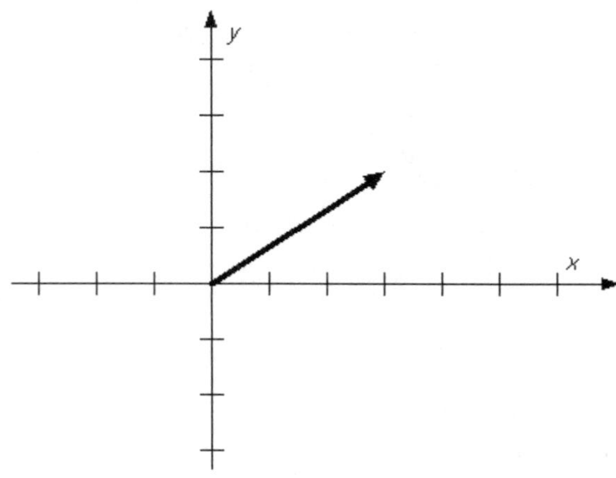

Tutor: Where can we start?
Student: I'm adding vectors, so I'll need their components.
Tutor: What are the components of the first vector?
Student: The x component is $+3$ and the y components is $+2$.
Tutor: What are the components of the second vector, the one we're trying to find?
Student: Don't I need an angle to find those?
Tutor: You need something. Do you know anything about the vector you'll get when you add them?
Student: It's horizontal.
Tutor: What are the components of that vector?
Student: I don't know how long it is, so how can I find the components?
Tutor: What is the y component of a horizontal vector?
Student: Zero. It doesn't go up or down.
Tutor: The y component of the first vector is $+2$, and the y component of the sum is zero, so...
Student: ...the y component of the second vector is -2.
Tutor: Now you have a vector with a length of 4 and a y component of -2. What is the x component?
Student: I can use the Pythagorean theorem.

$$4 = \sqrt{x^2 + (-2)^2} \longrightarrow x = \sqrt{4^2 - (-2)^2} = 3.46$$

Tutor: A square root could be positive or negative. Could your second vector be $(-3.46, -2)$?
Student: Maybe, what would that look like?
Tutor: Start at the end of the first vector, and draw a circle of length 4. The second vector has to end on this circle.

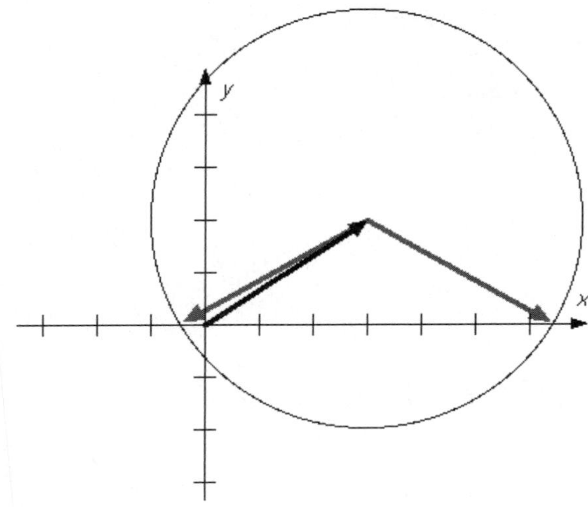

Student: I see, there are two points on the x axis where the vector could end. What if the first vector had been (3, 5)? Then there wouldn't be any point where the circle intersects with the x axis.
Tutor: And there wouldn't be any vector of length 4 that you could add to the first and get a horizontal vector. The y component would have to be -5, so the second vector would have to be at least 5 long.

Chapter 4

Motion in Two and Three Dimensions

In this chapter we learn how to use vectors. Many things in physics have direction, and we use vectors to describe the directions.

To demonstrate and practice vectors, we look at two basic techniques: projectile motion and relative motion. Here we connect constant acceleration (and constant velocity) with vectors in two dimensions.

When everything lines up along a single line (any line), then we have a one-dimensional problem. When everything lines up along a single plane, then we have a two-dimensional problem. If we can't find a plane in which everything happens, then we have a three-dimensional problem.

To deal with a one-dimensional problem, we choose one direction as the positive direction. Anything in the opposite direction is negative. With a two-dimensional problem, we need to choose two axes. The one limitation is that the axes need to be perpendicular to each other. Any set of axes will do, but there is often one set that works best. By "works best" I mean that it gives us much easier equations to solve.

Once we have a set of perpendicular axes, **we can treat the motion along each axis separately**.

While it is mathematically correct to use the $\hat{\imath}$ and $\hat{\jmath}$ notation of the previous chapter, it is not done so much in practice. Instead, it is easier to describe motion as being in the x or y direction.

EXAMPLE

A boy throws a ball toward a wall. The ball leaves his hand 1.4 m above the ground, with a speed of 15 m/s at 42° above horizontal. How far above the ground does the ball hits the wall, 5 m away?

Tutor: Is this a one-dimensional problem?
Student: The ball starts going up and over, but the acceleration of gravity is downward. Since not everything is along a single line, no.
Tutor: Is this a two-dimensional problem?
Student: Everything can be drawn in the plane of the page, with nothing going into the page or coming out of the page, so yes.
Tutor: For a two-dimensional problem, we need two axes.
Student: Gravity is vertical, so I choose x horizontal toward the wall and y vertically up.
Tutor: Those will work well, but not because the acceleration is vertical. They work because the wall is vertical.
Student: How does the wall come into it?
Tutor: We will want to know when the ball hits the wall, and it is much easier if this happens in one dimension.

Student: So I want to line up one axis with the "finish line"?
Tutor: Yes. Now what is happening along the x axis?
Student: Constant acceleration.
Tutor: What is happening along the y axis?
Student: Constant acceleration.
Tutor: How do we solve constant acceleration problems?
Student: We write down the five symbols and identify what we know.
Tutor: But now we need two lists, one each for the x and y axes.

$$\begin{aligned} \Delta x &= & \Delta y &= \\ v_{x0} &= & v_{y0} &= \\ v_x &= & v_y &= \\ a_x &= & a_y &= \\ t &= & t &= \end{aligned}$$

Tutor: What is the displacement in the x direction?
Student: +5 m
Tutor: What is the displacement in the y direction?
Student: We don't know; that's what we're trying to find.
Tutor: Is it really what we're trying to find?
Student: We want to find how high it is when the x displacement is +5 m, so we need to find the y displacement and add +1.4 m to it.
Tutor: What is the initial velocity in the x direction?
Student: The x component is adjacent to the 42° angle, so it is cosine: $15\ \text{m/s} \cos 42°$.
Tutor: What is the initial velocity in the y direction?
Student: The x component is opposite to the 42° angle, so it is sine: $15\ \text{m/s} \sin 42°$.
Tutor: Are both of them positive?
Student: The x component is toward the wall, which is our $+x$ direction. The y component is upward, which is our $+y$ direction, so both are positive.
Tutor: What is the final velocity in the x direction?
Student: There is no horizontal acceleration, so it should be the same as the initial velocity in the x direction.
Tutor: Yes, but unfortunately this is the exception to the three-of-five rule: when the acceleration is zero and it becomes a constant velocity problem, knowing two velocities (which are the same) only counts as one. What is the final velocity in the y direction?
Student: We don't know.
Tutor: What is the acceleration in the x direction?
Student: The acceleration is parallel to the y axis, so it has no horizontal component. Zero.
Tutor: What is the acceleration in the y direction?
Student: g downward, so $-9.8\ \text{m/s}^2$.
Tutor: What is the time in the x direction?
Student: Time has direction?
Tutor: No.
Student: So the time is the same for the x and y problems.
Tutor: Do we know it?
Student: No.

$$\begin{aligned} \Delta x &= +5\ \text{m} & \Delta y &= \ ? \\ v_{x0} &= +(15\ \text{m/s})\cos 42° & v_{y0} &= +(15\ \text{m/s})\sin 42° \\ v_x &= & v_y &= \\ a_x &= 0 & a_y &= -g \\ t &= \quad \longleftrightarrow & t &= \end{aligned}$$

Tutor: Can we solve the y problem to find the y displacement?
Student: No, we only know two of the five.
Tutor: Can we solve the x problem?
Student: Yes, we know three of the five, so we could solve for the final velocity or the time in the x direction.
Tutor: How does that help us with the y problem?
Student: The time in the x problem is the same as the time in the y problem. Then we'll be able to solve the y problem.

$$\Delta x = v_{x0}t + \frac{1}{2}a_x t^2$$

$$(5\text{ m}) = ((15\text{ m/s})\cos 42^\circ)\,t + \frac{1}{2}(0)t^2$$

$$t = \frac{(5\text{ m})}{(15\text{ m/s})\cos 42^\circ}$$

$$t = 0.448\text{ s}$$

$$\Delta x = v_0 t + \frac{1}{2}at^2 \rightarrow \Delta y = v_{y0}t + \frac{1}{2}a_y t^2$$

$$\Delta y = ((15\text{ m/s})\sin 42^\circ)\,(0.448\text{ s}) + \frac{1}{2}(-9.8\text{ m/s}^2)(0.448\text{ s})^2$$

$$\Delta y = 3.52\text{ m}$$

Student: The y displacement is 3.52 m.
Tutor: But that wasn't the question. What does the y displacement mean?
Student: The ball hits 3.52 m above where he threw it, or 1.4 m + 3.52 m = 4.92 m above the ground.
Tutor: If the y displacement had been less than −1.4 m, so that the height above the ground was negative, what would that have meant?
Student: That the ball hit the wall below the ground?
Tutor: Yes, what would that mean?
Student: That the ball hit the ground before reaching the wall.

EXAMPLE

A cannon fires a cannonball from the top of a cliff to the level ground 180 m below. The cannonball leaves the cannon at 100 m/s at 53° above horizontal. How far from the base of the cliff does the cannonball land?

Tutor: How many dimensions are there in this problem?
Student: The initial velocity and the acceleration are not along the same line, so more than one. They are in the same plane, so two is enough.
Tutor: For a two-dimensional problem, we need two axes.
Student: I choose x horizontal, parallel to the ground, and y perpendicular to x.
Tutor: Now what is happening along the x axis?
Student: Constant acceleration.
Tutor: What is happening along the y axis?
Student: Constant acceleration.
Tutor: How do we solve constant acceleration problems?
Student: We write down the five symbols and identify what we know.

$$\begin{array}{rclcrcl} \Delta x & = & & & \Delta y & = & \\ v_{x0} & = & & & v_{y0} & = & \\ v_x & = & & & v_y & = & \\ a_x & = & & & a_y & = & \\ t & = & & & t & = & \end{array}$$

Tutor: What is the displacement in the x direction?
Student: We don't know; that's what we're trying to find.
Tutor: What is the displacement in the y direction?
Student: -180 m.
Tutor: What is the initial velocity in the x direction?
Student: The x component is adjacent to the 53° angle, so it is cosine: $+100\text{ m/s}\cos 53°$.
Tutor: What is the initial velocity in the y direction?
Student: The x component is opposite to the 53° angle, so it is sine: $+100\text{ m/s}\sin 53°$.
Tutor: What is the final velocity in the x direction?
Student: There is no horizontal acceleration, so it should be the same as the initial velocity in the x direction, but it doesn't help us.
Tutor: What is the final velocity in the y direction?
Student: We don't know.
Tutor: What is the acceleration in the x direction?
Student: The acceleration is parallel to the y axis, so it has no horizontal component. Zero.
Tutor: What is the acceleration in the y direction?
Student: g downward, so -9.8 m/s^2.
Tutor: What is the time?
Student: We don't know, but it is the same for the x and y problems.

$$\begin{array}{rclcrcl} \Delta x & = & ? & & \Delta y & = & -180\text{ m} \\ v_{x0} & = & +(100\text{ m/s})\cos 53° & & v_{y0} & = & +(100\text{ m/s})\sin 53° \\ v_x & = & & & v_y & = & \\ a_x & = & 0 & & a_y & = & -g \\ t & = & & \longleftrightarrow & t & = & \end{array}$$

Tutor: Can we solve the x problem to find the x displacement?
Student: No, we only know two of the five.
Tutor: Can we solve the y problem?
Student: Yes, we know three of the five, so we could solve for the time in the y problem, then put that into the x problem.

$$\Delta x = v_0 t + \frac{1}{2}at^2 \rightarrow \Delta y = v_{y0}t + \frac{1}{2}a_y t^2$$

$$(-180\text{ m}) = (+80\text{ m/s})t + \frac{1}{2}(-9.8\text{ m/s}^2)t^2$$

$$(4.9\text{ m/s}^2)t^2 + (-80\text{ m/s})t + (-180\text{ m}) = 0$$

Student: So we have to solve a quadratic equation?
Tutor: We could solve the quadratic...

$$t = \frac{-b \pm \sqrt{b^2 - 4ac}}{2a}$$

$$t = \frac{-(-80\text{ m/s}) \pm \sqrt{(-80\text{ m/s})^2 - 4(4.9\text{ m/s}^2)(-180\text{ m})}}{2(4.9\text{ m/s}^2)} = \frac{(80\text{ m/s}) \pm (99.6\text{ m/s})}{(9.8\text{ m/s}^2)}$$

$$t = -2.0\text{ s or } 18.3\text{ s}$$

Tutor: ...or we could solve the y problem for the final velocity.
Student: How does that help us? We need the time.
Tutor: If we knew the final velocity we would have four of the five and would have our choice of equations. We could solve for the time without a quadratic.

$$v^2 - v_0^2 = 2a\ \Delta x \rightarrow (v_y)^2 - v_{y0}^2 = 2a_y\ \Delta y$$

$$(v_y)^2 = v_{y0}^2 + 2a_y\ \Delta y$$

$$v_y = \sqrt{(80\text{ m/s})^2 + 2(-9.8\text{ m/s}^2)(-180\text{ m})}$$

$$v_y = \pm 99.6\text{ m/s}$$

Tutor: The square root could be positive or negative. Which is it?
Student: What difference does it make?
Tutor: If the final velocity is positive then the cannonball is going upward as it lands.
Student: No, $v_y = -99.6$ m/s.

$$v - v_0 = at \rightarrow v_y - v_{y0} = a_y t$$

$$t = \frac{v_y - v_{y0}}{a_y}$$

$$t = \frac{(-99.6\text{ m/s}) - (80\text{ m/s})}{(-9.8\text{ m/s}^2)}$$

$$t = 18.3\text{ s}$$

Student: Now we can do the x problem.

$$x = v_{x0}t + \frac{1}{2}at^2$$

$$x = (60\text{ m/s})(18.3\text{ s}) + \frac{1}{2}(0)(18.3\text{ s})^2$$

$$x = 1100\text{ m}$$

As an object flies through the air, its path forms a parabola. The vertical motion is like $\Delta y = v_0 t - \frac{1}{2}gt^2$. The horizontal motion is constant, so that t is proportional to Δx. In the photo, the beanbags fly in a parabolic arc after they leave the airgun.

Imagine that you are driving down the road at 60 mph. If you look only at yourself, you don't appear to be moving. If you look out the window at the tree beside the road, the tree appears to be coming at you at 60 mph. You think of the car just ahead of you as going 60 mph, but it doesn't get any further from you. Likewise, a car going the other way appears to be coming at you at 120 mph.

This is the idea behind relative motion. We can measure the position, the velocity, or the acceleration of anything compared to any other thing. We often intuitively measure everything compared to the ground, and think of the ground as "stationary," but we don't have to approach things that way. (When we get to relativity, it's important to *not* approach things that way.)

The math behind relative motion is straightforward. If you have any two objects A and B, then

$$\vec{x}_{\text{AB}} = \vec{x}_{\text{AC}} + \vec{x}_{\text{CB}}$$

where C is any third person or object. $\vec{x}_{\text{AB}}$ is read as "the velocity of A as measured by B." Also,

$$\vec{x}_{\text{AB}} = -\vec{x}_{\text{BA}}$$

This says that whatever I see you doing, you see me doing the same thing in the other direction.

We can take the derivative of this, so

$$\vec{v}_{\text{AB}} = \vec{v}_{\text{AC}} + \vec{v}_{\text{CB}}$$

and

$$\vec{v}_{\text{AB}} = -\vec{v}_{\text{BA}}$$

The same is true for acceleration.

It is important that everyone agree on the axes. For displacement, we can think of each person carrying their axes with them, but they have to point in the same direction. These equations fail if I think north is positive and you think south is positive.

EXAMPLE

As a bus drives north at 60 mph, a passenger on the bus walks toward the rear at 2 mph, compared to the bus. How fast is the passenger moving, as determined by someone standing on the side of the road?

Tutor: How are you going to approach this problem?
Student: Because we're talking about relative motion, I'm going to use the relative motion equations.
Tutor: Not all problems are labelled with the technique needed to solve them. How might you identify that relative motion is involved in this problem?
Student: It talks about a measurement "compared to" and "as determined by."
Tutor: Yes. It involves measurements made in two different frames of reference, by two different people.
Student: So I'll start with the equation.

$$\vec{v}_{\mathrm{AB}} = \vec{v}_{\mathrm{AC}} + \vec{v}_{\mathrm{CB}}$$

Student: What are A, B, and C?
Tutor: What are the three people, objects, or frames of reference mentioned in the problem?
Student: The bus, the passenger, and the guy by the side of the road.
Tutor: It doesn't matter which one you make which letter, so long as you do it consistently.
Student: Because I want the velocity of the passenger as measured by the guy, I'll use A as the passenger and B as the guy.

$$\vec{v}_{\mathrm{PG}} = \vec{v}_{\mathrm{PB}} + \vec{v}_{\mathrm{BG}}$$

Tutor: What is the velocity $\vec{v}_{\mathrm{BG}}$ of the bus as measured by the guy?
Student: 60 mph.
Tutor: Is it positive or negative?
Student: It doesn't say.
Tutor: You need to choose an axis. Pick either north or south as positive and stick with your choice.
Student: I'll take north as positive, so it's +60 mph.
Tutor: What is the velocity $\vec{v}_{\mathrm{PB}}$ of the passenger as measured by the bus?
Student: 2 mph. He's walking toward the rear of the bus, or southward compared to the bus, so it's −2 mph.

$$\vec{v}_{\mathrm{PG}} = (-2\text{ mph}) + (+60\text{ mph}) = +58\text{ mph}$$

Student: The guy by the side of the road sees the passenger moving northward at 58 mph.

EXAMPLE

An airplane is pointed southward with an airspeed of 200 mph. The wind is blowing northwest at 30 mph. What is the speed of the plane compared to the ground?

Student: It says "compared to," so I'm going to use relative motion.
Tutor: Good.
Student: All I need to do is figure out whether I add or subtract the 200 and the 30.
Tutor: You're trying to skip steps, and that's a good way to make a mistake. South and northwest aren't parallel to each other.
Student: So when I add or subtract I'll need to do components. Darn.
Tutor: Harder but not impossible. What are you going to use for your axes?
Student: x and y, of course.
Tutor: The directions in the problem are south and northwest. Think about how x and y relate to the

directions in the problem, or you could just use some combination of north, south, east, west, or even northwest.

Student: The plane is going south, so I'll use south as one of my axes. The wind is more west than east, so I'll use west as the other. That way the axes are perpendicular and my results will probably be positive.

Tutor: Now figure out how to add the vectors.

Student: Using relative motion:

$$\vec{v}_{\mathrm{AB}} = \vec{v}_{\mathrm{AC}} + \vec{v}_{\mathrm{CB}}$$

Tutor: What are the three things in the problem?

Student: The airplane and the ground. That's two, so there must be a third thing moving. . . oh, the wind or air.

Tutor: Which is A, B, and C?

Student: I want the plane compared to the ground, so the Plane is A, the Ground is B, and the third thing C is the Air.

$$\vec{v}_{\mathrm{PG}} = \vec{v}_{\mathrm{PA}} + \vec{v}_{\mathrm{AG}}$$

Tutor: What are the components of $\vec{v}_{\mathrm{PA}}$, the velocity of the plane compared to the air?

Student: It's headed south at 200 mph, so +200 mph south and 0 west.

Tutor: What are the components of $\vec{v}_{\mathrm{AG}}$, the velocity of the air compared to the ground?

Student: That's the wind, so I draw my triangle.

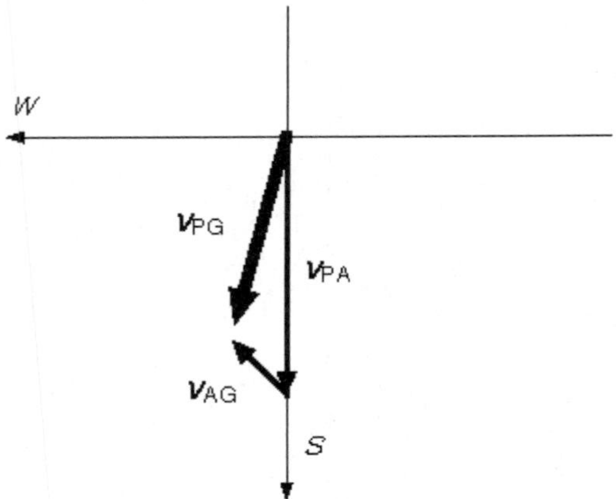

Student: Because it's 45°, I can use either angle, and the components are the same.

$$v_{\mathrm{AG},S} = v_{\mathrm{AG},W} = (30 \text{ mph}) \cos 45^\circ = 21 \text{ mph} \quad ?$$

Tutor: The wind is going *north*west, so the south component is negative (trying to skip steps again). Now you can do the vector addition.

Student: First I'll add the south components.

$$v_{\mathrm{PG},S} = v_{\mathrm{PA},S} + v_{\mathrm{AG},S} = (200 \text{ mph}) + (-21 \text{ mph}) = 179 \text{ mph}$$

$$v_{\mathrm{PG},W} = v_{\mathrm{PA},W} + v_{\mathrm{AG},W} = (0 \text{ mph}) + (21 \text{ mph}) = 21 \text{ mph}$$

Tutor: The south and west components are perpendicular, so you can't just add them.

Student: I need to use the Pythagorean theorem again.

$$v_{\mathrm{PG}} = \sqrt{v_{\mathrm{PG},S}^2 + v_{\mathrm{PG},W}^2} = \sqrt{(179 \text{ mph})^2 + (21 \text{ mph})^2} = 180 \text{ mph}$$

EXAMPLE

An airplane has an airspeed of 200 mph. The wind is blowing south at 30 mph. The airplane needs to fly directly southeast compared to the ground, so that it flies parallel to the runway. In what direction should the pilot point the airplane?

Student: It says compared to, so I'm going to try relative motion.

$$\vec{v}_{\rm PG} = \vec{v}_{\rm PA} + \vec{v}_{\rm AG}$$

Student: The directions are not all parallel, so I'll need axes to add the vectors.
Tutor: What are you going to use for your axes?
Student: How about south and west again?
Tutor: Any set of perpendicular axes will work, of course, but the goal is to get the plane flying southeast. The math will be easier if you pick the goal (southeast) as one of your axes.
Student: Really? Okay, and I'll use southwest for the other axis. $\vec{v}_{\rm PG}$ equals +200 mph.
Tutor: Is it clear that the plane is moving 200 mph compared to the ground? In the last problem, the groundspeed $|v_{\rm PG}|$ and the airspeed $|v_{\rm PG}|$ were different.
Student: I guess not. But the southwest component of the groundspeed is zero, so that the plane goes southeast compared to the ground.
Tutor: Yes. What are the components of the wind $\vec{v}_{\rm AG}$?
Student: It's a 45° angle again, and southward means the southwest and southeast components are both positive. Drawing the triangle isn't as easy because the axes seem strange. Did you pick the strange directions just so I could practice using unusual axes?
Tutor: Yes. If it seems strange, you can rotate the paper.

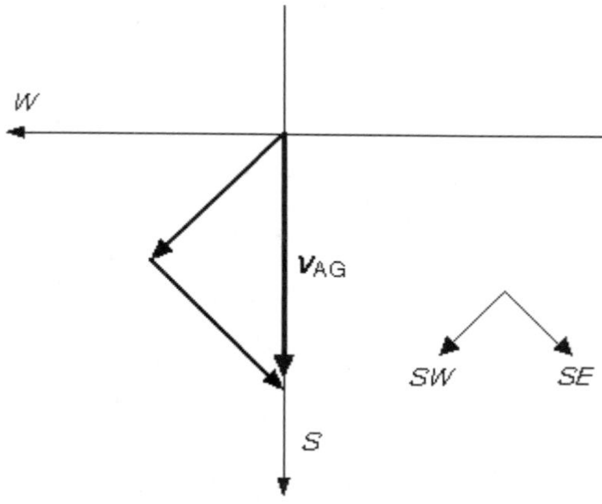

$$v_{{\rm AG},SW} = v_{{\rm AG},SE} = (30\text{ mph})\cos 45° = +21\text{ mph}$$

Tutor: What components do you have now?

	SE	SW
$\vec{v}_{\rm PA}$	?	?
$\vec{v}_{\rm AG}$	+21	+21
$\vec{v}_{\rm PG}$	?	0

Student: I see. I can find $\vec{v}_{\text{PA},SW} = -21$. But I still don't know the angle.
Tutor: You have one of the components of $\vec{v}_{\text{PA}}$, and you know the magnitude v_{PA}, so you can find the other component.
Student: Using the Pythagorean theorem backwards.

$$v_{\text{PA}}^2 = v_{\text{PA},SE}^2 + v_{\text{PA},SW}^2$$

$$(200 \text{ mph})^2 = v_{\text{PA},SE}^2 + (-21 \text{ mph})^2$$

$$v_{\text{PA},SE} = \sqrt{(200 \text{ mph})^2 - (-21 \text{ mph})^2} = 199 \text{ mph}$$

Student: It's a square root, so is it positive or negative?
Tutor: If it's negative then the plane is going northward.
Student: Of course. Then the angle is

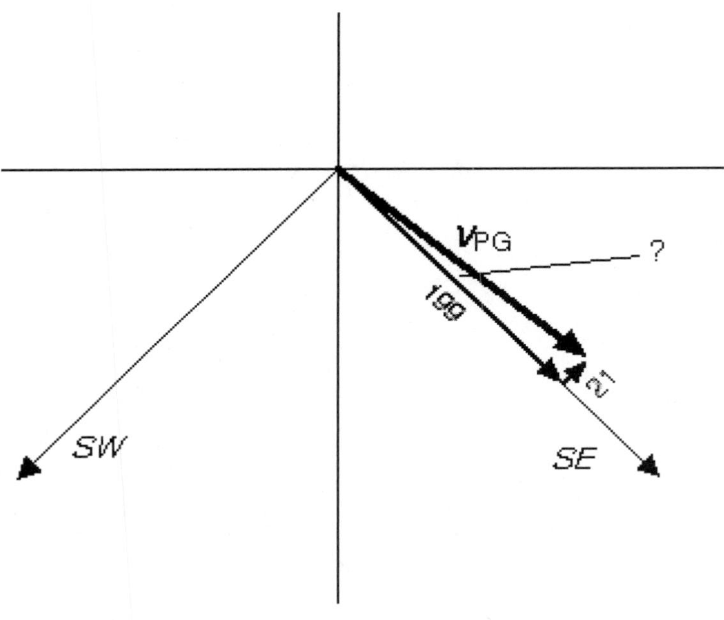

$$\arctan \frac{\text{opposite}}{\text{adjacent}} = \arctan \frac{21}{199} = 6°$$

Student: That's 6° measured eastward from southeast.

Chapter 5

Force and Motion — I

In this chapter we cover the single most important technique to learn in physics. If you don't get this, get help.

Every problem involving forces starts with the same basic steps:

- Draw a free-body diagram.
- Choose axes.
- Use the diagram to write Newton's second law equations along each axis.
- Solve for the unknown.

Forces are things that push or pull on an object. Anything that touches an object can apply a force to it. Also, there are two types of forces that can occur without touching: gravity and electromagnetic forces. We won't have any electromagnetic forces for a while, so we only need to deal with gravity and things in contact with our object. Also, there is no friction until the next chapter.

For the moment, we need to worry about three forces.

- gravity
- normal forces
- tension forces

Gravity is easy. Every object has a mass m, and a weight (force) of mg, where g is the same acceleration of gravity as in the earlier chapters. $g = 9.8\ \text{m/s}^2$ on the surface of the Earth, and g is *never* negative. g is the magnitude of the gravity force, and does not contain direction information.

Any time two objects are in contact, there could be a force where they meet. This force is to keep the objects from occupying the same place at the same time. This force is always perpendicular to the surfaces, and is just hard enough to keep the objects apart and no harder. There is a word in mathematics that means "perpendicular to the surface," and that word is normal. So the normal force is the perpendicular to the surface force.

You don't know the magnitude of the normal force. One common mistake is to assume that you know the normal force, but you don't know it. The only time that you know the normal force is when someone is standing on a scale. The purpose of the scale is to measure the normal force, so if it says 200 pounds, then the normal force is 200 pounds. If you don't have a scale, you don't know the normal force.

We pull on an object with a string or rope. Usually in physics problems we use massless ropes (purchased at the theoretical physics store). Tension forces are away from the object in the direction of the rope.

You don't know the magnitude of the tension force. One common mistake is to assume that you know the tension force, but you don't know it. The only time that you know the tension force is when the tension force is applied by a spring scale.

The last important point about forces is dealing with paired forces. Newton's third law says that if A puts a force on B, then B puts a force on A, and that the two forces are equal in magnitude and have opposite direction. Note that because one of these forces is on A and the other on B, these forces never occur on the same object.

The purpose of the free-body diagram is to get a complete list of the forces, get the direction right, and attach a name or label to each force.

EXAMPLE

A 3 kg box sits on the floor. The box is pulled to the right by a rope that is 35° above horizontal. Draw a free-body diagram for the box.

Tutor: Are there any forces on the box?
Student: It has mass, so it has weight mg downward.
Tutor: Are there any other forces acting on the box?
Student: There is a rope pulling on it, so there is a tension force.
Tutor: Do we know how big the tension force is?
Student: Uh, no?
Tutor: Correct. Later we're going to want to put the magnitude of the tension force into an equation, so we need a symbol to represent the magnitude of the tension force.
Student: I choose T.
Tutor: Good. What is the direction of the tension force?
Student: 35° from horizontal right toward upward.
Tutor: Are there any other forces acting on the box?
Student: The box is in contact with the floor, so there is a normal force.
Tutor: What is the direction of the normal force?
Student: Perpendicularly out of the floor, so upward.
Tutor: What is the size of the normal force?
Student: The normal force is the same as the weight.
Tutor: Not necessarily. The tension will pull upward on the box, so if the normal force is the same as the weight, the box will lift off of the floor.
Student: So the normal force isn't the same as the weight?
Tutor: Sometimes it is, but a normal force is only as big as it needs to be to keep the objects from moving into each other.
Student: So how big is the normal force?
Tutor: Do you know?
Student: No.
Tutor: So pick a variable name or symbol to represent the magnitude of the normal force.
Student: I choose N.
Tutor: F_{N} and n are also common choices. Are there any other forces acting on the box?
Student: Friction.
Tutor: We're going to save friction until the next chapter.
Student: Then that's all of the forces.
Tutor: How do you know?

Student: Because we did the weight and we did everything in contact with the box.

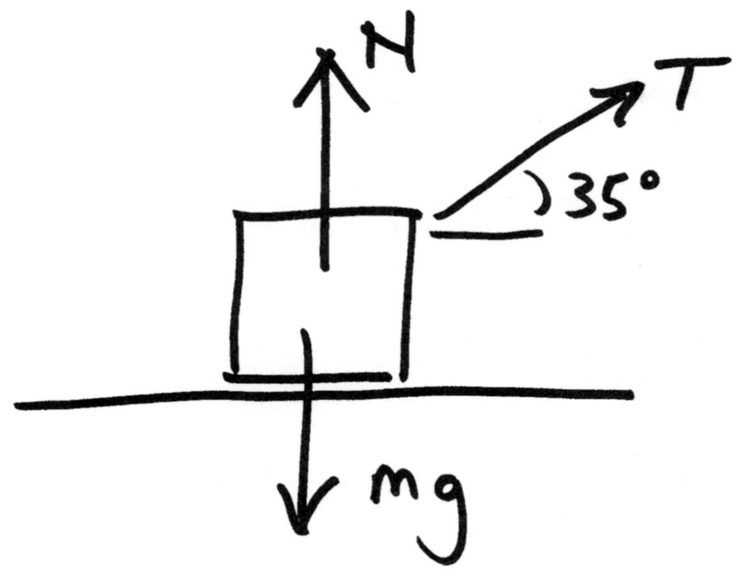

EXAMPLE

A 2 kg book sits on a 7 kg table. Draw a free-body diagram for the book and one for the table.

Tutor: Are there any forces on the book?
Student: It has mass, so it has weight mg downward.
Tutor: Are there any other forces acting on the book?
Student: It is in contact with the table, so there is a normal force upward.
Tutor: How big is the normal force?
Student: It is the same as the weight, because those are the only forces on the book.
Tutor: What if the table is in an elevator that is accelerating upward?
Student: Who would put a book on a table in an elevator?
Tutor: The point is that just having two forces on an object doesn't make them equal. How big is the normal force that the table puts on the book?
Student: I don't know, so I'll give it a symbol N.
Tutor: Are there any other forces acting on the book?
Student: I did the weight, and the only thing touching the book is the table, so those are the only forces on the book.
Tutor: What forces are there on the table?
Student: There is the weight of the table and the weight of the book and a normal force up from the floor.
Tutor: The weight of the book doesn't act on the table.
Student: Surely the book affects the table.
Tutor: Yes, but the gravity force of the book is the Earth pulling it down, and Newton's third law says that the gravity force of the book pulling up on the Earth is equal and opposite. The table doesn't come into the gravity force on the book.
Student: Then how does the book affect the table?
Tutor: They are in contact with each other.
Student: So there is a normal force perpendicularly out of the book acting on the table, pushing the table downward.

Tutor: Yes. How big is this normal force?
Student: Equal to the weight of the book?
Tutor: Not necessarily. Newton's third law says that the force that the table puts on the book is equal to and in the opposite direction as the force that the book puts on the table. How big is the force that the table puts on the book?
Student: It was N, so the normal force of the book pushing down on the table is also N.
Tutor: How big is the normal force that the floor puts on the table?
Student: It's a normal force, so N.
Tutor: Is the normal force that the floor puts on the table the same size as the normal force that the book puts on the table?
Student: Not necessarily, so I need to use a different symbol for the floor-table force, so I choose N_{floor}.
Tutor: Are the two weights the same?
Student: No, so I'll use mg for the weight of the book and Mg for the weight of the table.
Tutor: Are there any more forces acting on the table?
Student: We did the weight of the table, and it is in contact with two things so there are two normal forces. We're done.

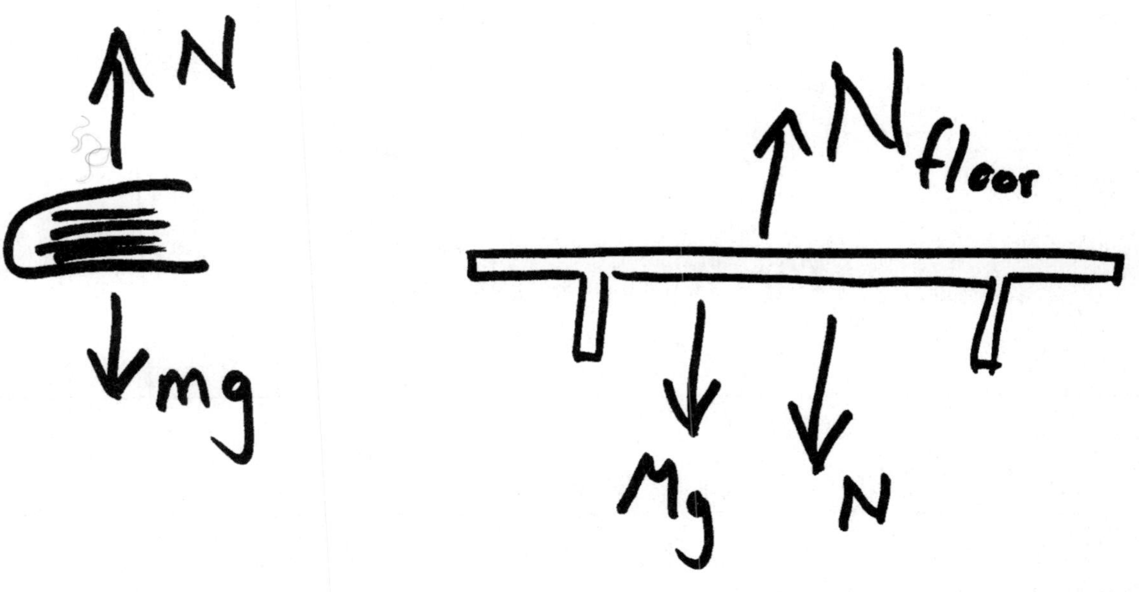

EXAMPLE

A 6 N box and a 12 N box hang from a rope over a frictionless, massless pulley. Draw a free-body diagram for each box.

Tutor: What forces are there on the 6 N box?
Student: It has mass, so it has weight mg downward. There is a tension force T upward.
Tutor: What forces are there on the 12 N box?
Student: It also has mass, so it has weight downward. There is a also tension force upward.
Tutor: Are the two masses the same?
Student: No, so I'll use m_6g for the 6 N box and $m_{12}g$ for the 12 N box.
Tutor: Newtons is a unit of force, not mass, so 6 N is mg for the lighter box.
Student: So the mass is 6 N/9.8 m/s^2?
Tutor: Yes, and likewise for the 12 N box.

Student: Are the tensions the same or different?
Tutor: As long as it is the same massless rope, and the pulley is massless and frictionless, then the tensions are the same.
Student: So I use the same symbol T in each free-body diagram. Isn't the tension equal to the weight of the lighter box?
Tutor: If it is, then the light box has zero net force on it and doesn't accelerate, but the net force on the heavy box is downward and it does accelerate.
Student: And then the rope would have to become longer. So I don't know the size of the tension force.

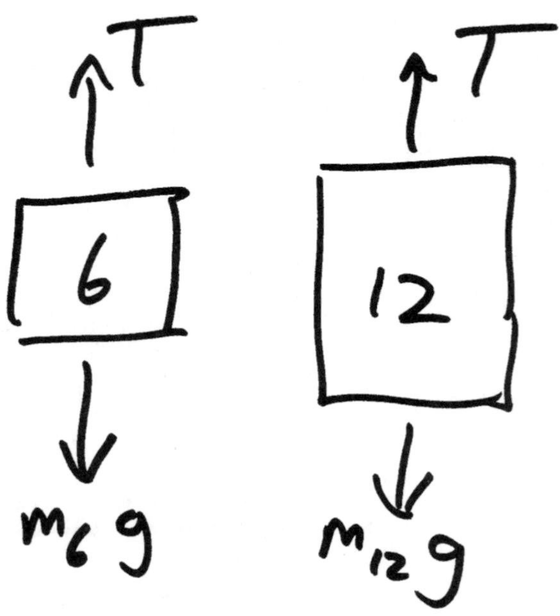

EXAMPLE

A 3 kg box hangs from the ceiling by a rope, and a 4 kg box hangs from the 3 kg box from another rope. Draw a free-body diagram for each box.

Tutor: What forces are there on the 3 kg box?
Student: It has mass, so it has weight mg downward. There is a tension force upward and a tension force downward.
Tutor: Are the tension forces the same?
Student: If they were, they would add to zero and only the weight would remain, so the box would fall. They must be different, so T_1 up and T_2 down.
Tutor: They could be the same, if the boxes were falling with acceleration g. What forces are there on the 4 kg box?
Student: It also has mass, so it has weight downward. There is also a tension force upward.
Tutor: Are the two masses the same?
Student: No, so I'll use $m_3 g$ for the 3 kg box and $m_3 g$ for the 4 kg box.
Tutor: Is the tension pulling up on the 4 kg box equal to T_1 or T_2?
Student: It is the same rope as the one that pulls down on the 3 kg, so it is T_2.
Tutor: As long as the rope is massless.
Student: What if it had mass?
Tutor: We would need to draw an additional free-body diagram for the rope. If it is massless, then we

have the tension at top and bottom and they add to the mass, zero, times the acceleration of the rope.
Student: Which is zero no matter what the acceleration is, so the tension is the same everywhere on the rope.

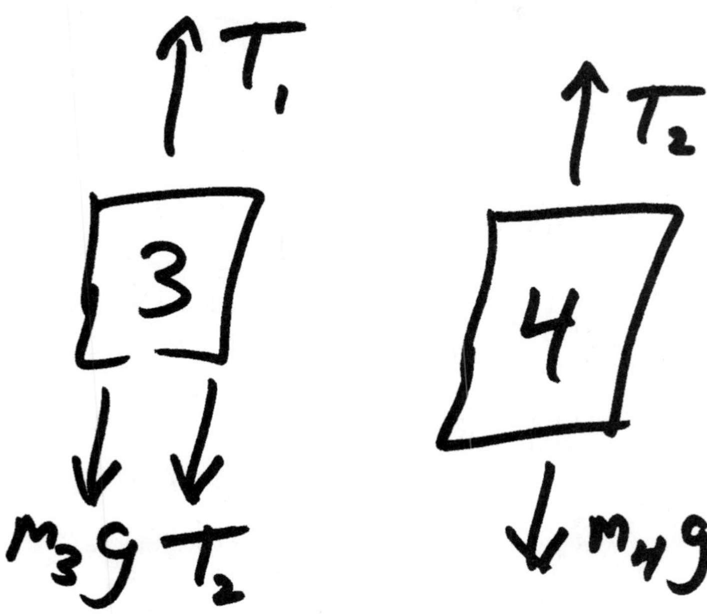

Newton says that if we add all of the forces on an object we get the mass times the acceleration of that object. The forces are vectors, and often the forces are not colinear (along a single line), so we need to divide them into components to add them. To do this we need axes (usually two, sometimes three). Any set of axes will work, so long as they are perpendicular to each other. If you **choose one of the axes to be parallel to the acceleration**, then the math will be much easier to do. If the acceleration is zero, then choose any axes you like.

Once you have your axes, you can add the force vectors. Take one axis, and go through the forces one at a time. Find the component of that force along the axis. Add all of the components and set that total equal to the mass times the acceleration along that axis.

Don't worry if you don't know values for all of the symbols in the equation. We've got to be missing one variable, or we would have nothing to solve for. It is common to be missing two or more, so we'll need additional equations in order to get a solution. Do each axis, then think about solving the equations.

EXAMPLE

A 3 kg box sits on the floor. The box is pulled to the right by a rope that is 35° above horizontal and has a force of 26 N. Find the acceleration of the box. Find the normal force that the floor exerts on the box.

Tutor: How do we begin?
Student: We draw a free-body diagram for the box. We already did that for this problem, with the weight mg downward, the normal force N upward, and the tension force T up and to the right.

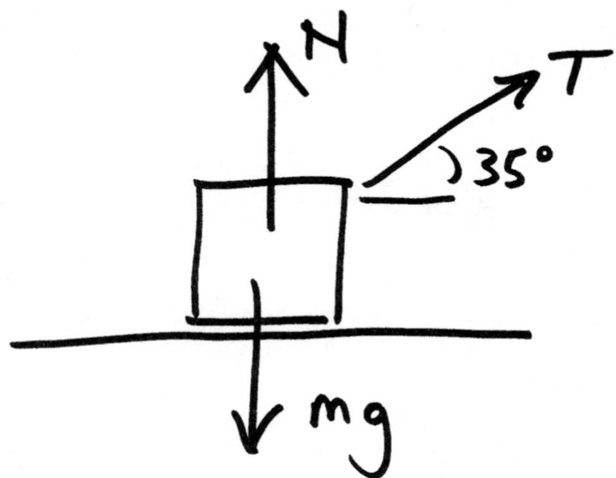

Tutor: What is the direction of the acceleration?
Student: The box will slide along the floor, so horizontal and to the right.
Tutor: Choose one axis parallel to the acceleration and the second axis perpendicular to the first.
Student: I choose x to the right and y upward.
Tutor: Newton says that if we add the force vectors together, the result will be equal to the mass of the box times its acceleration.
Student: To add the forces, we need to divide the vectors into components.
Tutor: Yes, then we can apply Newton's second law for each axis.

$$\Sigma F_x = ma_x \qquad \text{and} \qquad \Sigma F_y = ma_y$$

Tutor: What is the x component of the weight?
Student: The weight is perpendicular to the x axis, so the x component is zero.
Tutor: What is the x component of the normal force?
Student: The normal force is also perpendicular to the x axis, so the x component is zero.
Tutor: What is the x component of the tension?
Student: The x component is adjacent to the 35° angle, so it's cosine.

$$T \cos 35° = ma_x$$

Tutor: Let's do the y components. What is the y component of the weight?
Student: The weight is parallel to the y axis, so all of it. It's down, so it's negative.
Tutor: But the weight is negative because it is opposite to the y axis. Had you chosen the y axis downward, the weight would be positive.
Student: Yes, so the weight is negative, and the normal force is also parallel to the y axis, so the y component is $+N$.
Tutor: What is the y component of the tension?
Student: The x component is opposite to the 35° angle, so it's sine. It's in the same direction as the y axis, so it's positive.

$$-mg + N + T \sin 35° = ma_y$$

Tutor: Can we solve these?
Student: The tension T is 26 N, so I can solve the first equation and find a_x.

$$(26 \text{ N}) \cos 35° = (3 \text{ kg})a_x$$

$$a_x = 7.1 \text{ m/s}^2$$

Student: I don't know the normal force so I can't find a_y.
Tutor: a_y is the acceleration in the y direction, or vertical. We chose the axes so that all of the acceleration

would be in the x direction.
Student: So $a_y = 0$.

$$-(3\text{ kg})(9.8\text{ m/s}^2) + N + (26\text{ N})\sin 35^\circ = (3\text{ kg})(0)$$

Student: Now I can solve for the normal force N.

$$-(29.4\text{ N}) + N + (14.9\text{ N}) = 0$$

$$N = 14.5\text{ N}$$

Student: That looks strange, with N on each side.
Tutor: The N on the left is in italics, which indicates that it's a variable. The N on the right is in plaintype, so it is a unit. It can be confusing, so some people try to choose other symbols for the normal force, like F_N, n, or η. What is the direction of the normal force?
Student: It's positive so it's upward.
Tutor: N is the magnitude of the normal force, so it must be positive no matter what direction the normal force is.
Student: If I had chosen down as the y axis, then shouldn't N be negative?
Tutor: No, then the y component of the normal force would have been $-N$, and because N is positive, this would be negative. We use the drawing to get the direction right, and the symbol is equal to the magnitude.
Student: So to see if the normal force is up or down, I have to look back at the drawing.

EXAMPLE

A man pushes on a box that is on a ramp. The box has a mass of 24 kg and the ramp is 28° compared to the horizontal. He pushes with a force of 100 N at 10° above horizontal. What is the acceleration of the box?

Tutor: How do we begin?
Student: We draw a free-body diagram for the box. The weight is mg downward. There is a normal force N upward.
Tutor: Is the normal force really upward? The surface isn't horizontal.
Student: Right, the normal force is perpendicular to the surface, but we still don't know how big it is. Also, the man is pushing on the box, so there is a force that I'll call P, 10° above horizontal.
Tutor: Are there any other forces acting on the box?
Student: The only things touching it are the man and the ramp surface, and we did those and gravity, so we have them all.

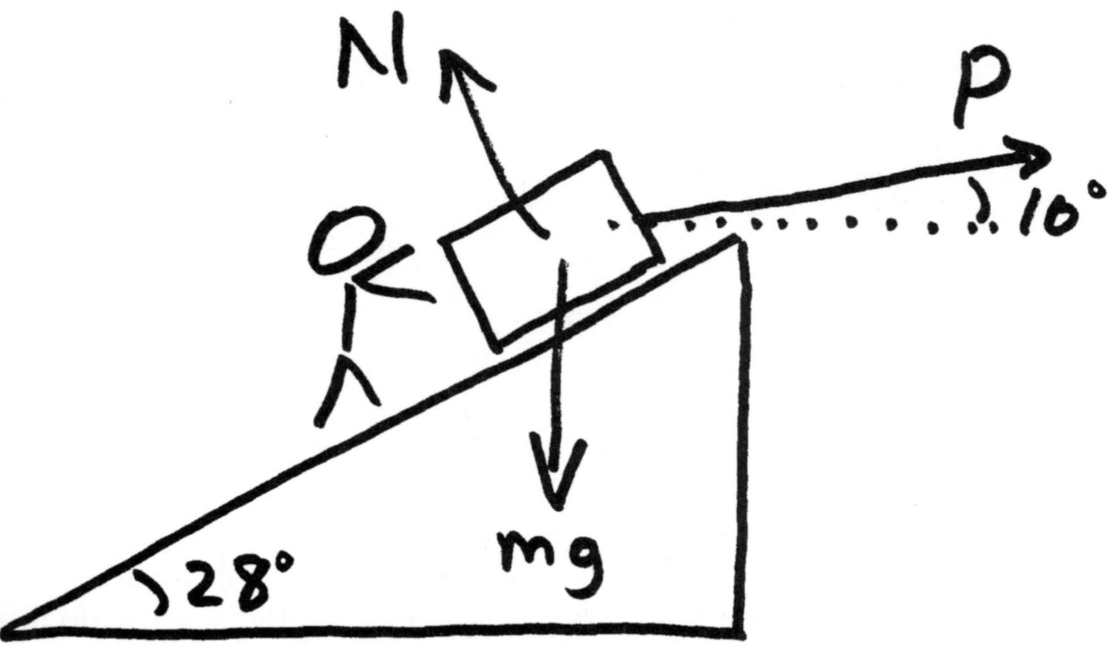

Tutor: The vectors are not all colinear, so to add the vectors we need to have two axes. The math will be much easier if one axis is parallel to the acceleration.
Student: Why is that? Why can't I use x right and y up?
Tutor: You could, but then you'd have a_x and a_y, and you wouldn't know either. Because the box slides along the ramp, you'd have a third equation $a_y/a_x = \tan 28°$, and you'd have to solve simultaneous equations. If $a_y = 0$, then you have two only equations and often you can solve each one by itself; much easier.
Student: Okay, x is up the ramp and y is perpendicular to it, up and to the left.
Tutor: Now we can write Newton's second law.
Student: We have to do it in both the x and y direction.

$$\Sigma F_x = ma_x \qquad \text{and} \qquad \Sigma F_y = ma_y$$

Tutor: What is the x component of the normal force?
Student: The normal force is perpendicular to the x axis, so the x component is zero.
Tutor: What is the x component of the weight?
Student: Part of the weight is parallel to the x axis, but how do we find the angle?
Tutor: Consider the triangle formed by the ramp and the weight force. The angle at the top is $90° - 28° = 62°$. That angle and the angle between the weight and the $-y$ axis form a right angle, so it's $90° - 62° = 28°$.
Student: So the x component of the weight is opposite to the $28°$ angle, so it's sine. It goes in the opposite direction as the x axis, so it's negative.
Tutor: What is the x component of the push force?
Student: The push is $18°$ from the x axis.

$$-mg \sin 28° + P \cos 18° = ma_x$$

Tutor: What is the y component of the normal force?
Student: The normal force is parallel to the y axis, so the y component is $+N$.
Tutor: What is the y component of the weight?
Student: The y component of the weight is adjacent to the $28°$ angle, so it's cosine. It goes in the opposite direction as the y axis, so it's negative.

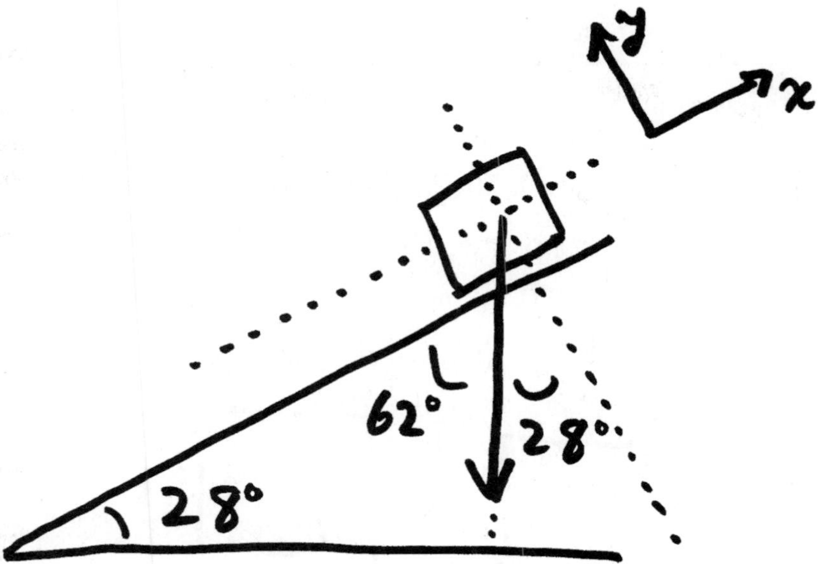

Tutor: What is the y component of the push force?
Student: The y component is opposite to the 18°, so it's sine. It's opposite to the y axis, so it's negative.

$$N - mg\cos 28° - P\sin 18° = m\cancelto{0}{a_y}$$

Tutor: Now that we've applied Newton's second law, can we solve for the acceleration?
Student: We want the acceleration up the ramp a_x, and we know everything else in the x equation, so we can solve.

$$-(24 \text{ kg})(9.8 \text{ m/s}^2)\sin 28° + (100 \text{ N})\cos 18° = (24 \text{ kg})a_x$$
$$-(110 \text{ N}) + (95 \text{ N}) = (24 \text{ kg})a_x$$
$$a_x = \frac{-15 \text{ N}}{24 \text{ kg}} = -0.64 \text{ m/s}^2$$

Student: Doesn't a_x have to be positive?
Tutor: a_x is a component, not a magnitude, so it could be negative. What does a negative a_x mean?
Student: A negative a_x means that the acceleration is down the ramp. He isn't pushing hard enough to move the box up the ramp.
Tutor: Which way is the box moving?
Student: The acceleration is down the ramp, so the box is moving down the ramp.
Tutor: Perhaps the box was moving up the ramp, because he was pushing harder earlier. Does the acceleration tell us which way something is moving?
Student: The acceleration tells us how the velocity is changing. The box could be moving up the ramp but slowing. How can we tell?
Tutor: We can't, not from the information we have.

EXAMPLE

In an Atwood's machine, a 6 N box and a 12 N box hang from a rope over a frictionless, massless pulley. Find the acceleration of the boxes.

Student: We start by drawing the free-body diagram. We already did this one.

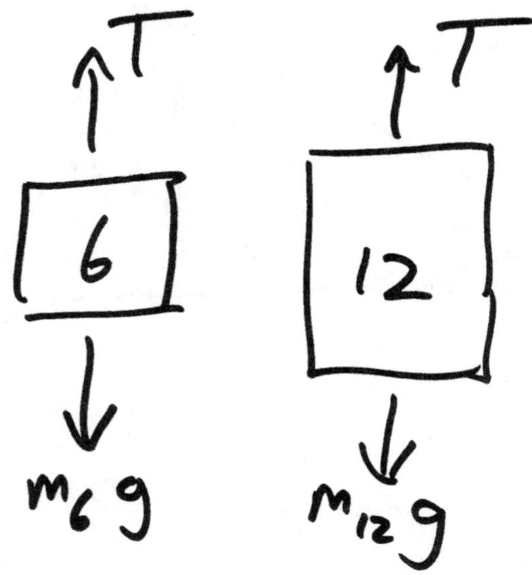

Tutor: Newton says that if we add the forces, we get the mass times the acceleration. How do we add forces?
Student: They're vectors, but they are all up and down, so I only need one axis. I choose y upward.
Tutor: Will both boxes accelerate upward?
Student: One will move up when the other moves down.
Tutor: How will their accelerations compare?
Student: They will be the same, except one will be positive and the other will be negative.
Tutor: It would be handy if they were the same. To do this, we use separate axes for the two objects.
Student: We can do that?
Tutor: Yes, each object can have its own set of axes. But you still need to be consistent in using the correct axes for each object. This will come in handy when the accelerations aren't parallel, like if one of the boxes was on a ramp.
Student: Okay. I think that the heavier box will move down, so down is positive for the 12 N box and up is positive for the 6 N box.
Tutor: You mean that the heavier box will accelerate down, since we could apply an additional force so that the lighter one moves down.
Student: Of course. I'll add the forces on the 12 N box. The tension is equal to the weight of the lighter box.

$$m_{12}g - m_6 g = m_{12}a \quad ?$$

Tutor: If the tension is equal to the weight of the lighter box, then the total force on the lighter box is zero and it doesn't move but the heavier one does.
Student: You mean doesn't accelerate. That would be a problem.
Tutor: For the lighter box to accelerate up, the tension must be more than its weight. For the heavier box to accelerate down, the tension must be less than its weight.
Student: So the tension is between $m_6 g$ and $m_{12}g$.
Tutor: Yes. How big is the tension force?

Student: We don't know, that's why we made it a variable.
Tutor: So apply Newton's second law to each box.

$$m_{12}g - T = m_{12}a$$
$$T - m_6 g = m_6 a$$

Student: Each equation has two variables, so I can't solve either of them.
Tutor: But they have the same two variables, so you have two equations and two variables, and you can solve them together.
Student: I want the acceleration and not the tension, so I'll add the two equations.

$$m_{12}g - m_6 g = m_{12}a + m_6 a$$
$$(m_{12} - m_6)g = (m_{12} + m_6)a$$
$$(12\ \text{N} - 6\ \text{N})(9.8\ \text{m/s}^2) = (12\ \text{N} + 6\ \text{N})a$$
$$a = \frac{(6\ \cancel{\text{N}})(9.8\ \text{m/s}^2)}{(18\ \cancel{\text{N}})} = 3.3\ \text{m/s}^2$$

Tutor: The result is positive. What does that mean?
Student: That the boxes are moving — no, accelerating — in the directions I chose as positive.

Chapter 6

Force and Motion — II

In this chapter we add two things to what we did last chapter. This first is friction forces, and the second is things moving in a circle.

Of all of the things that you think you intuitively know, the one you are most likely to get backwards is friction. You may think that friction opposes motion, and many physics books say so, but **friction opposes sliding of surfaces. Friction will even cause motion in order to prevent sliding.**

Imagine that you are driving a car that is currently at rest. If you push on the accelerator, the engine turns the tires. If there was no friction, the tires would rotate but the car wouldn't move, so the tires would slip on the ground. To prevent this slipping, friction pushes the car forward. The same thing works when walking, where friction pushes your shoe forward, causing motion.

Friction comes in two types, static and kinetic. Kinetic friction is when the surfaces are already sliding against each other, and static friction is when they aren't sliding yet. Kinetic friction tries to stop the sliding, and static friction tries to prevent sliding.

EXAMPLE

Consider a tablecloth on a table, and a bottle sitting on the tablecloth. A physics teacher slowly pulls the tablecloth to the left. As he does so, the bottle slides with the tablecloth. Draw free-body diagrams for the bottle and the tablecloth.

Tutor: What are the forces acting on the bottle?
Student: There is gravity mg down, of course, and a normal force N_1 perpendicular to the surface, or up. There is also a friction force f_1 against the motion, or right.
Tutor: The bottle accelerates to the left. What force causes it to move to the left?
Student: The teacher is pulling on it.
Tutor: The teacher is pulling on the tablecloth, but he isn't touching the bottle. He doesn't exert a force on the bottle.
Student: It just moves because the tablecloth is moving.
Tutor: Newton says that the bottle can't accelerate to the left without a force to the left. What force could be to the left?
Student: Gravity is down and the normal force is up, so it can't be those. The only force remaining is friction, and how can that be to the left?
Tutor: If there was no friction, what would happen to the bottle?
Student: There would be no horizontal forces, so I guess it wouldn't move.
Tutor: Correct, but the tablecloth would move to the left, so the bottle and tablecloth would be sliding against each other. Friction tries to prevent sliding.

Student: So friction pushes the bottle to the left so that it doesn't slide?
Tutor: Yes, friction moves the bottle to keep it from sliding on the tablecloth. What type of friction is it?
Student: I was going to say "kinetic" because the bottle is moving, but I'll think again. The bottle isn't sliding on the tablecloth, so it's static friction.

Tutor: Correct. What are the forces acting on the tablecloth?
Student: It has gravity Mg downward, and a normal force N_2 from the table upward. Because there is a normal force, there could be a friction force f_2 from the table.
Tutor: Which way does the friction force from the table act?
Student: The tablecloth is sliding to the left, so it's kinetic friction and it pushes the tablecloth right.
Tutor: Good. There are also forces from the bottle.
Student: The tablecloth pushes the bottle up, so the bottle exerts a normal force N_1 down.
Tutor: The tablecloth also exerts a friction force on the bottle, so the bottle exerts one on the tablecloth.
Student: So the bottle pushes the tablecloth to the right with an equal and opposite force f_1, even though there isn't any sliding between the bottle and the tablecloth?
Tutor: One way to prevent sliding is to push the bottle to the left, but the other way is to push the tablecloth right.

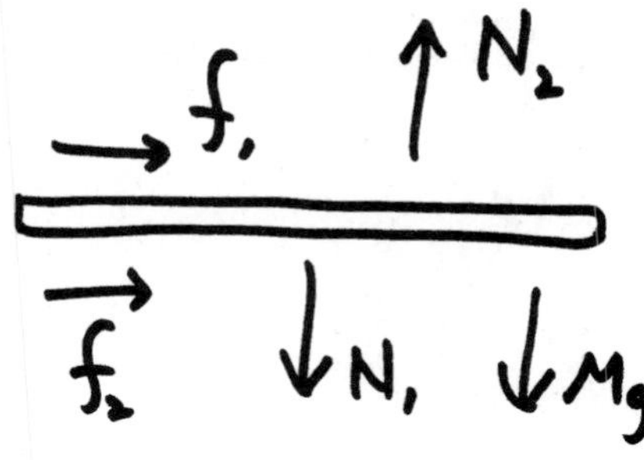

Just because two surfaces are in contact does not mean that there is a friction force. Imagine a book sitting on a table. There are no horizontal forces on the book, so it sits motionless on the table. If there was a friction force, it would cause the book to move. Since there doesn't need to be any friction to prevent sliding, there is no friction force.

When drawing your free-body diagram, **always put the static friction forces in last**. This is because the static friction can be in any direction needed to keep sliding from happening. Until you know in which direction something would slide if there was no friction, you don't know which way the static friction will be. Do all other forces first, then ask, "in which direction would it slide if there was no friction?" Add the static friction in to prevent this sliding.

The magnitude of the kinetic friction force is $F_\mathrm{k} = \mu_\mathrm{k} N$, where μ_k is the coefficient of friction and N is the normal force between the surfaces. The magnitude of the static friction force is anything it needs to be to prevent sliding, up to a maximum of $F_{\mathrm{s,\ max}} = \mu_\mathrm{k} N$, where μ_s is the coefficient of friction and N is the normal force between the surfaces.

EXAMPLE

A box sits on a table. The box has a mass of 8 kg and the coefficients of friction between the box and the table are $\mu_\mathrm{s} = 0.7$ and $\mu_\mathrm{k} = 0.4$. A horizontal force of 45 N pushes the box to the right. What is the friction force on the box?

Tutor: Where do we start?
Student: Can't we skip the free-body diagram?
Tutor: If you have more than one force, then you need a free-body diagram. I still make mistakes if I skip the diagram with only two forces.
Student: Okay. Gravity mg goes down, and a normal force N pushes up. Because there is a normal force, there could be a friction force $\mathcal{F}$, but we'll do that last. Someone is pushing to the right with a force $P = 45$ N.
Tutor: What would happen if there were no friction force?
Student: The box would accelerate to the right, sliding across the floor. Friction point to the left to oppose the sliding.

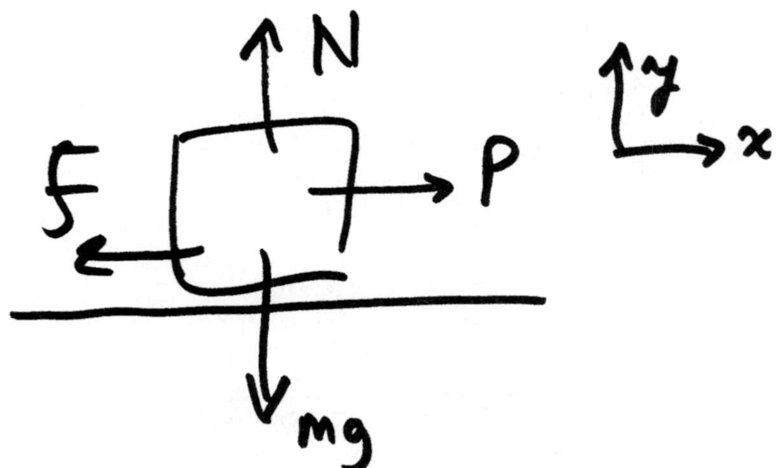

Tutor: Good. Is it static or kinetic friction?
Student: Kinetic, because the box is sliding.
Tutor: How do you know that the box is sliding?
Student: Because someone pushes it.
Tutor: Have you ever pushed something and then it didn't move?
Student: Okay, how do I check?
Tutor: How big could the static friction force be?
Student: Static friction could be as big as $\mathcal{F}_\mathrm{max} = \mu_\mathrm{s} N$. I need to know the normal force.

Tutor: To get the normal force, add the vertical forces:

$$N - mg = m\cancelto{0}{a_y}$$

$$N = mg$$

Student: Static friction could be as big as

$$\mathcal{F}_{\text{max}} = \mu_s N = \mu_s mg = (0.7)(8\text{ kg})(9.8\text{ m/s}^2) = 55\text{ N}$$

Tutor: How big is the friction force?
Student: It's 55 N.
Tutor: The static friction force can be as big as 55 N, but it doesn't have to be 55 N. What happens if the friction force is 55 N?
Student: The total force is to the left.
Tutor: Yes, he pushes to the right and the box accelerates to the left. Does that make sense?
Student: No, so the friction force will only be 45 N, and the box doesn't move.
Tutor: What if the box were already moving?
Student: Then it would be sliding, and the friction would be kinetic friction.

$$\mathcal{F}_{\text{k}} = \mu_s N = \mu_s mg = (0.4)(8\text{ kg})(9.8\text{ m/s}^2) = 31\text{ N}$$

EXAMPLE

A boy pulls on a box with a rope that is 23° above horizontal to the right. He pulls with a force of 50 N and the mass of the box is 8 kg. The coefficients of friction between the box and the floor are $\mu_s = 0.56$ and $\mu_k = 0.47$. What is the acceleration of the box?

Student: First we draw the free-body diagram. Gravity mg is down, and the normal force N is up. There is a tension $T = 50$ N up and to the right, and there is a friction force $\mathcal{F}$ to the left.

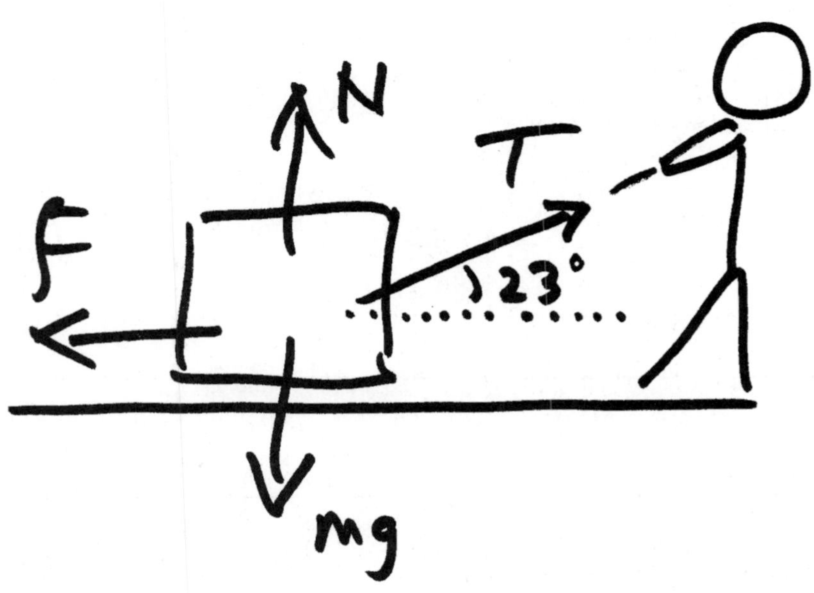

Tutor: Is it static friction or kinetic friction?
Student: That depends on whether the box is sliding across the floor.
Tutor: Good. How do we check?
Student: We see if the static friction would be enough to keep it from sliding. If it is, then the acceleration is zero. If it isn't enough, then the box slides and there is kinetic friction.
Tutor: To do that, we need to write down the Newton's second law equations.

$$\Sigma F_x = ma_x \qquad \text{and} \qquad \Sigma F_y = ma_y$$

Student: If the box accelerates, it will do so to the right. I'll use that as the x axis and up as the y axis.

$$T\cos 23° - \mathcal{F} = ma_x$$

$$T\sin 23° + N - mg = m\cancelto{0}{a_y}$$

Student: a_y is zero because all of the acceleration is in the x direction.
Tutor: Can you do anything with these equations?
Student: I can solve the second one to find N.
Tutor: Does that help you?
Student: Yes, because $\mathcal{F}_{\text{s,max}} = \mu_{\text{s}}N$.

$$(50\ \text{N})\sin 23° + N - (8\ \text{kg})(9.8\ \text{m/s}^2) = 0$$

$$(19.5\ \text{N}) + N - (78.4\ \text{N}) = 0$$

$$N = 58.9\ \text{N}$$

$$\mathcal{F}_{\text{s,max}} = \mu_{\text{s}}N = (0.56)(58.9\ \text{N}) = 33.0\ \text{N}$$

Tutor: What are you going to do with the maximum static friction force?
Student: I'll put it in the x equation and see if the box moves.

$$(50\ \text{N})\cos 23° - (33.0\ \text{N}) = (8\ \text{kg})a_x$$

$$(46.0\ \text{N}) - (33.0\ \text{N}) = (8\ \text{kg})a_x$$

Student: The total x force is positive, or to the right, so the box does move. Now I can find the acceleration.
Tutor: Since the box moves, is the friction still static.
Student: The friction is really kinetic, because the box is sliding. That means I have to start over with kinetic friction.
Tutor: You only need to go back to the x equation. Nothing about the y equation has changed.
Student: So the normal force is still 58.9 N.

$$(50\ \text{N})\cos 23° - (0.47)(58.9\ \text{N}) = (8\ \text{kg})a_x$$

$$(18.3\ \text{N}) = (8\ \text{kg})a_x$$

$$a_x = \frac{(18.3\ \text{N})}{(8\ \text{kg})} = 2.3\ \text{m/s}^2$$

EXAMPLE

A 6 kg box is sliding up a 34° ramp at 14 m/s. The coefficient of kinetic friction between the box and the ramp is $\mu_{\text{k}} = 0.36$. What is the acceleration of the box?

Tutor: Where do we begin?
Student: With the free-body diagram, of course. Gravity mg is down, and the normal force N is perpendicular to the surface. Because there is a normal force, there could be a friction force $\mathcal{F}$. Also there is a force P pushing the box up the ramp.
Tutor: The problem doesn't mention a force pushing the box up the ramp. Where did that come from?
Student: Something had to push the box or it wouldn't go up the ramp.
Tutor: Something did push the box to get it going up the ramp, but that something is done pushing now. Can something be moving up without an upward force on it?
Student: Yes, if there was a force in the past. I see. There is no pushing force.
Tutor: The friction force has to be parallel to the ramp surface. Will it point up or down the ramp?
Student: Won't friction always be down the ramp?
Tutor: Friction opposes sliding, so if the box is sliding up the ramp, the friction force will be down the ramp.
Student: Of course. And since it's already sliding, it will be kinetic friction.

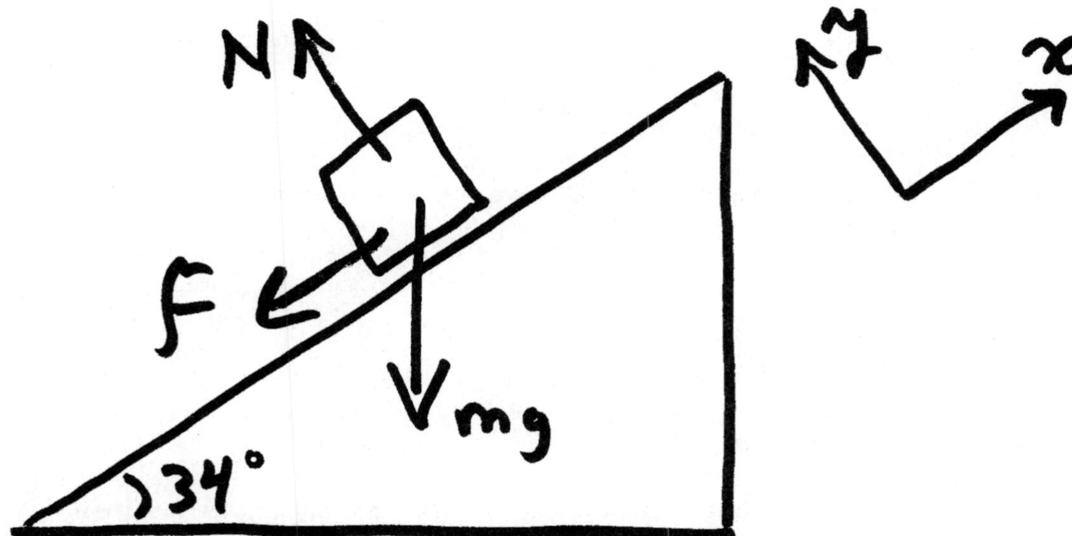

Tutor: Now you can choose axes and apply Newton's second law. Remember to pick one axis parallel to the acceleration.
Student: The box is going up the ramp, but friction and gravity are both pushing it down the ramp. Which way should I pick as the axis?
Tutor: It really doesn't matter whether up or down the ramp is positive, but if you choose x horizontal the math will get messy.
Student: I choose x up the ramp, in the direction of the initial velocity. Just to be different, I'll take y down into the ramp.

$$-\,mg\sin 34^\circ - \mathcal{F} = ma_x$$

$$+\,N - mg\cos 34^\circ = m\cancelto{0}{a_y}$$

Tutor: Can you solve these equations to find the acceleration?
Student: To find a_x I need to know $\mathcal{F}$. It's kinetic friction, so $\mathcal{F} = \mu_k N$.
Tutor: How will you get the normal force?
Student: I can solve the y equation to find the normal force N, then use that to solve the x equation for the acceleration.

$$N = mg\cos 34^\circ = (6\text{ kg})(9.8\text{ m/s}^2)\cos 34^\circ = 48.7\text{ N}$$

$$-mg\sin 34^\circ - \mu_k N = ma_x$$

$$-(6\text{ kg})(9.8\text{ m/s}^2)\sin 34^\circ - (0.36)(48.7\text{ N}) = (6\text{ kg})a_x$$

$$a_x = -8.4\text{ m/s}^2$$

Student: As we expected, the acceleration is down the ramp, in the negative direction. Is the normal force always $N = mg\cos\theta$?
Tutor: Not always. You need to go through the steps.

When something goes in a circle, its velocity is always changing. Even if it travels at a constant speed, the velocity is changing because the direction is changing. The **acceleration of something going in a circle is toward the middle** and is

$$a_c = \frac{v^2}{r}$$

where v is the speed and r is the radius of the circle.

It is important to realize that **centripetal force is not a real force and never goes in the free-body diagram**. The name "centripetal force" refers to the total force when the object is going in a circle. There will be forces that you already know that add up to equal the centripetal force mv^2/r.

Imagine that you are in a car when the car turns to the left. You might think that you feel a force pushing you to the right, toward the right of the car. A stationary observer sees that you are continuing straight while the car turns underneath you. The real force on you is friction with the seat pushing you to the left. The perceived force to the right is similar to a perceived force backwards when the car accelerates forward — if the seat didn't push you forward you would stay there while the car moved out from under you.

If something is going in a circle, treat the problem like you would any other. Draw the free-body diagram, choose axes, and apply Newton's second law. The only difference is that the total force will be toward the middle of the circular path, and the acceleration will be v^2/r.

EXAMPLE

A roller coaster goes over a hill of radius 30 m. What speed is needed to achieve weightlessness?

Student: As the coaster goes over the hill, it is going in a circle, so the acceleration is downward, toward the middle of the circle.
Tutor: Good. Where do we begin?
Student: With the free-body diagram, as always. Gravity mg is going down and the normal force N is going up. The centripetal force F_c is going down, toward the middle of the circle.
Tutor: There is no centripetal force, and it never goes in the free-body diagram.
Student: Then how can the coaster accelerate downward?
Tutor: The forces that do exist must add up to go downward. Is the normal force equal to the weight?
Student: Not necessarily.
Tutor: Are there any other forces besides weight and normal force?
Student: We did the weight, and the only thing in contact with the coaster is the track, so that's all of them.

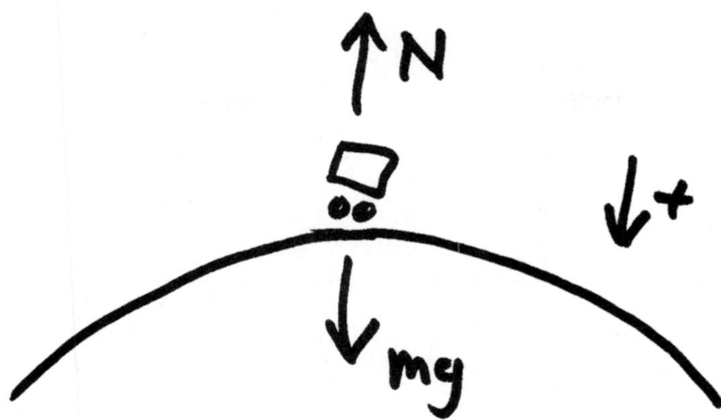

Tutor: Is it possible for the forces to add to a total that points downward?
Student: If the weight is greater than the normal force, then the total force will be down.
Tutor: Apply Newton's second law. Choose an axis and add the forces.
Student: I choose up to be positive.
Tutor: Remember that the acceleration is downward. If up is positive, then the acceleration will be negative.
Student: Okay, I choose down to be positive so that the acceleration will be positive.

$$\Sigma F = ma$$

$$mg - N = m\frac{v^2}{r}$$

Tutor: When something is going in a circle, we sometimes call the total force mv^2/r the "centripetal force." Which of the forces in the problem is the centripetal force?
Student: Neither, it's the sum of the two.
Tutor: Yes, the *total* force is equal to the centripetal force. What happens as the coaster goes faster?
Student: As v increases, the right side gets larger, so the left side gets larger too.
Tutor: Does the weight increase as the coaster goes faster?
Student: No, so the normal force must get smaller.
Tutor: We call it "weightlessness" when $N = 0$. The weight hasn't changed, but you don't "feel" a force. Perhaps we should call it normalforcelessness.
Student: But how can there be no normal force? Won't the coaster fall?
Tutor: Yes, it does fall. But to go over the hill, in an arc, it needs to accelerate downward, and the weight provides the force to accelerate it downward.
Student: And all of this happens when

$$mg - \cancelto{0}{N} = m\frac{v^2}{r}$$

$$\cancel{m}g = \cancel{m}\frac{v^2}{r}$$

$$v = \sqrt{rg} = \sqrt{(30\text{ m})(9.8\text{ m/s}^2)} = 17\text{ m/s}$$

Tutor: Right. The coaster "falls" at exactly the same curvature that the track has. The rider feels the normal force from the seat momentarily disappear. What happens if the coaster goes even faster?
Student: Then the curvature of the track is greater than the path of the coaster, and the coaster comes off of the track. That would be dangerous.
Tutor: Modern coasters have wheels under the track, and restraints for the riders, so that the normal force can pull downward too. When the coaster goes too fast you feel yourself being jerked downward, like on the Magnum at Cedar Point in Ohio.
Student: So would I put in a negative value for N? I thought N was a magnitude and couldn't be

negative.
Tutor: A negative value for N indicates that the force goes in the opposite direction as drawn in the diagram. Or you could draw a new diagram with the normal force down.

EXAMPLE

A small coin is placed on a turntable (like an old record player), 14 cm from the center. The coefficients of friction between the coin and the turntable are $\mu_s = 0.62$ and $\mu_k = 0.48$. How fast can the coin go before it slips? If it slips, in what direction will it go?

Student: We start with the free-body diagram. The coin has weight mg downward, and a normal force N upward. We don't know how big N is. Because there is a normal force, there could be a friction force $\mathcal{F}$.
Tutor: In which direction is the friction force?
Student: Well, it has to be parallel to the surface, and it doesn't need to be opposite to the motion. The coin isn't sliding yet, so it's static friction.
Tutor: Correct on all counts. Which way is the coin accelerating?
Student: It's going in a circle, so in toward the middle.
Tutor: What force is pushing it that way?
Student: Centripetal force. No, wait, there is no such thing. There isn't any force in that direction. Friction must be toward the middle of the turntable.
Tutor: Correct. How big can friction be?
Student: Up to $\mu_s N$.
Tutor: Since the problem asks for the fastest that the coin can go, we want the maximum static friction force.
Student: Do we always want the maximum friction force?
Tutor: When we want a limit, like when does the coin start to slip, then we want the maximum friction force.
Student: Okay, I pick axes. x is in toward the center and parallel to the acceleration, and y is up. But when the coin has gone halfway around, x will be away from the center.
Tutor: We can move the axes with the coin, so that x is always pointing from the coin into the center of the circle.

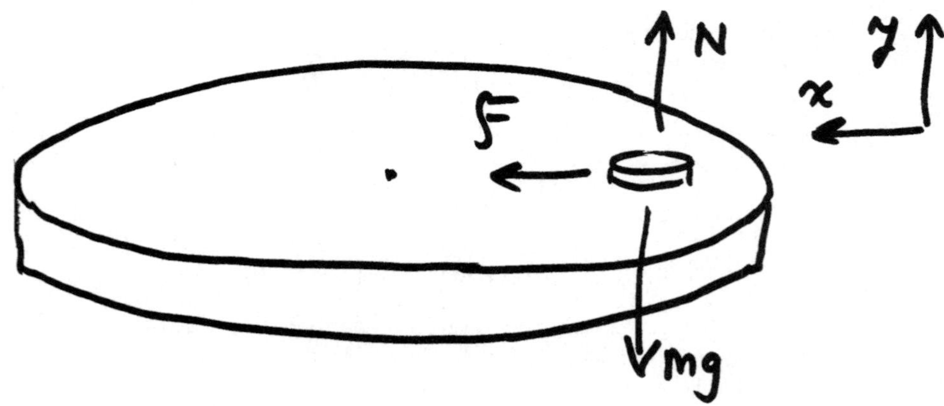

Student: Now I apply Newton's second law.

$$\mathcal{F} = ma_x$$

$$+N - mg = \cancelto{0}{a_y}$$

Student: I can solve the second equation for N, then find $\mathcal{F}$, and then find a_x.

$$N = mg$$

Student: I need to know the mass m.
Tutor: Keep going, maybe it'll cancel.

$$\mu_s N = m\frac{v^2}{r}$$

$$(0.62)\cancel{m}(9.8\ \text{m/s}^2) = \cancel{m}\frac{v^2}{(0.14\ \text{m})}$$

$$v = \sqrt{(0.62)(9.8\ \text{m/s}^2)(0.14\ \text{m})} = 0.92\ \text{m/s}$$

Tutor: What if the coin goes faster than this?
Student: It flies off of the turntable, away from the center.
Tutor: Does it go directly away from the center? It was moving around the center. What are the forces after it slips?
Student: Once it slips, the friction will be kinetic, which is smaller. Once the coin is off of the turntable, there will be no friction, so the coin will move in a straight line.

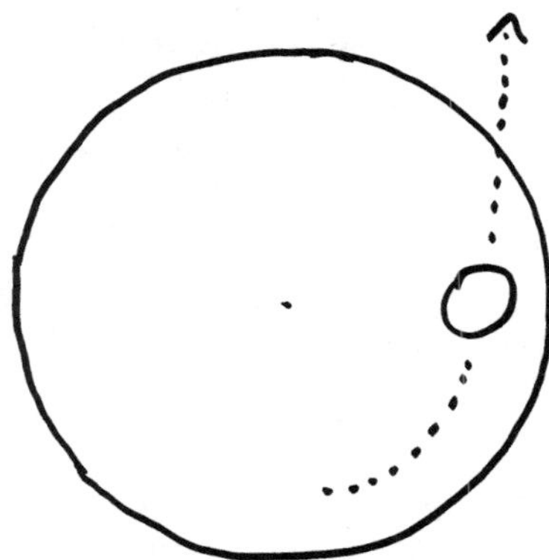

EXAMPLE

A 1500 kg car goes around a curve of radius 200 m. The curve is banked at 11° and the coefficient of static friction between the car and the road is 0.48. What is the maximum speed of the car around the curve?

Tutor: We start by drawing the free-body diagram, without which all hope is lost.
Student: There is a gravity force mg pulling it down. There is a normal force N perpendicular to the surface of the road. Because there is a normal force there could also be a friction force f. Nothing else is in contact with the car, so there are no other forces.
Tutor: What is the direction of the friction force?
Student: It isn't necessarily opposite to the motion. Which way would the car slide without friction?
Tutor: Very good. There is a speed at which the horizontal component of the normal force is just enough force so that the car goes around the curve. Our car is going even faster, so it needs even more force in toward the middle.
Student: So friction points in toward the middle? Doesn't it have to be parallel to the road surface?

Tutor: Yes, but of the choices up and down the ramp, which is in toward the middle of the turn?
Student: Down the ramp. So friction points down the ramp.
Tutor: Yes. Imagine trying to go around a turn on an icy day. There isn't enough friction and the car goes straight, or doesn't turn enough. Friction pushes the car in toward the turn.
Student: Okay, now I need axes. The acceleration is also down the ramp.
Tutor: The acceleration is toward the middle of the circle. Is the circle that the car is traveling completely horizontal or parallel to the ramp?
Student: It is horizontal, so the acceleration is horizontal. The x axis needs to be horizontal, toward the middle of the circle, and the y axis needs to be perpendicular to that, so up.

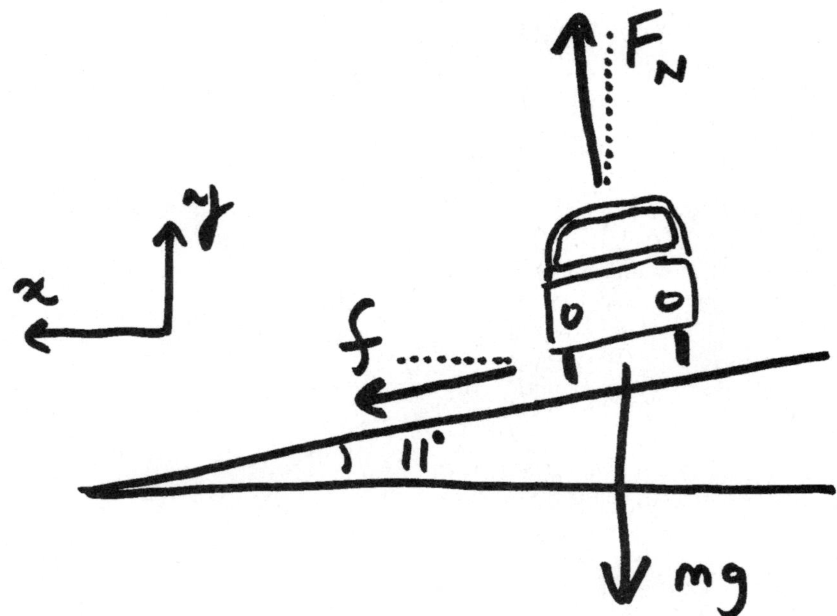

Student: Now I can apply Newton's second law.

$$+N\sin 11^\circ + f\cos 11^\circ = ma_x = m\frac{v^2}{r}$$

$$+N\cos 11^\circ - f\sin 11^\circ - mg = m\cancel{a_y}^{\,0}$$

Tutor: Good. Can you do anything with these equations?
Student: I have three unknowns, N, f, and v. I want to find v. I need another piece of information.
Tutor: Yes, you need the relation between the friction force f and the normal force N.
Student: Is $f = \mu N$, or $f \leq \mu N$ this time?
Tutor: We want the *fastest* speed, so the static friction force is as big as it can be, $f = \mu N$. If the speed was some arbitrary value, we couldn't do this.
Student: Why is it static friction? Isn't the car moving compared to the road?
Tutor: The tires are rolling, rather than slipping. Since there is no slipping, it's static friction.

$$+N\sin 11^\circ + \mu N\cos 11^\circ = m\frac{v^2}{r}$$

$$+N\cos 11^\circ - \mu N\sin 11^\circ - mg = 0$$

Student: I know everything in the second equation except N. I can find N and use it in the top equation to find v.

$$N\left(\cos 11^\circ - \mu\sin 11^\circ\right) = mg$$

$$N = \frac{mg}{(\cos 11° - \mu \sin 11°)}$$

$$\left(\frac{mg}{(\cos 11° - \mu \sin 11°)}\right)(\sin 11° + \mu \cos 11°) = m\frac{v^2}{r}$$

$$v = \sqrt{Rg\frac{(\sin 11° + \mu \cos 11°)}{(\cos 11° - \mu \sin 11°)}}$$

$$v = \sqrt{(200 \text{ m})(9.8 \text{ m/s}^2)\frac{(\sin 11° + (0.48)\cos 11°)}{(\cos 11° - (0.48)\sin 11°)}}$$

$$v = 38.2 \text{ m/s} = 85 \text{ mph}$$

Tutor: What if we were going *slower* than the magic speed where no friction is needed?
Student: Then the horizontal component of the normal force would be too much, and friction would have to point the other way.

Chapter 7

Kinetic Energy and Work

Energy is the power to make something move. That might not sound terribly interesting, but it includes a lot of stuff. Sound is moving air molecules, so energy is the power to make tunes. Lots of things use energy.

It is a principle of physics that **energy is conserved**. This means that energy is neither created nor destroyed, but only turned from one form of energy into another. Conservation laws are a very powerful technique for solving problems. Conservation of energy is one of those laws, and is the important technique of this chapter.

Just because something is "conserved" does not mean it doesn't change. What it does mean is that we can account for the change. For example, chairs in a chair factory are conserved. By this we mean that chairs do not suddenly appear or disappear, but are made and shipped. If there are more chairs now than there were an hour ago, then someone made some more. If there are fewer chairs, then some must have been loaded onto a truck and sent to the customer. If we take the number of chairs at the beginning of the day, add the number made and subtract the number shipped, we get the number of chairs in the factory now.

How do we use conservation? What if the truck left the factory and no one counted how many chairs were on board. We could count how many chairs were still in the factory, and use the equation from the last paragraph to determine how many chairs had been loaded. If we know all of the terms in the equation except one, we can solve for the one we don't know.

The equation for energy is

$$K_i + W = K_f$$

or, the initial kinetic energy plus the work is the final kinetic energy. Doing work is how we change the energy.

Objects have kinetic energy when they are moving. The kinetic energy of a moving object is

$$KE = \frac{1}{2}mv^2$$

Note that the kinetic energy is positive even if the velocity is in the negative direction.

The work done on an object is

$$W = \vec{F} \cdot \vec{d} = Fd\cos\phi$$

If the force $\vec{F}$ is in the same direction as the motion $\vec{d}$, the work is positive, the energy increases, and the speed increases. If the force is in the opposite direction as the motion, the work is negative, the energy decreases, and the speed decreases. **If the force is perpendicular to the motion, then the work is zero**, the energy doesn't change, and the speed stays the same but the direction of motion may change.

The units of work and energy are the same — they need to be so that we can add them. A newton times a meter is a joule.

$$1 \text{ N} \times 1 \text{ m} = 1 \text{ J}$$

EXAMPLE

A 3 kg box is pulled 6 m across the floor. The box is pulled to the right by a rope that is 35° above horizontal and has a force of 26 N. The coefficient of kinetic friction between the box and the floor is 0.15. Find the work that each force does on the box, and find the speed of the box after it has been pulled 6 m.

Student: Do we still start with a free-body diagram, even though we're using energy?
Tutor: Yes. To find the work done by each force we need to know what they are.
Student: Gravity mg is down, normal force N is up, because there is a normal force there could be a friction force f, and tension T is up and to the right.
Tutor: Is it kinetic or static friction?
Student: The box is sliding, so it's kinetic. As the box slides to the right, the friction on the box will be to the left.

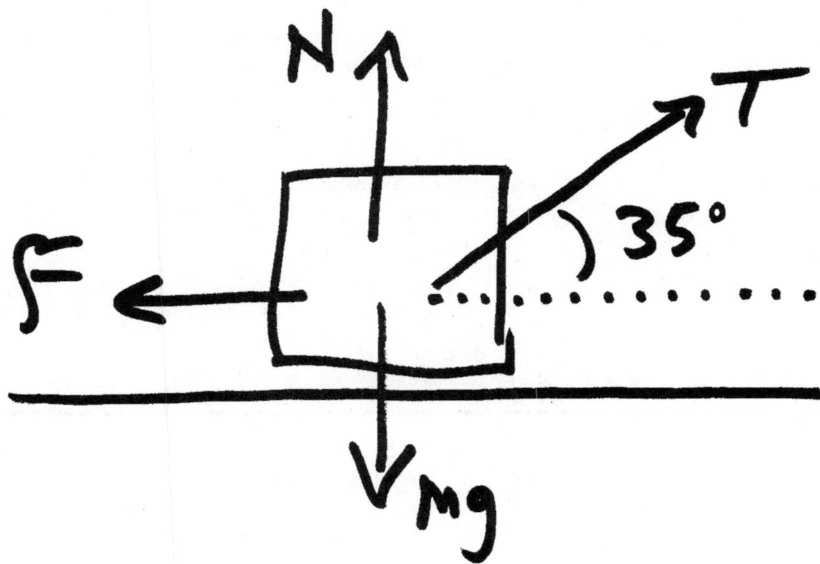

Tutor: What is the work that the tension does on the box?
Student: The tension is 26 N and at a 35° angle.

$$W_T = Fd\cos\phi = (26 \text{ N})(6 \text{ m})\cos 35^\circ = 128 \text{ J}$$

Tutor: Good. How much work is done by gravity?
Student: Gravity is up, and the box moves sideways.

$$W_G = Fd\cos\phi = \left((3 \text{ N})(9.8 \text{ m/s}^2)\right)(6 \text{ m})\cos 90^\circ = 0 \text{ J}$$

Student: Does that make sense? Gravity really does no work?
Tutor: Does gravity cause the box to speed up as it moves, gaining kinetic energy?
Student: Not on a level floor.
Tutor: So if gravity doesn't change the kinetic energy, it does no work. Any force that is perpendicular to the motion does zero work.

Student: That's a good thing to remember. The normal force is also perpendicular to the motion, so

$$W_N = Fd\cos\phi = N(6\text{ m})\cos 90^\circ = 0\text{ J}$$

Student: And I didn't even need to figure out what the normal force is.
Tutor: How much work does friction do?
Student: Friction doesn't cause the box to speed up, so it does no work.
Tutor: Friction could cause the box to slow down, which would be taking energy out, or doing negative work.
Student: Negative work? Then I need the friction force μN, so I need the normal force after all.

$$+N + T\sin 35^\circ - mg = m\cancelto{0}{a_y}$$

$$N = mg - T\sin 35^\circ = (3\text{ kg})(9.8\text{ m/s}^2) - (26\text{ N})\sin 35^\circ = 14.5\text{ N}$$

Student: I just thought of something. What if T was so big that the normal force was negative?
Tutor: Could the floor be pulling the box down? What would happen if you pulled *really* hard?
Student: No. The box would lift off of the floor, so a_y wouldn't be zero. The work done by friction is

$$W_f = Fd\cos\phi = [(0.15)(14.5\text{ N})]\,(6\text{ m})\cos 180^\circ = -13\text{ J}$$

Student: Now I can apply Newton's second law in the x direction and find the acceleration.
Tutor: You could do that. Many physics students try to use the first thing they learned for everything, even if a newer technique works better. Try using conservation of energy.

$$K_i + W = K_f$$

Student: The box isn't moving to start, so $K_i = 0$. We do $128\text{ J} + 0 + 0 + (-13\text{ J}) = 115\text{ J}$ of work, so

$$0 + (115\text{ J}) = \frac{1}{2}mv^2$$

$$v = \sqrt{\frac{2(115\text{ J})}{3\text{ kg}}} = 8.8\text{ m/s}$$

At this point we introduce springs. The force from a spring is

$$F_{\text{spring}} = -kx$$

The minus sign generally does not go into an equation. It is there to remind us that the force of the spring is opposite to x, where x is how much the spring is stretched or compressed. k is the "spring constant," a constant for any particular spring (though a different spring will have a different k). The units of k are newtons of force for each meter that the spring is compressed (newtons/meter).

The reason to introduce springs here, when talking about work and energy, is that work and energy is the easiest way to deal with springs. Because the force from a spring will change if it is compressed, the acceleration will not be constant. **Conservation of energy works well when the acceleration is not constant**. The work needed to compress or stretch a spring a distance x from its normal length is

$$W_{\text{spring}} = \frac{1}{2}kx^2$$

When first compressing a spring, the force is zero so no work is needed. But the more you compress it the greater the force you need, and each cm takes more work than the previous one. The work is equal to the average force $\frac{1}{2}kx$ times the distance x.

EXAMPLE

A wall is tilted in by 15°, and a 2 kg block is placed against the wall. The block is held in place with a spring. The coefficients of friction between the block and the wall are $\mu_s = 0.61$ and $\mu_k = 0.49$. Holding the block in place requires that the spring be compressed by 12 cm. What is the spring constant of the spring?

Student: We start with the free-body diagram. The weight mg is down. The normal force N is perpendicularly out of the surface. Because there is a normal force, there could be a friction force. There is also a spring force. The friction force pushes up the wall, to keep the block from sliding down the wall.
Tutor: Doing well. What is the direction of the spring force?
Student: It doesn't say, exactly. Pushing in toward the wall, opposite the normal force?
Tutor: Probably. Let's assume that this is the case.

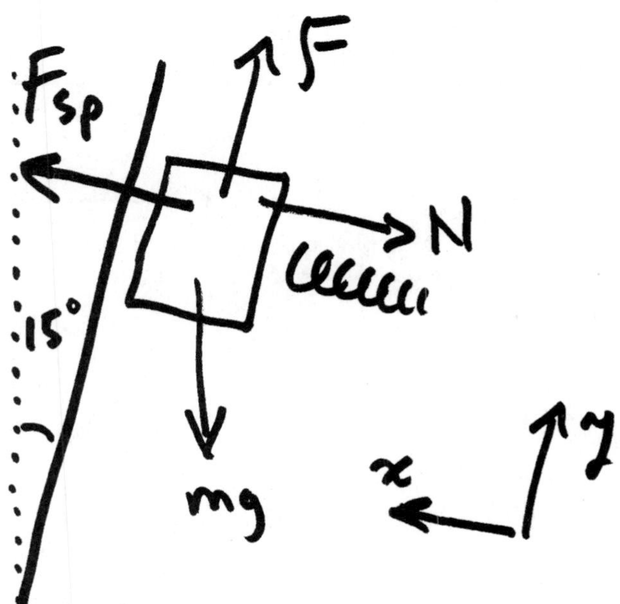

Student: So the spring force is equal to the normal force, and friction is equal to the weight, and...
Tutor: Not all of the weight is parallel to the friction force.
Student: Ah, yes. I choose axes, but there is no acceleration.
Tutor: So you can use any pair of perpendicular axes.
Student: I choose parallel to the spring force as x, and parallel to friction as y.

$$-N + F_{\text{spring}} - mg\sin 15° = m\cancelto{0}{a_x}$$

$$+\mathcal{F} - mg\cos 15° = m\cancelto{0}{a_y}$$

Tutor: How many unknowns do you have?
Student: I have N, F_{spring}, and $\mathcal{F}$. Three unknowns and only two equations. I need one more piece of information. I can use $F_{\text{spring}} = -kx$.
Tutor: F_{spring} is the magnitude of the spring force, and you have the direction in your equation already. Do you really want $F_{\text{spring}} = -kx$?
Student: No, I want $F_{\text{spring}} = kx$.
Tutor: But then you introduce the unknown k. This is fine, since you want to solve for k eventually, but you still have three unknowns.
Student: It's static friction because it isn't sliding. Can we say that the static friction force is equal to

the maximum static friction force?
Tutor: If 12 cm is the smallest compression of the spring that will hold the block, then we are at the limit and we can say that.

$$-N + kx - mg\sin 15^\circ = 0$$
$$+\mu_s N - mg\cos 15^\circ = 0$$

Student: I can solve the bottom equation for N, then put it in the top equation and solve for k.

$$N = \frac{mg\cos 15^\circ}{\mu_s}$$

$$kx = mg\sin 15^\circ + \frac{mg\cos 15^\circ}{\mu_s} = mg\left(\sin 15^\circ + \frac{\cos 15^\circ}{\mu_s}\right)$$

$$k = \frac{mg}{x}\left(\sin 15^\circ + \frac{\cos 15^\circ}{\mu_s}\right) = \frac{(2\text{ kg})(9.8\text{ m/s}^2)}{0.12\text{ m}}\left(\sin 15^\circ + \frac{\cos 15^\circ}{0.61}\right) = 301\text{ N/m}$$

EXAMPLE

A spring with spring constant $k = 1600$ N/m is placed on the ground and pressed down 6 cm. A 2 kg block is placed on the spring, and the spring is released. How far above its initial position does the block go?

Tutor: First we draw the free-body diagram. What forces act on the block?
Student: Gravity mg points down, and a normal force N is upward.
Tutor: It really is a normal force, since it is a force between surfaces, but we call it a spring force.
Student: Okay, a spring force F_{sp} points upward.

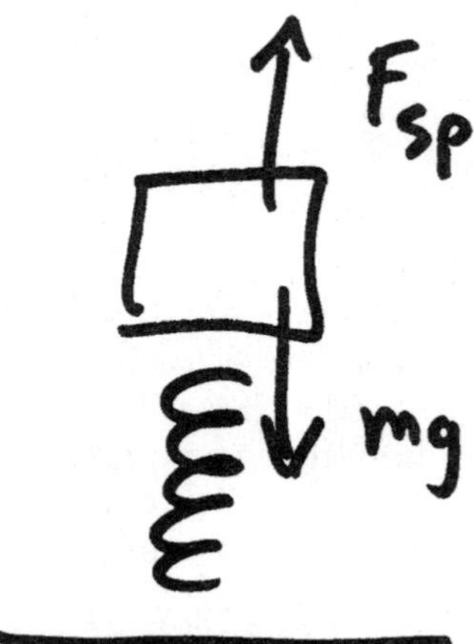

Tutor: Is the spring force constant?
Student: No, as the spring uncompresses, the spring force will decrease.
Tutor: Is the acceleration constant?
Student: No, as the spring force changes, the total force changes, so the acceleration changes.
Tutor: When the acceleration is not constant, it's almost always easier to use a conservation law, like conservation of energy.
Student: So whenever a spring is involved, use energy?
Tutor: There was a spring in the last problem, but we didn't use energy. More precisely, whenever a spring

is changing its length, that's a tip-off to try energy first.

$$K_i + W = K_f$$

Student: The block starts at rest, so K_i is zero.
Tutor: How fast is the block moving when it reaches its maximum height?
Student: It's at rest then too, so K_f is also zero. Then the work is zero and everything is zero and we can't solve for anything.
Tutor: Don't panic. How much work does gravity do on the block?
Student: The force mg times the distance d times the cosine of 180°, since it moves up and gravity points down.
Tutor: Good. Of course, we don't know d, but that's what we're trying to find, so this is how it gets into our equation. How much work does the spring do on the block?
Student: The work of a spring is $\frac{1}{2}kd^2$.
Tutor: Is the initial compression of the spring equal to the distance that the block moves up?
Student: I guess not. That means we need another variable.
Tutor: How much is the spring initially compressed?
Student: By 6 cm.
Tutor: How much is it compressed after the block flies up?
Student: Zero. Ah, the spring compression changes by 6 cm.
Tutor: Because the equation for the work of a spring is nonlinear, we can't just use the change in compression. We need to do both the initial and final compression. Does the spring do positive or negative work?
Student: The spring pushes up, and the block moves up, and since the directions are the same it's positive work.

$$K_i + (W_{\text{gr}} + W_{\text{gr}}) = K_f$$

$$0 + \left(mgd\,\cancelto{-1}{\cos 180^\circ} + \frac{1}{2}k(0.06\text{ m})^2\,\cancelto{+1}{\cos 0^\circ} \right) = 0$$

$$mgd = \frac{1}{2}k(0.06\text{ m})^2$$

$$(2\text{ kg})(9.8\text{ m/s}^2)d = \frac{1}{2}(1600\text{ N/m})(0.06\text{ m})^2$$

$$d = 0.147\text{ m} = 14.7\text{ cm}$$

Student: The block moves up 14.7 cm.
Tutor: As part of our solution, we assumed that the spring fully uncompressed. Is that consistent with our answer?
Student: The spring needed to uncompress by 6 cm, and the block moved more than that, so we're okay.
Tutor: Good.
Student: What if we hadn't converted the 6 cm to meters, but had left it as 6 cm?
Tutor: Then we would have had

$$d = \frac{(1600\text{ N/m})(6\text{ cm})^2}{2(2\text{ kg})(9.8\text{ m/s}^2)} = \frac{(1600\text{ kg m/s}^2\text{/m})(6\text{ cm})^2}{2(2\text{ kg})(9.8\text{ m/s}^2)} = 1470\text{ cm}^2\text{/m}$$

Tutor: It's still a unit of length, but not one that we're really familiar with.

An important distinction in science is the difference between the total amount and the rate at which it is done. Consider the following conversation:

Student: I drove from New York to Boston yesterday.
Tutor: How fast did you go?
Student: 217 miles.

Most of us would immediately recognize that the response answered a different question, that the correct response would be something like "65 miles per hour." Now consider another conversation, about energy but exactly the same as the previous conversation.

Student: I carried a heavy box up the stairs yesterday.
Tutor: How much power did you exert?
Student: 5000 joules.

Joules measure energy or work, so 5000 joules is the amount of work done. **Power measures how fast you do work, not the amount done**. To do 5000 joules of work quickly requires a large power. To do 5000 joules of work slowly requires a small power. In each case the amount of work done is still 5000 joules.

To determine how fast the work was done, we divide the work by the time.

$$P = \frac{E \text{ or } W}{t}$$

Power is measured in watts, so one watt is one joule per second. To do 5000 joules of work in 10 seconds requires 500 joules per second or 500 watts or 500 W. Such a power would be quite a feat for a human (1 horsepower = 746 W). To do 5000 joules of work is not difficult for a human — just don't try to do it so fast.

Humans tire quickly if they try to exert more than about 70% of their maximum power. Sprinting, for example, will tire a person much quicker than jogging. Since you can jog much longer (time) than you can sprint, you can jog further — doing more work — than you can sprint, even though you sprint faster. The time you can jog is much longer.

EXAMPLE

A 1600 kg car drives 1 mile up a 5% grade at a constant 60 mph. How much power does this take?

Student: We need to know the forces, so we start where we always do. Gravity mg points down, there is a normal force N perpendicular to the surface, and because there is a normal force there could be a friction force f. Nothing else is in contact with the car, so that's all of the forces.
Tutor: Which way is the friction force acting?
Student: The only way for the car to go up the hill is if friction pushes it up the hill. If the tires are rolling without slipping, then it's static friction.
Tutor: As the engine turns the tires, what would happen without friction?
Student: The tires would slip on the road, so static friction pushes the car forward and up the hill to prevent slipping.

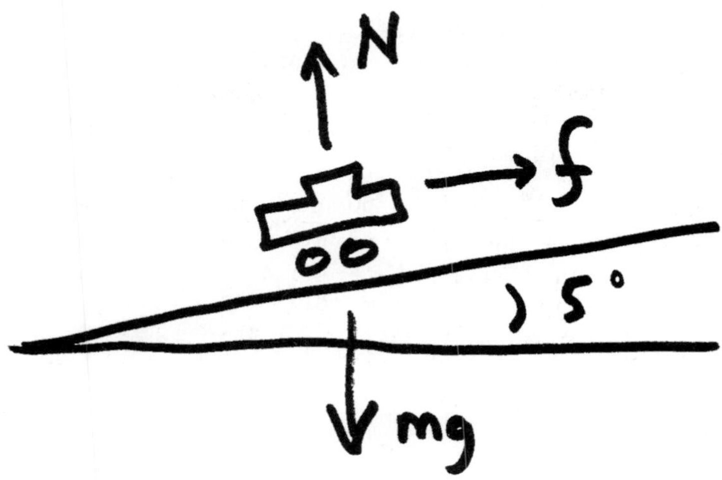

Tutor: We want to find the power of the friction force. Does the friction force do positive or negative work?
Student: Friction points up the hill, and the motion is up the hill. Since the force is in the same direction as the motion, friction does positive work.
Tutor: Yes. How much work does the normal force do?
Student: The normal force is always perpendicular to the motion, so it doesn't do any work. Does the normal force ever do any work?
Tutor: Yes, if the surface moves. If you push on an object and the object moves, your hand moves so the normal force of your hand does work. Or the floor in an elevator does work because the floor surface moves.
Student: But the road surface doesn't move, so the normal force doesn't do any work.
Tutor: How much work does gravity do?
Student: We don't need to know that to do the problem.
Tutor: True, but try applying conservation of energy anyway.
Student: If the final speed of the car is the same as the initial speed, then the total work is zero. The normal force doesn't do any work, so gravity must do the same amount of work as friction, except negative.
Tutor: Excellent. Does it make sense that gravity does negative work?
Student: Gravity is down, and the motion of the car is up and over. Since gravity is opposite to the motion, I guess it has to do negative work.
Tutor: Yes. Also, gravity by itself would tend to slow the car down, reducing the kinetic energy, so it is doing negative work.
Student: That makes sense. So I can find the work done by gravity instead of the work done by friction.
Tutor: That would work, so long as you keep track of the signs. Which is easier?
Student: I already know the force of gravity, so I'll find the work done by gravity.

$$W_{\text{mg}} = Fd\cos\phi = (mg)d\cos\phi$$

Student: What is a 5% grade?
Tutor: That's when the rise over the run is 5%, or when $\tan\theta = 0.05$.
Student: But if the "run" is a mile, then the road would be longer than a mile.
Tutor: True, but not by much. For a 5% grade the difference is about one-eighth of 1%, so there is not much error in pretending that they are the same.
Student: That's tricky. Okay, the angle of the roadway is

$$\tan\theta = 0.05 \quad\longrightarrow\quad \theta = 2.86^\circ$$

Student: One mile is 1610 meters, so

$$W_{\text{mg}} = (mg)d\cos\phi = (1600\text{ kg})(9.8\text{ m/s}^2)(1610\text{ m})\cos 87.14^\circ = 1.26\times10^6\text{ J} \quad ?$$

Tutor: Remember that the work of gravity is negative.
Student: Oh yeah, I should have used 92.86° instead. But the work by friction is the same but positive, and it is 1.26×10^6 J.
Tutor: Yes, now what is the power?
Student: I divide the work by the time.

$$t = \frac{d}{v} = \frac{1 \text{ mile}}{60 \text{ mi/hr}} \times \frac{60 \text{ min}}{1 \text{ hr}} \times \frac{60 \text{ s}}{1 \text{ min}} = 60 \text{ s}$$

$$P = \frac{W}{t} = \frac{1.26 \times 10^6 \text{ J}}{60 \text{ s}} = 21 \times 10^3 \text{ J/s or W}$$

Student: Is that a lot?
Tutor: Try converting it into horsepower.
Student: Okay. 1 horsepower = 746 W.

$$(21 \times 10^3 \text{ W}) \times \left(\frac{1 \text{ hp}}{746 \text{ W}}\right) = 28 \text{ hp}$$

Student: Any car should be able to do that easily.
Tutor: Of course we left out air resistance and friction in the engine, so it's not quite that simple.
Student: Why do trucks go uphill so slowly?
Tutor: A truck can have 40 times the mass of the car, so to go up the hill at 60 mph would take over 1100 hp.
Student: And by doing it slower, they still do the same amount of work but with less power.
Tutor: Yep.

Chapter 8

Potential Energy and Conservation of Energy

Some types of work can be expressed as potential energy. **Potential energy is work done in the past that could be turned into work again.** If I have lifted a rock above my head, the rock has potential energy, and by dropping the rock I can turn the potential energy into kinetic energy. If I push a box across a floor, pushing to keep the box moving despite friction, then the work I do goes into heating the floor and the box (through friction), and it is very difficult to turn this energy into some other useful form of energy. Therefore we treat gravity with potential energy, but we don't treat friction this way. There are three forces that we commonly treat with potential energy: gravity, springs, and forces from electric fields (to be covered later).

What makes potential energy useful is that the potential energy only depends on where something is, and not how it got there. This makes **potential energy easier to calculate than work.**

$$\text{PE} = U = W_{\text{we did}} = \int F_{\text{we applied}}$$

While this may appear intimidating, in practice it is easy. Consider gravity. The force of gravity is a constant mg, so

$$\int_0^h (mg)\,dx = [mgx]_0^h = mgh - mg(0) = mgh$$

This says that the *difference* in potential energy between "zero" and "h" is mgh. This means that it takes mgh amount of work to move a mass m from wherever is "zero" to the spot "h", a distance h high.

The potential energy from gravity is

$$U_{\text{gravity}} = mgh$$

What is the height? We can choose *anywhere* to be zero height, from which all heights are measured. Every time we use conservation of energy in an equation, the potential energy before and the potential energy after both show up, so that only the change in potential energy matters. Therefore, we can choose any spot to be zero height, but once we choose we must stick with that spot for the whole problem (like a coordinate axis).

The potential energy from a spring is

$$U_{\text{spring}} = \frac{1}{2}kx^2$$

Here x is not the length of the spring, but the change in length from the unstretched, uncompressed length. Zero potential energy occurs when the spring is neither stretched nor compressed. Then the work of a spring from the last chapter becomes the potential energy stored in the spring.

If we use potential energy to express the work done by a force, then we don't include the work by this force when we calculate the work. We use an expanded conservation of energy equation

$$KE_i + PE_i + W = KE_f + PE_f$$

where W is the work done by forces not already treated with conservation of energy. It is possible to include E_th and E_int, but these are dealt with by W (heat generated by friction is the same as work done by friction).

EXAMPLE

A spring ($k = 900$ N/m) is compressed by 31 cm and a 6 kg block is placed in front of it. When the spring is released the block slides across the floor. The coefficients of friction between the block and the floor are $\mu_\mathrm{s} = 0.61$ and $\mu_\mathrm{k} = 0.46$. How far does the block slide across the floor?

Tutor: How would you like to attack the problem?
Student: I want to draw a free-body diagram, find the acceleration, and use the constant acceleration technique to find how far it goes. Of course, there is a different acceleration when the spring is pushing on the block.
Tutor: As the spring decompresses, the force of the spring changes, so the acceleration changes.
Student: Does that mean that I can't use Newton's laws?
Tutor: Newton's laws are still true, but to use the acceleration would mean setting up differential equations.
Student: Yuck.
Tutor: On the other hand, if we use energy techniques, we won't need any differential equations.
Student: Okay, I'll try energy, and I'll use potential energy for gravity and the spring.

$$KE_i + PE_i + W = KE_f + PE_f$$

Tutor: What is the kinetic energy of the block just as the spring is released?
Student: It isn't moving yet, so the speed is zero and the kinetic energy is zero.
Tutor: What is the kinetic energy of the block just as it stops?
Student: Wait. Shouldn't I do the kinetic energy just as the block leaves the spring?
Tutor: We could use that instant as the final time, and then use constant acceleration to see how far the block slides. But we can also do the whole thing in one step.
Student: Sounds good. The kinetic energy when the block stops is zero.
Tutor: It's not unusual for two or three of the terms in the energy equation to be zero. What is the potential energy when the block is released?
Student: Do you want the potential energy of gravity or the spring?
Tutor: Both. To do gravity you have to pick a spot as height equals zero.
Student: I'll pick the starting spot as height equals zero. Then the potential energy of gravity is zero. The potential energy of the spring is $\frac{1}{2}kx^2$.
Tutor: What is the potential energy when the block stops?
Student: Then the potential energy of gravity is still zero, because the block is at the same height. The potential energy of the spring is zero, because the spring is uncompressed.

$$\cancel{KE_i} + PE_i + W = \cancel{KE_f} + PE_f$$
$$\frac{1}{2}kx^2 + W = 0$$

Tutor: W is the work done by all forces other than gravity and the spring. What other forces are there?
Student: There is a normal force upward, and a friction force against the sliding.
Tutor: How much work does the normal force do?
Student: The normal force is perpendicular to the motion, so it doesn't do any work.
Tutor: How much work does the friction force do?

Student: The work is the force times the displacement times the cosine of the angle. The displacement is what I'm looking for, so it's a variable x.
Tutor: Is the displacement of the block the same as the initial compression of the spring?
Student: Not necessarily. I guess I need a different variable for the displacement. I'll use y. The angle is $180°$, and cosine of the angle is -1, because the friction force is in the opposite direction as the sliding.
Tutor: Good. What is the friction force.
Student: Isn't it just μmg?
Tutor: Don't skip steps now. How do you find it?
Student: I need the normal force, so I add vertical forces. The normal force is up and gravity is down, and the vertical acceleration is zero, so the normal force *is* equal to the weight, and the friction force is μmg.
Tutor: This is the special case where there are exactly two forces and they need to add to zero. We've seen before that the normal force is not always mg.
Student: Okay, so I can look for the special case of two forces and zero acceleration.

$$\frac{1}{2}kx^2 + (\mu mg)(y)(-1) = 0$$

$$y = \frac{kx^2}{2\mu mg} = \frac{(900\text{ N/m})(0.31\text{ m})^2}{2(0.46)(6\text{ kg})(9.8\text{ m/s}^2)} = 1.60\text{ m}$$

Tutor: By setting the final spring energy to zero, you assumed that the spring would fully uncompress. If friction were really strong, then the spring wouldn't get the block moving.
Student: How do we check?
Tutor: The distance the block slid is way past where the spring uncompresses, so we're safe. We should have checked first that the spring force was greater than the maximum static friction force.

EXAMPLE

A 2 kg block slides without friction down a track. It leaves the track at a $35°$ angle above horizontal at a point 1.3 m below its starting point. After leaving the track, how far up does the block fly?

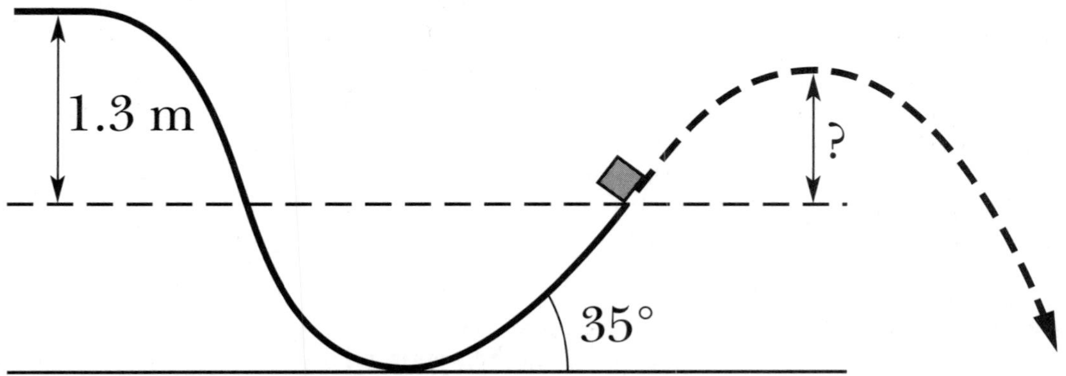

Tutor: How will you attack this problem?
Student: It's the energy chapter, so I'll use energy.
Tutor: If you weren't reading the problem in a book, but instead on a test or somewhere else, why would you use energy?
Student: Because the acceleration isn't constant, and we don't care about the time.

$$KE_i + PE_i + W = KE_f + PE_f$$

Tutor: Good. What is the kinetic energy when the block starts?
Student: It doesn't appear to be moving yet, so zero.
Tutor: What is the potential energy when the block starts?
Student: I need to pick somewhere as height equals zero. Where's the bottom of the ramp?
Tutor: You don't need to use the lowest point as height equals zero. Try using the top.
Student: Okay, the block starts at $h = 0$, so $mgh = 0$.
Tutor: What is the potential energy at the "final" position?
Student: At the maximum height, it isn't moving, so $KE_f = 0$.
Tutor: We have two issues here. The block leaves the ramp at an angle, with some horizontal velocity, and it will still have that horizontal velocity when it reaches maximum height. Therefore $KE_f \neq 0$. You need to rethink what your "final" position is, and how you are going to get to the answer.
Student: So what I need to do is find the speed of the block as it leaves the ramp, then use projectile motion and constant acceleration to see how high it goes.
Tutor: And where is your "final" position?
Student: Where the block leaves the ramp. The potential energy there is $mg(-1.3\text{ m})$. Can potential energy be negative?
Tutor: Potential energy is how much work you need to do to move the object from "zero" to that spot. A negative potential energy means that the object will go there on its own, that you could even get work out of the process. The block will slide down the ramp on its own, gaining speed. It's possible to extract that energy later.

$$\cancel{KE_i} + \cancel{PE_i} + W = KE_f + mg(-1.3\text{ m})$$

Tutor: What is the final kinetic energy?
Student: $\frac{1}{2}mv_f^2$. We don't know v, but we hope to solve for it.
Tutor: What is the work done by all forces *other* than gravity?
Student: Why not gravity?
Tutor: Because we already took care of gravity by using potential energy.
Student: Right. The other forces are the normal force and friction. The normal force is perpendicular to the motion, so it doesn't do any work. And there's no friction on the track.

$$\cancel{KE_i} + \cancel{PE_i} + \cancel{W} = \frac{1}{2}mv^2 + mg(-1.3\text{ m})$$

$$\frac{1}{2}\cancel{m}v^2 = \cancel{m}g(1.3\text{ m})$$

$$v = \sqrt{(2)(9.8\text{ m/s}^2)(1.3\text{ m})} = 5.05\text{ m/s}$$

Student: When the block leaves the track it is moving at 5.05 m/s.
Tutor: How much of that is in the horizontal direction?
Student: The horizontal is adjacent to the angle, so

$$v_x = (5.05\text{ m/s})\cos 35° = 4.13\text{ m/s}$$

$$v_y = (5.05\text{ m/s})\sin 35° = 2.90\text{ m/s}$$

Tutor: How fast is the block going when it reaches maximum height?
Student: You mentioned that it would still be moving horizontally. At maximum height it is going 4.13 m/s horizontally and it's not moving vertically.
Tutor: And because we don't care about the time, we can use energy to see how high it goes.
Student: But the acceleration is constant, so we don't need to use energy.
Tutor: But we *can* use energy, even if we have a constant acceleration. If the acceleration isn't constant, then we may *need* to use energy.
Student: Okay. Taking the start when it leaves the track, calling that the new height equals zero, and the finish when it reaches maximum height

$$KE_i + PE_i + W = KE_f + PE_f$$

$$\frac{1}{2}\cancel{m}(5.05\text{ m/s})^2 + \cancel{m}g(0) + (0) = \frac{1}{2}\cancel{m}(4.13\text{ m/s})^2 + \cancel{m}gh$$

$$\frac{1}{2}(5.05\text{ m/s})^2 + = \frac{1}{2}(4.13\text{ m/s})^2 + gh$$

$$h = \frac{1}{2(9.8\text{ m/s}^2)}\left((5.05\text{ m/s})^2 - (4.13\text{ m/s})^2\right) = 0.43\text{ m}$$

EXAMPLE

A 65 kg man bungee-jumps off of a platform. The bungee cord is 20 m long and has a spring constant of 60 N/m. What is the greatest distance below the platform that he reaches after leaping from the platform?

Student: The bungee cord acts like a spring, so the acceleration won't be constant and we'll use conservation of energy.
Tutor: Good. First you need to pick "initial" and "final" positions.
Student: The start should be at the top, when he first jumps. The end is where he reaches his lowest point.
Tutor: If we want to know where his lowest point is, then it helps if that is one of the points. What are you going to use as height equals zero?
Student: I'm going to use the top, where he jumps from. That means that his potential energy at the end will be negative, but that's okay.
Tutor: You mean the potential energy from gravity; there's also the spring involved. What is the kinetic energy at the start?
Student: He isn't moving yet, so it's zero.
Tutor: What is his kinetic energy at the end?
Student: The lowest point is when he isn't moving, so that's also zero.

$$\cancel{KE_i} + PE_i + W = \cancel{KE_f} + PE_f$$

Tutor: Correct. If he were still going down, he wouldn't be at the lowest point yet. If he were going up, he would be past it. What is the potential energy at the top?
Student: The spring hasn't been stretched yet, and he is at height equals zero, so both types of potential energy are zero.

$$\cancel{PE_i} + W = PE_f$$

Tutor: What is the work done by forces other than gravity and the bungee cord?
Student: There aren't any other forces, because he's not in contact with anything, right?
Tutor: Right, so the work is zero.

$$\cancel{W} = PE_f$$

Student: That means that PE_f is zero. How can that be?
Tutor: There are two types of potential energy, and at the end they must add up to zero. What is the potential energy at the end?
Student: We don't know how far he's fallen, so let's call it x. The height at the end is negative, so the gravity potential energy is $mg(-x)$. The spring potential energy is $\frac{1}{2}kx^2$.
Tutor: Does the bungee cord stretch the same distance that he falls? Does it begin stretching immediately?
Student: No, he falls 20 m before it starts stretching. The bungee cord stretches $x - (20\text{ m})$.

$$0 = [mg(-x)] + \left[\frac{1}{2}k\left(x - (20\text{ m})\right)^2\right]$$

$$mg(x) = \frac{1}{2}k\Big(x - (20\ \text{m})\Big)^2$$

$$\frac{2mgx}{k} = x^2 - 2(20\ \text{m})x + (20\ \text{m})^2$$

$$x^2 + \left[-2(20\ \text{m}) - \frac{2mg}{k}\right] x + (20\ \text{m})^2 = 0$$

$$x^2 + \left[-2(20\ \text{m}) - \frac{2(65\ \text{kg})(9.8\ \text{m/s}^2)}{(60\ \text{N/m})}\right] x + (20\ \text{m})^2 = 0$$

$$x^2 + (-61.2\ \text{m})\,x + (400\ \text{m}^2) = 0$$

$$x = \frac{-b \pm \sqrt{b^2 - 4ac}}{2a} = \frac{-(-61.2\ \text{m}) \pm \sqrt{(-61.2\ \text{m})^2 - 4(1)(400\ \text{m}^2)}}{2(1)}$$

$$x = 7.4 \text{ or } 53.8\ \text{m}$$

Student: How can there be two answers?
Tutor: Which one do you want?
Student: The bungee cord is 20 m long, so it has to be 53.8 m.
Tutor: The other answer would occur if the bungee cord could be compressed.
Student: So if he bounced back up, and the bungee cord compressed instead of going slack, then he would reach a spot 7.4 meters high.

Chapter 9

Center of Mass and Linear Momentum

This chapter is about a second conservation law, conservation of momentum. In order to explain momentum, we begin by explaining center of mass.

So far, all objects have been "point" particles. They act as if all of their mass is at a single point. Most real objects are not like this — their mass is spread out. For many purposes it is possible to treat objects as point particles — as if all of their mass is at a single point. When this happens, that point is the center of mass.

To calculate the location of the center of mass, we do a "weighted average." Divide the object into pieces, and for each piece, multiple the mass of that piece by its coordinate. Sum the product for all of the pieces, and divide by the total mass.

$$x_{\text{CM}} = \frac{\Sigma_i x_i m_i}{\Sigma_i m_i}$$

Some things to keep in mind:

- You get to decide how to divide the pieces.
- For each piece, you use the coordinate of the center of mass of that piece.
- You get to choose the coordinate axes, but as always you have to keep the same axes for the whole calculation.
- Wise use of symmetry can simplify the calculation considerably.

How do you use symmetry? Consider a uniform square. "Uniform" means that the mass of the square is spread evenly across the square, the same everywhere. Pick a point, not the center, and let's use symmetry to check to see if that point is the center of mass. Draw an imaginary line vertically down the center of the square, and rotate the square 180° around this line. The square looks the same, so the center of mass *must* be in the same place as it was before. If the center of mass had moved, then we would have two centers of mass for the same object. If the point you picked was on the vertical line, then it is in the same place as it was and it could be the center of mass. If the point you picked was not on the vertical line, then it is not in the same place as it was, and it could not be the center of mass. By using symmetry, we have determined that the x coordinate of the center of mass is in the middle, without doing a calculation.

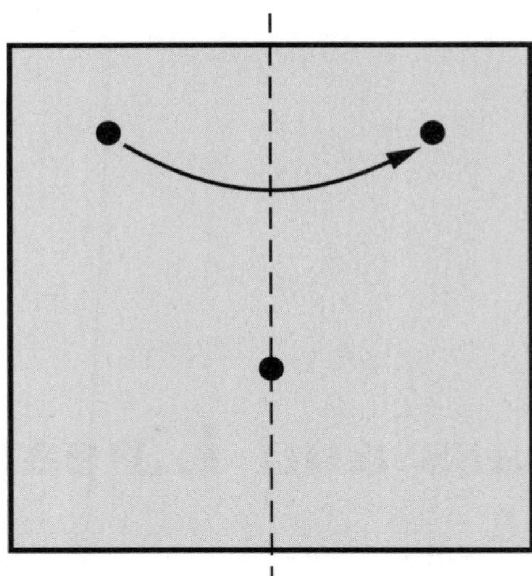

EXAMPLE

Find the center of mass of the four objects.

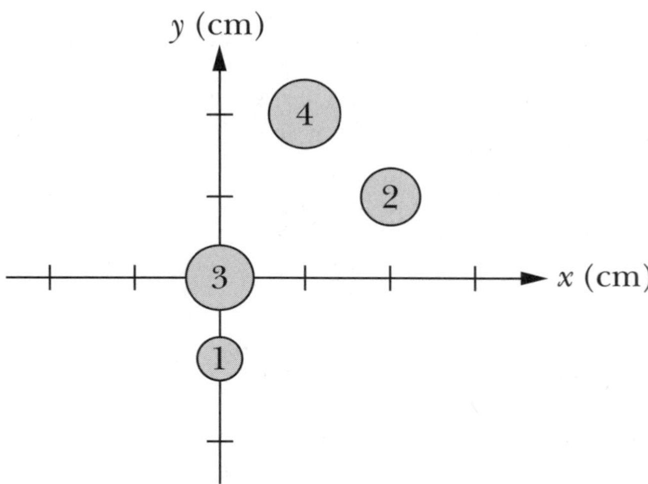

Student: So I take each object and add it up?

$$1 + 3 + 4 + 2 = 10 \quad ?$$

Tutor: For the denominator, you do that. In the numerator you multiply each mass by its coordinate.
Student: Where are the axes?
Tutor: You get to pick them. What would you like?
Student: I'll pick the biggest mass to be the origin, with x to the right and y up.
Tutor: Because you have two axes, you have to do them separately. First do the x axis and then do the y axis.
Student: So I multiply the mass of 3 by the x coordinate of 0, and I get zero?
Tutor: Yes. If that was the only mass, then the numerator would be zero, so the center of mass would be at zero.

Student: Which is where the object is. But I have to do that for all of them.

$$x_{\text{CM}} = \frac{(1\text{ kg})(0\text{ cm}) + (3\text{ kg})(0\text{ cm}) + (4\text{ kg})(1\text{ cm}) + (2\text{ kg})(2\text{ cm})}{(1\text{ kg}) + (3\text{ kg}) + (4\text{ kg}) + (2\text{ kg})} = \frac{8\text{ cm kg}}{10\text{ kg}} = 0.8\text{ cm}$$

Tutor: Now do the y coordinate.
Student: For the 1 kg object, do I use +1 or −1 m?
Tutor: If the object were at (0,+1), would the location of the center of mass be different?
Student: Yes, so I need the negative.

$$y_{\text{CM}} = \frac{(1\text{ kg})(-1\text{ cm}) + (3\text{ kg})(0\text{ cm}) + (4\text{ kg})(2\text{ cm}) + (2\text{ kg})(1\text{ cm})}{(1\text{ kg}) + (3\text{ kg}) + (4\text{ kg}) + (2\text{ kg})} = \frac{9\text{ cm kg}}{10\text{ kg}} = 0.9\text{ cm}$$

Tutor: Which object is located at the center of mass?
Student: None of them.
Tutor: The center of mass doesn't have to correspond to one of the objects.

EXAMPLE

Find the center of mass of the uniform triangle.

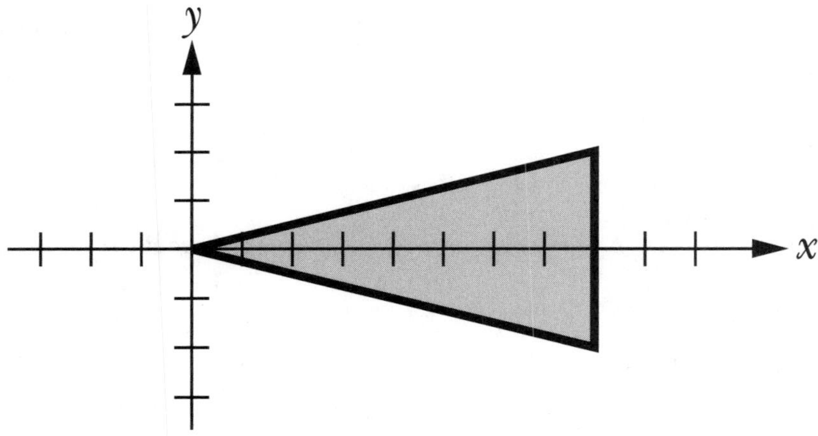

Student: It's just one piece! How do I do the equation for each piece when there's only one piece?
Tutor: Can you use symmetry to find the center of mass?
Student: If I flip it top-to-bottom, it looks the same, so the center of mass has to be on the horizontal axis.
Tutor: What if you flip it left-to-right?
Student: It doesn't look the same. That means that the center of mass isn't halfway along the horizontal axis.
Tutor: It means that the center of mass doesn't *have* to be there. It could be there but there's no reason to think that it is.
Student: So the y component of the center of mass is at $y = 0$, but we don't know where the x component is, and it's only one piece.
Tutor: You need to divide it into pieces yourself, where you can determine the mass and location of each piece.

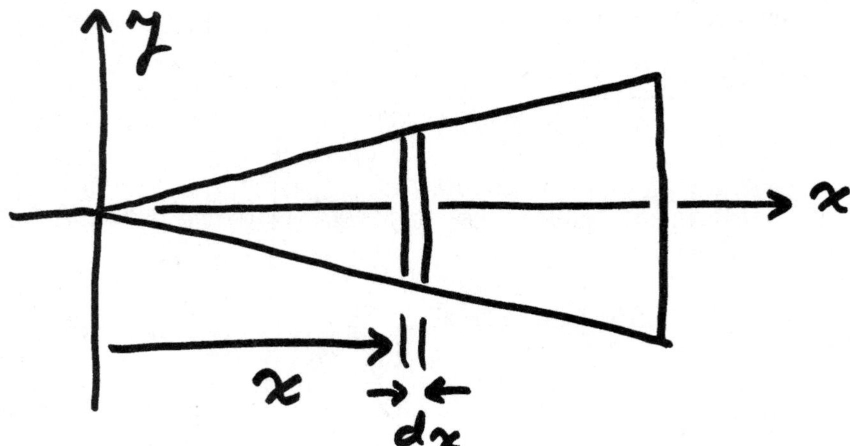

Tutor: Divide it into a lot of pieces, and then each piece is small enough that we can call it a rectangle. Then we add up all of the pieces.
Student: That sounds like an integral!
Tutor: Yep. What is the width of each piece?
Student: Each piece is dx wide.
Tutor: Yes. What is the height of each piece?
Student: They aren't all the same height.
Tutor: That's right, they aren't. Given that a piece is located at x, how high is it?
Student: Well, the bigger x is, the greater the height of the piece.
Tutor: Yes, also it's linear, so that the height is proportional to x. At $x = 0$, the height is zero, and at $x = 8$, the height is 2.
Student: If it's linear, then $h = Cx$, where C is a constant. If the height is 2 at $x = 8$, then that constant needs to be $\frac{1}{4}$, so the height of each piece is $\frac{1}{4}x$.
Tutor: Good. To find the mass of a piece, we multiply the area times the density. That is, we multiply the square meters by the kilograms per square meters, and get kilograms.
Student: What is this density? It isn't given.
Tutor: True. Hopefully it will cancel, or we won't be able to do the problem. Make up a variable for it. Physicists tend to use δ or ρ for densities.
Student: Okay, the density is δ. Then the mass of a piece is the area times the density, or $(\frac{1}{4}x \times dx)(\delta)$.

$$x_{\text{CM}} = \frac{\int_0^8 (\frac{1}{4}x \times dx)(\delta) \cdot x}{\int_0^8 (\frac{1}{4}x \times dx)(\delta)} = \frac{\frac{1}{4}\delta \int_0^8 x^2\, dx}{\frac{1}{4}\delta \int_0^8 x\, dx} = \frac{\int_0^8 x^2\, dx}{\int_0^8 x\, dx} = \frac{\left[\frac{1}{3}x^3\right]_0^8}{\left[\frac{1}{2}x^2\right]_0^8} = \frac{\left[\frac{1}{3}(8)^3\right]}{\left[\frac{1}{2}(8)^2\right]} = \frac{\left[\frac{1}{3}(8)\right]}{\left[\frac{1}{2}\right]} = \frac{16}{3}$$

Tutor: Does the result make sense?
Student: How should I know?
Tutor: Well, could the answer be more than 8?
Student: No, all of the mass is to the left of 8.
Tutor: Could the answer be less than 4?
Student: No, more of the mass is on the right side, and further to the right.
Tutor: Is the answer between 4 and 8?
Student: Yes, it's about 5. So it works.
Tutor: At least it passes the sanity test.

Just as energy is conserved, so is momentum. When a large truck collides head-on with a small car, we expect to see the car bounce backwards while the truck continues on. This is not because the force of the truck on the car is larger than the force of the car on the truck — the forces are the same. Instead, this is because the truck has more momentum than the car.

Conservation of momentum looks very similar to conservation of energy. The equations have the same general form and the techniques are the same. Momentum is the product of mass and velocity.

$$\vec{P} = m\vec{v}$$

Impulse is the name given to the change in the momentum, which thus has the same units as momentum (kg m/s or N s). The basic equation is

$$\vec{P_i} + \vec{J} = \vec{P_f}$$

$$\vec{J} = \vec{F}t$$

where $\vec{J}$ is the "impulse," or change in momentum.

One difference between momentum and energy is that momentum is a vector, so it has direction. In the car – truck head-on collision, one of the vehicles has a positive momentum and the other has a negative. The truck has more momentum, so the total momentum is in the direction that the truck is going. After the collision, the total momentum is also in the direction that the truck was going.

When we find the momentum, we can either calculate it for one object or for a group of objects, often called a "system." Consider a system of two objects. If they create a force on each other, then the forces are equal and opposite. So the impulses that they exert on each other are equal and opposite. Whatever that impulse is, the impulse of A on B plus the impulse of B on A add to zero. So any forces between the two bodies of our system don't change the momentum of the system. (The center of mass of the system keeps moving with the same velocity.) The result of this logic is that **conservation of momentum works particularly well for collisions**.

EXAMPLE

A 2500 kg train car moves down the track at 5 m/s. It hits and couples to a second (stationary) train car of 3500 kg. What is the speed of the train cars afterward, and what was the impulse that the second car gave to the first one?

Tutor: How are you going to solve this problem?
Student: There's a collision, which is a tip-off to try conservation of momentum.
Tutor: Good. Are you going to use the momentum of one of the train cars or of the two together?
Student: What's the advantage of doing the two together?
Tutor: If the train cars exert forces on each other, these forces will cancel if we do them together. Do they exert forces on each other?
Student: They certainly do. So I want to do them together.

$$\vec{P_i} + \vec{J} = \vec{P_f}$$

Tutor: What forces act on the train cars?
Student: They push on each other, of course. Also, there is gravity and a normal force on each. These might be equal.
Tutor: We could try to figure out if they're equal. Instead, what are the directions of gravity and the normal force?
Student: They are both vertical.
Tutor: So if we only consider the horizontal component of the momentum, then we don't have to worry about them at all.
Student: Cool. All we need is the forces they create on each other.
Tutor: Correct, and those forces are equal, so they add to zero.

$$P_{1i} + P_{2i} + (\cancel{F_{1\text{ on }2}t} - \cancel{F_{2\text{ on }1}t}) = P_{1f} + P_{2f}$$

$$P_{1i} + P_{2i} = P_{1f} + P_{2f}$$

Tutor: What is the momentum of the first car before the collision?
Student: The momentum is mv, so (2500 kg)(5 m/s).
Tutor: What is the momentum of the second car before the collision?
Student: It wasn't moving, so its momentum is zero.
Tutor: What is the momentum of the first car after the collision?
Student: Momentum is mv, but we don't know the velocity. I'll use v_1 for the velocity of the train car afterward.

$$(2500 \text{ kg})(5 \text{ m/s}) + 0 = (2500 \text{ kg})v_1 + P_{2f}$$

Tutor: What is the momentum of the second car after the collision?
Student: We don't know its velocity either, so I'll use v_2 for that.
Tutor: The two train cars couple together. Does that mean anything?
Student: They have to have the same velocity afterward?
Tutor: Yes.
Student: So $v_2 = v_1$, and I only have one variable.

$$(2500 \text{ kg})(5 \text{ m/s}) = (2500 \text{ kg})v_1 + (3500 \text{ kg})v_1$$
$$(12500 \text{ kg m/s}) = (6000 \text{ kg})v_1$$
$$v_1 = \frac{(12500 \text{ kg m/s})}{(6000 \text{ kg})} = 2.08 \text{ m/s}$$

Tutor: Can you tell from your result which way the train cars go?
Student: Well, if the first car is originally moving to the right, shouldn't they move off to the right?
Tutor: Yes. Can you express that same thought, using the word "momentum?"
Student: If the initial momentum is to the right, and there is no impulse from outside forces, then the final momentum is to the right.
Tutor: Very good. Do your numbers say the same thing? Keep in mind that you used an axis but you didn't explicitly say what you were using.
Student: Oops. So when I said the velocity of the first car beforehand was positive, I picked that as the positive direction?
Tutor: Yep. And when the final velocity was positive...
Student: Then the train cars move in that same direction. How do I find the impulse?
Tutor: Repeat the conservation of momentum equation, but for just one car.
Student: So using the first car

$$P_{1i} + J_{2 \text{ on } 1} = P_{1f}$$
$$(2500 \text{ kg})(5 \text{ m/s}) + J_{2 \text{ on } 1} = (2500 \text{ kg})(2.08 \text{ m/s})$$
$$J_{2 \text{ on } 1} = -7292 \text{ kg m/s}$$

Student: Why is it negative?
Tutor: Which way is the force that the second car exerts on the first?
Student: Against the initial velocity, so it should be negative.
Tutor: And $J_{1 \text{ on } 2}$ is positive, so that the final momentum of the second car is positive.

EXAMPLE

A 72 lb boy on a skateboard throws a 16 lb bowling ball to the right at a speed of 8 m/s compared to him. What is the boy's velocity after he throws the ball?

Student: This doesn't sound like a "collision." Why do I want to use momentum for this?
Tutor: Do you know the force that the boy applies to the bowling ball?

Student: No, do I need to know it?
Tutor: By using conservation of momentum, you can eliminate forces that two objects put on each other. You won't need to find the force. It's true that momentum works well for collisions, but that's because using momentum gets the forces between colliding objects to cancel.
Student: So I want to use the combined momentum of the two objects, boy and bowling ball.

$$\vec{P}_i + \vec{J} = \vec{P}_f$$

Tutor: Correct. What are the forces acting on them?
Student: The boy pushes on the bowling ball, and the bowling ball pushes back. These forces are opposite and equal, so they cancel. There is gravity and a normal force on the boy. These are vertical, so we do only horizontal momentum and they disappear.

$$P_{1i} + P_{2i} = P_{1f} + P_{2f}$$

Tutor: What is the momentum of the boy beforehand?
Student: He isn't moving, so it's zero. The bowling ball isn't moving either.
Tutor: What is the momentum of the boy afterward?
Student: Momentum is mv. We don't know his velocity, so I'll use a variable.

$$0 + 0 = m_{\text{boy}} v_{\text{boy}} + m_{\text{ball}} v_{\text{ball}}$$
$$0 + 0 = (72 \text{ lb}) v_{\text{boy}} + (16 \text{ lb}) v_{\text{ball}}$$

Student: Do I need to convert pounds into kilograms?
Tutor: You can keep them as they are for now, and see how they work out. What is the velocity of the bowling ball after he throws it?
Student: 8 m/s to the right. I'll use to the right as the positive direction, so it's +8 m/s.
Tutor: The 8 m/s is with respect to the boy, or as measured by the boy.
Student: Ah, so I need to find the velocity compared to the ground.

$$v_{\text{AB}} = v_{\text{AC}} + v_{\text{CB}}$$
$$v_{\text{boy,gr}} = v_{\text{boy,ball}} + v_{\text{ball,gr}}$$
$$v_{\text{ball,boy}} = -v_{\text{boy,ball}}$$
$$v_{\text{boy,gr}} = -v_{\text{ball,boy}} + v_{\text{ball,gr}}$$
$$v_{\text{boy}} = -(8 \text{ m/s}) + v_{\text{ball}}$$
$$0 = (72 \text{ lb}) v_{\text{boy}} + (16 \text{ lb}) \Big(v_{\text{boy}} + (8 \text{ m/s}) \Big)$$

Tutor: Now we have an equation with a single unknown.
Student: So we can solve it.

$$0 = (72 \text{ lb}) v_{\text{boy}} + (16 \text{ lb}) v_{\text{boy}} + (16 \text{ lb})(8 \text{ m/s})$$
$$(88 \text{ lb}) v_{\text{boy}} = -(16 \text{ lb})(8 \text{ m/s})$$
$$v_{\text{boy}} = -\frac{(16 \cancel{\text{lb}})(8 \text{ m/s})}{(88 \cancel{\text{lb}})} = -1.45 \text{ m/s}$$

Tutor: What does the negative result mean?
Student: It means that the boy goes in the negative direction, so he goes to the left.
Tutor: Does that make sense? Remember that the initial momentum was zero.
Student: And the impulse was zero, so the final momentum needs to be zero. If the bowling ball goes to the right, then the boy needs to go to the left.

In some collisions, the kinetic energy is conserved. When two billiard balls strike, the total kinetic energy before the collision is the same as after the collision. When two cars strike, some of the kinetic energy goes into other forms of energy, such as heat or denting the cars. **An elastic collision is one where the kinetic energy is conserved**. An inelastic collision is one where some of the kinetic energy is turned into other forms of energy.

Because elastic collisions are somewhat common, and because of the unpleasantness of solving simultaneous equations (especially with squares in them), physicists do something they don't often do — work out the result and reuse it the next time. If object 1 collides elastically with object 2, then their velocities afterward will be

$$v_1' = \frac{m_1 - m_2}{m_1 + m_2} v_1 + \frac{2m_2}{m_1 + m_2} v_2$$

$$v_2' = \frac{m_2 - m_1}{m_1 + m_2} v_2 + \frac{2m_1}{m_1 + m_2} v_1$$

or, if object 2 is stationary before the collision,

$$v_1' = \frac{m_1 - m_2}{m_1 + m_2} v_1 \quad \text{and} \quad v_2' = \frac{2m_1}{m_1 + m_2} v_1$$

To use these equations, the **positive direction for each object must be the same**.

How do you tell if a collision is elastic or not?

- If the problem says that the collision is elastic, then the collision is elastic (really).
- If the two objects stick together, then the collision is not elastic.
- If one object passes through the other, then the collision is not elastic.
- If you can solve the problem using only conservation of momentum, then there is no need to worry whether the collision is elastic.
- After that, use your experience with the objects in question: are they likely to bounce off of each other (elastic) or is some of the energy likely to be turned into heat (not elastic)?

EXAMPLE

A 4000 kg truck is coming straight at Joe at 14 m/s. He throws a 0.3 kg rubber ball at the truck with a speed of 10 m/s, and it bounces off of the truck elastically. After the ball bounces off of the truck, how fast is the ball going?

Tutor: How are you going to attack this problem?
Student: There's a collision, so I'm going to use momentum. The ball and the truck put forces on each other, but if I use the momentum of them together those forces cancel out. All of the other forces are vertical, and by looking at the horizontal component, they go away.

$$P_{1i} + P_{2i} = P_{1f} + P_{2f}$$

$$m_{\text{ball}} v_{\text{ball}} + m_{\text{truck}} v_{\text{truck}} = m_{\text{ball}} v_{\text{ball}}' + m_{\text{truck}} v_{\text{truck}}'$$

$$(0.3\text{ kg})(10\text{ m/s}) + (4000\text{ kg})(14\text{ m/s}) = (0.3\text{ kg}) v_{\text{ball}}' + (4000\text{ kg}) v_{\text{truck}}' \quad ?$$

Tutor: Before the collision, are the ball and the truck going the same direction?
Student: No, they have a head-on collision.

Tutor: So one of them has to have a negative initial velocity.
Student: I forgot to choose an axis! I'll take the direction that the ball is thrown as positive.

$$(0.3\text{ kg})(10\text{ m/s}) + (4000\text{ kg})(-14\text{ m/s}) = (0.3\text{ kg})v'_{\text{ball}} + (4000\text{ kg})v'_{\text{truck}}$$

Tutor: Can you solve the equation?
Student: I guess not; I have one equation and two variables. I need another piece of information.
Tutor: The piece you are missing is that the collision is elastic.
Student: That means that I can write down the kinetic energy before and after, set them equal, and get another equation.
Tutor: Yes, that is what it means. To avoid all of that work, you could just use the elastic collision equations.
Student: Yes, that would be easier.

$$v'_1 = \frac{m_1 - m_2}{m_1 + m_2}v_1 + \frac{2m_2}{m_1 + m_2}v_2$$

$$v'_{\text{ball}} = \frac{m_{\text{ball}} - m_{\text{truck}}}{m_{\text{ball}} + m_{\text{truck}}}v_{\text{ball}} + \frac{2m_{\text{truck}}}{m_{\text{ball}} + m_{\text{truck}}}v_{\text{truck}}$$

$$v'_{\text{ball}} = \frac{(0.3\text{ kg}) - (4000\text{ kg})}{(0.3\text{ kg}) + (4000\text{ kg})}(10\text{ m/s}) + \frac{2(4000\text{ kg})}{(0.3\text{ kg}) + (4000\text{ kg})}(-14\text{ m/s})$$

$$v'_{\text{ball}} = \frac{(-3999.7\text{ kg})}{(4000.3\text{ kg})}(10\text{ m/s}) + \frac{(8000\text{ kg})}{(4000.3\text{ kg})}(-14\text{ m/s})$$

$$v'_{\text{ball}} = \frac{(-3999.7\text{ kg})}{(4000.3\text{ kg})}(10\text{ m/s}) + \frac{(8000\text{ kg})}{(4000.3\text{ kg})}(-14\text{ m/s})$$

$$v'_{\text{ball}} = -37.996\text{ m/s}$$

Tutor: What does the minus sign mean?
Student: It means that the ball bounces in the direction that the truck was originally going. Why is the speed so much faster than before?
Tutor: Ah, it's dangerous to throw a rubber ball at an oncoming truck. Look at the problem from the truck's point of view. What does the truck driver see?
Student: He sees a ball coming at him at, oh, 24 m/s.
Tutor: Yes, and how much does the ball change the truck's momentum?
Student: None at all.
Tutor: If the truck puts a force on the ball, then the...
Student: ...ball puts a force on the truck, equal but opposite. The ball accelerates more because it has less mass. The truck's momentum must change, but not very much.
Tutor: Correct. So he sees the ball bounce off at about 24 m/s. If he's coming at you at 14 m/s and he thinks the ball is coming at you at 24 m/s, then what do you see?
Student: I see the ball coming at me at 38 m/s.
Tutor: Mathematically what we do is switch everything to the "center of mass" frame of reference. The truck's frame is almost the center of mass frame here. The idea here is the same one used to "slingshot" space probes around planets to gain speed.

EXAMPLE

A 0.85 kg ball on the end of a 35–cm–long string falls and hits a 0.56 kg block of putty. The block sticks to the ball. What is the speed of the putty after it is hit by the ball?

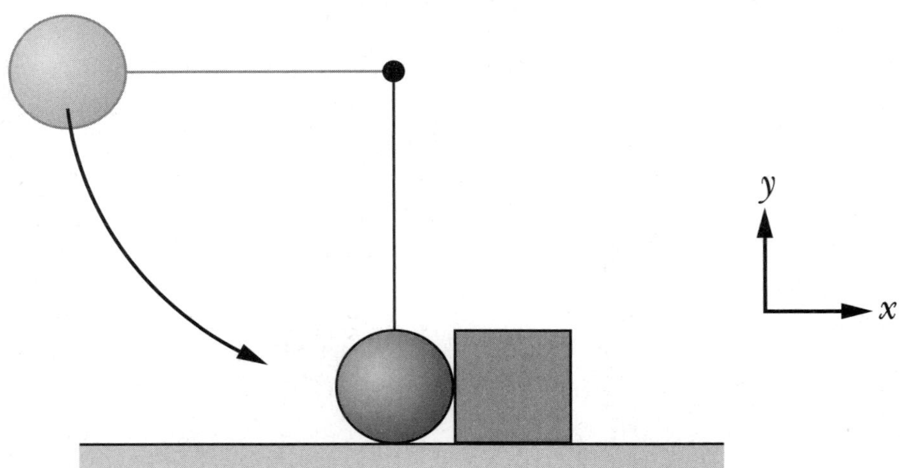

Student: There's a collision, so we need conservation of momentum. The putty and ball stick together, so it's not an elastic collision. Other than the forces that the ball and the putty put on each other, all other forces are vertical, so I'm going to use horizontal momentum.

$$P_{1i} + P_{2i} = P_{1f} + P_{2f}$$

$$m_{\text{ball}} v_{\text{ball}} + m_{\text{putty}} v_{\text{putty}} = (m_{\text{ball}} + m_{\text{putty}})\, v'_{\text{ball+putty}}$$

Tutor: All well and good. What is the velocity of the putty before the collision?
Student: It isn't moving, so $v_{\text{putty}} = 0$.

$$m_{\text{ball}} v_{\text{ball}} + m_{\text{putty}} \cancel{v_{\text{putty}}} = (m_{\text{ball}} + m_{\text{putty}})\, v'_{\text{ball+putty}}$$

Tutor: What is the velocity of the ball before the collision?
Student: The problem doesn't say. That means I have one equation but two unknowns.
Tutor: What do you know about the ball?
Student: It starts 35 cm high and falls in an arc until it hits the putty.
Tutor: How can we find the speed of the ball after it falls?
Student: Conservation of momentum, I assume?
Tutor: What are the forces on the ball?
Student: Gravity and tension.
Tutor: The tension force changes as the ball falls, both in magnitude and direction, and we need to know the time it takes to fall.
Student: That looks hard.
Tutor: It is hard. How much work does the tension do?
Student: *Work?* I thought work was in the last chapter.
Tutor: It was, but we can still use it. We're allowed to use all of the techniques we've learned. Just because it's the momentum chapter doesn't mean that we can only use momentum.
Student: Okay, the work from the tension is zero because the tension is perpendicular to the motion.

$$KE_{\text{before}} + PE_{\text{before}} + \cancel{W} = KE_{\text{after}} + PE_{\text{after}}$$

$$\frac{1}{2} m_{\text{ball}} (0)^2 + m_{\text{ball}} g h = \frac{1}{2} m_{\text{ball}} (v)^2 + m_{\text{ball}} g (0)$$

$$\cancel{m_{\text{ball}}} gh = \frac{1}{2}\cancel{m_{\text{ball}}}(v)^2$$

$$gR = \frac{1}{2}(v_{\text{ball}})^2$$

$$v_{\text{ball}} = \sqrt{2gR} = \sqrt{2(9.8\ \text{m/s}^2)(0.35\ \text{m})} = 2.62\ \text{m/s}$$

Tutor: Now you have the speed of the ball just before the collision.
Student: I'll choose the direction of the ball as positive, stick this in the momentum equation, and solve for the final velocity.

$$m_{\text{ball}} v_{\text{ball}} = (m_{\text{ball}} + m_{\text{putty}})\, v'_{\text{ball+putty}}$$

$$(0.85\ \text{kg})(2.62\ \text{m/s}) = ((0.85\ \text{kg}) + (0.56\ \text{kg}))\, v'_{\text{ball+putty}}$$

$$(2.23\ \text{kg m/s}) = (1.41\ \text{kg})v'$$

$$v' = \frac{(2.23\ \text{kg m/s})}{(1.41\ \text{kg})} = 1.58\ \text{m/s}$$

Chapter 10

Rotation

Sometimes things move, but sometimes things rotate (and talented objects can do both at the same time). Everything we've done so far also applies for rotation, using the same equations, but with different variables.

x	$\rightarrow$	θ (Greek "theta")	angular displacement
v	$\rightarrow$	ω (Greek "omega")	angular velocity
a	$\rightarrow$	α (Greek "alpha")	angular acceleration
m	$\rightarrow$	I	angular mass = moment of inertia
F	$\rightarrow$	τ (Greek "tau")	angular force = torque
P	$\rightarrow$	L	angular momentum

Every equation we've had will also have an angular equivalent. For example, force equals mass times acceleration ($F = ma$), so angular force equals angular mass times angular acceleration ($\tau = I\alpha$).

An object can have a velocity, which is the change in its position. It can have an acceleration, which is how fast the velocity is changing. It can have an acceleration even if the velocity is zero: throw something (other than this book) straight up, and at the instant that it is at its peak height the velocity is zero but changing, so that there is an acceleration. We've done all this before.

It's all true for rotation as well. An object can have an angular velocity, which is the change in its angular position or angular displacement, or how fast it is spinning. It can have an angular acceleration, which is how fast the angular velocity is changing — is the rotation speeding up or slowing down? It can have an angular acceleration even if the angular velocity is zero. Roll a basketball or other round object up a ramp, and at the instant that it is at its peak the angular velocity is zero but changing: it was rotating one way, is not rotating now, and is about to rotate the other way. There is an angular acceleration.

If the acceleration was constant then we could solve a number of problems with a few short equations. We could solve problems where the acceleration wasn't constant, but that was harder. All of the equations we used for constant acceleration work for constant angular acceleration as well — just replace the variable with the angular equivalent. Just like constant acceleration, if we know any three of the five rotational variables, then we can solve for the other two.

$$
\begin{aligned}
v - v_0 = at \quad &\rightarrow \quad \omega - \omega_0 = \alpha t \\
\Delta x = v_0 t + \tfrac{1}{2} a t^2 \quad &\rightarrow \quad \Delta\theta = \omega_0 t + \tfrac{1}{2}\alpha t^2 \\
\Delta x = \tfrac{1}{2}(v_0 + v)\, t \quad &\rightarrow \quad \Delta\theta = \tfrac{1}{2}(\omega_0 + \omega)\, t \\
v^2 - v_0^2 = 2a\, \Delta x \quad &\rightarrow \quad \omega^2 - \omega_0^2 = 2\alpha\, \Delta\theta
\end{aligned}
$$

We could measure lengths in many different units, such as meters, inches, or miles. We can also measure angles in many different units, with the most common being degrees, radians, or cycles. Once around is one cycle, 360 degrees, or 2π radians. With lengths, it didn't matter what unit we used, so long as we kept track of the units. Sometimes it became necessary to use meters, because a newton is a kilogram meter per second^2. With angles, it doesn't matter which unit we use, so long as we keep track of the units. Sometimes, though, it becomes necessary to use radians (more on this later).

EXAMPLE

A turntable starting from rest rotates 5 times while accelerating to 48 rpm (rotations per minute). What is the angular acceleration of the turntable, in radians/sec^2?

Tutor: What is happening here?
Student: Constant *angular* acceleration.
Tutor: How do we solve constant acceleration problems?
Student: If we know three of the five variables, we can solve for the other two.
Tutor: Start by writing down the five variables.

$$\begin{array}{ccccc} \Delta x & \longrightarrow & \Delta\theta & = & \\ v_0 & \longrightarrow & \omega_0 & = & \\ v & \longrightarrow & \omega & = & \\ a & \longrightarrow & \alpha & = & \\ t & \longrightarrow & t & = & \end{array}$$

Tutor: What is the angular displacement $\Delta\theta$?
Student: 5 rotations, or cycles. Do I need to convert that to radians?
Tutor: Not necessarily. Let's keep it and see what happens. What is the initial angular velocity?
Student: It starts at rest, so zero.
Tutor: What is the final angular velocity?
Student: 48 cycles per minute. We know three, so we can solve.
Tutor: Is it positive 48 or negative 48?
Student: Uh-oh, we didn't choose an axis. We have to do that even in rotation?
Tutor: Yes. Typically we use clockwise or counterclockwise as positive. The problem doesn't say, so the question here is whether the final angular velocity is in the same direction as the angular displacement, and has the same sign.
Student: It has to be in the same direction, so ω is positive.
Tutor: More precisely, it has the same sign as $\Delta\theta$.

$$\begin{array}{ccccl} \Delta x & \longrightarrow & \Delta\theta & = & 5 \text{ rotations} \\ v_0 & \longrightarrow & \omega_0 & = & 0 \\ v & \longrightarrow & \omega & = & 48 \text{ rot/min} \\ a & \longrightarrow & \alpha & = & ? \\ t & \longrightarrow & t & = & \end{array}$$

Tutor: We want to find the angular acceleration, of course. Do we know the time?
Student: No, but that's okay. We don't care about the time, so we choose the equation that doesn't include the time.

$$v^2 - v_0^2 = 2a\ \Delta x \longrightarrow \omega^2 - \omega_0^2 = 2\alpha\ \Delta\theta$$

$$\alpha = \frac{\omega^2 - \omega_0^2}{2\,\Delta\theta}$$
$$\alpha = \frac{(48 \text{ rot/min})^2 - (0)^2}{2(5 \text{ rot})}$$
$$\alpha = 230 \text{ rot/min}^2$$

Student: We wanted the angular acceleration in radians/sec^2.
Tutor: So now we have to convert.
Student: Okay. There are 2π radians in a rotation, and 60 seconds in a minute.

$$\alpha = \left(230 \frac{\text{rot}}{\text{min}^2}\right)\left(\frac{2\pi \text{ rad}}{1 \text{ rot}}\right)\left(\frac{1 \text{ min}}{60 \text{ s}}\right)^2$$
$$\alpha = 0.40 \text{ rad/s}^2$$

Tutor: Should the angular acceleration be positive?
Student: It should be in the same direction as the angular displacement, so it should have the same sign.

Sometimes we need to connect the angular motion with the linear motion. Imagine, for example, that we put a small drop of paint on the wheel before it rotates. We might then ask the question "how fast is the drop of paint moving?" Note that the velocity of the paint is constantly changing, because its direction is changing, but if the angular velocity is constant then the speed of the paint drop is also constant.

The further out from the pivot point (or axle) the paint is, the further it must move in each rotation, and the faster we expect it to move. Since the distance covered is $2\pi R$ for each rotation, the speed is

$$v = 2\pi R \times (\text{rotations per second})$$

By choosing the unit for angles to be 2π radians per rotation, this becomes:

$$\Delta x = R\,\Delta\theta$$
$$v = R\,\omega$$
$$a = R\,\alpha$$

If we had chosen some other unit for measuring angles, then the above formulas would need to include an extra constant; by using radians we get beautiful, simple formulas. This means that we *must* use radians whenever we use one of the formulas above. Sometimes we use one of them without knowing it, so the general rule is that we **use radians whenever there is a length** in the problem — any length.

EXAMPLE

A 42–cm–diameter wheel is rotating with an angular speed of 300 rpm. It is slowing with an angular deceleration of 52 rad/s^2. What is the (linear) acceleration of a point on the edge of the wheel?

Tutor: What is happening in this problem?
Student: The point is moving in a circle. When something moves in a circle, there is an acceleration v^2/r.
Tutor: But above we saw that $a = r\alpha$.
Student: Yes, that must mean that the two are equal, so $v^2/r = r\alpha$, right?
Tutor: But if something moves in a circle at a constant speed, α is zero. Then the acceleration is zero, even though the velocity is changing?
Student: That can't be right. What's going on?
Tutor: As the wheel turns, a point on the edge moves. If the wheel turns faster, then point on the edge has to move faster. As the rotation accelerates, the point accelerates. This acceleration is in the direction

of the motion of the point, or tangent to the circle. The acceleration $a = v^2/r$ was toward the center and made the point go in a circle. The acceleration $a = r\alpha$ makes the point go in a circle *faster*.
Student: So the two are perpendicular to each other?
Tutor: Yes, always.

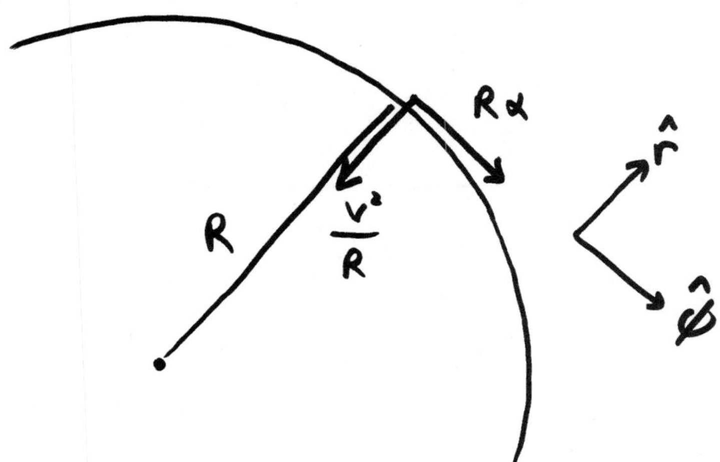

Student: If they are perpendicular to each other, how can they cancel out?
Tutor: They don't. Whenever the wheel is turning, the point on the edge has v^2/r. If the rotation rate is changing, so that the wheel spins faster or slower, then the point on the edge has $r\alpha$ as well.
Student: And to find the total acceleration we have to add them.
Tutor: Remember to add them as vectors.
Student: I see you've draw the coordinate axes. Why not x and y, and why can't the one be in so that the acceleration is positive?
Tutor: Because the point is moving, we rotate the axes so that one of them points along the radius. We call this the "radial" direction, and it is traditional to point it outward, in the radial direction. The other axis is tangent to the circle, and is traditionally called ϕ or sometimes θ.
Student: Okay. The "radial" acceleration is inward, so it's negative.

$$a_{\mathrm{R}} = -\frac{v^2}{r}$$

Tutor: We can substitute $v = r\omega$.

$$a_{\mathrm{R}} = -\frac{(r\omega)^2}{r} = -r\omega^2$$

Student: That's easier, because we have ω.

$$a_{\mathrm{R}} = -\left((21\ \mathrm{cm})\left(\frac{1\ \mathrm{m}}{100\ \mathrm{cm}}\right)\right)\left((300\ \mathrm{rot/min})\left(\frac{2\pi\ \mathrm{rad}}{1\ \mathrm{rot}}\right)\left(\frac{1\ \mathrm{min}}{60\ \mathrm{s}}\right)\right)^2$$

$$a_{\mathrm{R}} = -33\ \mathrm{m/s^2}$$

Student: Now we can do the "tangential" acceleration.

$$a_{\mathrm{T}} = R\alpha$$

$$a_{\mathrm{T}} = (0.21\ \mathrm{m})\,(52\ \mathrm{rad/s^2})$$

$$a_{\mathrm{T}} = 11\ \mathrm{m/s^2}$$

Tutor: We can express the acceleration vector using unit vectors.

$$\vec{a} = a_{\rm R}\vec{\hat{r}} + a_{\rm T}\vec{\hat{\phi}}$$

$$\vec{a} = (-33\ {\rm m/s^2})\vec{\hat{r}} + (11\ {\rm m/s^2})\vec{\hat{\phi}}$$

Student: The radial and tangential parts are perpendicular to each other. To find the magnitude of the vector I use the Pythagorean theorem.

$$|\vec{a}| = \sqrt{(-33\ {\rm m/s^2})^2 + (11\ {\rm m/s^2})^2}$$

$$|\vec{a}| = 35\ {\rm m/s^2}$$

The angular equivalent of force is angular force, called torque. Try opening a door by pushing near the hinge rather than at the handle. You will find that the door does not open as easily or quickly, or perhaps even at all. The further away from the pivot point the force is applied the greater the torque.

The torque a force creates is equal to the force times the moment arm times the sine of the angle between them:

$$\tau = FR\sin\phi$$

The "moment arm" is the line from the pivot point to the spot where the force is applied. If the force is parallel or antiparallel to the moment arm, it does not cause the object to spin and the torque is zero. There are variations on this method, and we'll explore them in the example below.

Torques can be positive or negative. Like velocities and accelerations, we get to pick the positive direction and a negative torque is one in the other direction. Once we choose the positive direction we need to stick with that direction for the whole problem. Physicists usually choose counterclockwise as the positive direction, but it's not necessary to do that.

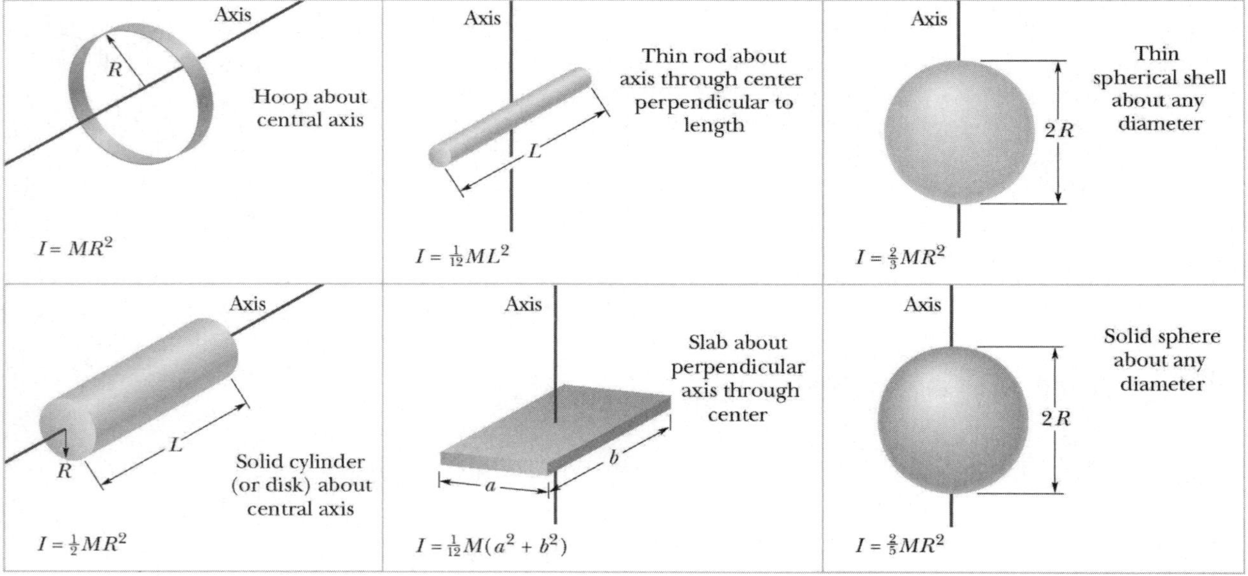

EXAMPLE

Find the total torque applied to the shape around the pivot point.

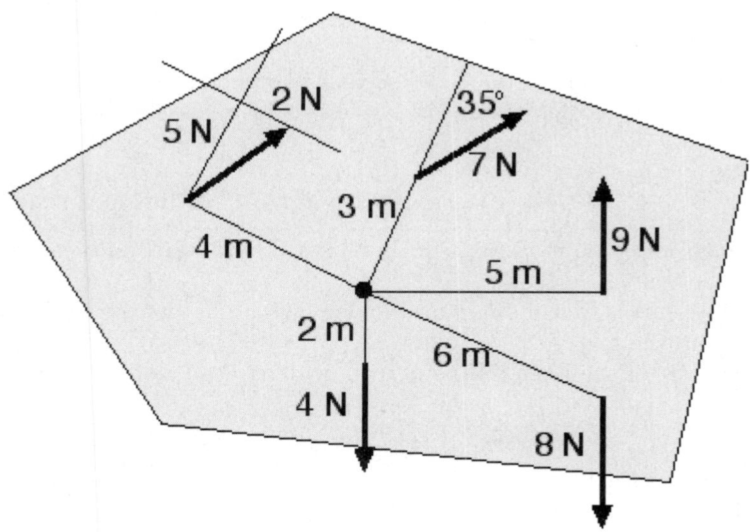

Student: How do we do torques?
Tutor: We find each torque, and add them like we would forces.
Student: Does that mean we need to find the components of each torque?
Tutor: In advanced problems that is needed, but for most problems we just have clockwise and counterclockwise. What is the torque that the 9 N force creates around the pivot point?
Student: There is a force of 9 N, it is applied 5 m from the pivot point, and there is a right angle between them.

$$\tau = FR\sin\phi = (9 \text{ N})(5 \text{ m})\sin 90^\circ = 45 \text{ N}\cdot\text{m}$$

Tutor: Is that torque in the clockwise or counterclockwise direction?
Student: How do I tell?
Tutor: Start at the force arrow and trace a circle that goes around the pivot point. Do you trace in the clockwise or counterclockwise direction?
Student: Counterclockwise.
Tutor: And is it a positive or a negative torque?
Student: I need to choose a positive direction. I'll pick counterclockwise as positive, so the 45 N·m is positive.
Tutor: Good. What is the torque created by the 7 N force?
Student: There is a 35° angle between the 3 m "moment arm" and the force, so

$$\tau = FR\sin\phi = (7 \text{ N})(3 \text{ m})\sin 35^\circ = 12 \text{ N}\cdot\text{m}$$

Student: If I start in the direction of the 7 N force arrow and go around the pivot point, I go clockwise, so it's a negative 12 N torque.
Tutor: What is the torque created by the 4 N force?
Student: The angle is 0°, so

$$\tau = FR\sin\phi = (4 \text{ N})(2 \text{ m})\sin 0^\circ = 0 \text{ N}\cdot\text{m}$$

Student: How can a force create no torque?
Tutor: If the 4 N force was the only force on the object, would it rotate around the pivot point?

Student: I guess it wouldn't.
Tutor: Right. It's like pulling a door outward from the hinges — it doesn't rotate. What about the 8 N force?
Student: The moment arm is 6 m, but there is no angle.
Tutor: The torque is the moment arm times the part of the force that is perpendicular to the moment arm. Or, the torque is the force times the part of the moment arm that is perpendicular to the force. Look at the 5 m line.
Student: So the 5 m line is the part of the 6 m line that is perpendicular to the 8 N force?
Tutor: Yes, the line from the pivot point that is perpendicular to the line of the force.
Student: So the torque is

$$\tau = F(R\sin\phi) = FR_{\perp} = (8\text{ N})(5\text{ m}) = 40\text{ N·m}$$

Student: And it goes clockwise, so it's −40 N·m. Is it okay to write mN for meter-newtons, or do we have to use N·m for Newton-meters?
Tutor: The problem with mN is that it could mean a force of a milli-newton. What is the torque from the force that is 4 m away from the pivot?
Student: We need the part of the force that is perpendicular to the moment arm, so

$$\tau = FR\sin\phi = F_{\perp}R = (5\text{ N})(4\text{ m}) = 20\text{ N·m}$$

Student: And it goes clockwise, so it's −20 N·m.
Tutor: What is the total torque about the pivot point?
Student: We just add the torques, so

$$\tau = 45\text{ N·m} - 12\text{ N·m} + 0\text{ N·m} - 40\text{ N·m} - 20\text{ N·m} = -27\text{ N·m}$$

Just as Newton's second law says $F = ma$, so the same equation holds with the angular equivalents: $\tau = I\alpha$ or angular force equals angular mass times angular acceleration. The angular mass is called <u>moment of inertia</u>, a combination of moment arm and inertia. The moment of inertia is equal to

$$I = \sum_i m_i r_i^2$$

We divide the object into tiny little pieces, then for each piece multiply its mass times the square of how far from the pivot point it is, then add all of those together. We need to use lots of tiny pieces so that each piece is all the same distance from the pivot point. Fortunately, many common shapes have simple moments of inertia, and it's much easier to look them up than to figure them out every time. See the table for common shapes.

Sometimes you might want to rotate one of these objects about a point other than the center. The moment of inertia of a circle (or disk) about its center is $I = \frac{1}{2}MR^2$, but what if it rotates about a point on its edge rather than its center? The parallel axis theorem tells us that

$$I = I_{\text{CM}} + Md^2$$

or the moment of inertia about any point on an object is equal to the moment of inertia about its center of mass plus its mass times the distance the pivot point is from the center of mass. So a disk rotated about a point on its edge would have a moment of inertia of $I = I_{\text{CM}} + Md^2 = \frac{1}{2}MR^2 + MR^2 = \frac{3}{2}MR^2$.

EXAMPLE

A boy leaving a store pushes on the door handle with a force of 18 N. The door is 0.78 m wide and has a mass of 7.2 kg. The boy pushes perpendicular to the surface of the door. How long does it take for the door to open (rotate to 90°)?

Tutor: If this were linear motion instead of rotation, how would you approach it?
Student: If I can find the acceleration, I can do a constant acceleration problem to find the velocity. To find the acceleration, I need the force and the mass.
Tutor: Good, but with rotation...
Student: To find the angular acceleration, I need the angular force (torque) and the angular mass (moment of inertia).
Tutor: What is the torque?
Student: There is only one force, and it is applied at a right angle at the edge, so

$$\tau = FR\sin\phi = F_{\perp}R = (18\text{ N})(0.78\text{ m}) = 14\text{ N·m}$$

Tutor: There are other forces, from gravity and from the pivot itself. The force from the pivot occurs at the pivot point, so the moment arm is zero. Gravity doesn't cause the door to open or close.
Student: So the other forces don't create any torque. Do I need a positive direction?
Tutor: Since everything is happening in the same direction, we can call that positive and not worry about it any more. If there was another torque, like friction at the pivot, then we would have to be careful about direction.
Student: From above the door looks like a rod, so the moment of inertia is

$$I_{\text{door}} = \frac{1}{12}ML^2 = \frac{1}{12}(7.2\text{ kg})(0.78\text{ m})^2 = 0.468\text{ kg m}^2 \quad ?$$

Tutor: That would be the moment of inertia if the door was being rotated about the center of the door, but it is being rotated about the edge.
Student: So I need the parallel axis thingee.

$$I = I_{\text{CM}} + Md^2$$
$$I = \frac{1}{12}ML^2 + M\left(L/2\right)^2 = \frac{1}{3}ML^2$$
$$I_{\text{door}} = \frac{1}{3}ML^2 = \frac{1}{3}(7.2\text{ kg})(0.78\text{ m})^2 = 1.46\text{ kg m}^2$$

Student: Now I can find the angular acceleration.

$$\tau = I\alpha$$
$$\alpha = \frac{\tau}{I} = \frac{14\text{ N·m}}{1.46\text{ kg m}^2} = 9.59\text{ rad/s}^2$$

Student: I see that the kg and the m^2 cancelled the same units in the numerator, and that the seconds squared came from the newtons, but where did radians come from?
Tutor: Radians aren't really a unit, not like other units. We can insert them and remove them at will. We can't do this with degrees or revolutions, because radians are assumed so that $v = r\omega$ works. It's really just a word used to remind us but not really a unit.
Student: Okay. Now I have the angular acceleration and I can do the constant angular velocity problem.
Tutor: Write down the five variables and identify the ones you know.
Student: Angular displacement is 90°, initial angular velocity ω_0 is zero, and we have the angular acceleration.
Tutor: The angular acceleration is in radians, so you'll want the angular displacement in radians. That's why we use the word even though it isn't a unit.
Student: 90° is one-fourth of a revolution, so it's $\frac{1}{2}\pi$ radians.

$$\begin{array}{ccccl} \Delta x & \longrightarrow & \Delta\theta & = & \pi/2 \\ v_0 & \longrightarrow & \omega_0 & = & 0 \\ v & \longrightarrow & \omega & = & \\ a & \longrightarrow & \alpha & = & 9.59 \text{ rad/s}^2 \\ t & \longrightarrow & t & = & ? \end{array}$$

$$\Delta\theta = \omega_0 t + \frac{1}{2}\alpha t^2$$

$$\left(\frac{\pi}{2}\text{ rad}\right) = (0)t + \frac{1}{2}\left(9.59 \text{ rad/s}^2\right) t^2$$

$$t = \sqrt{\frac{\pi \text{ rad}}{9.59 \text{ rad/s}^2}}$$

$$t = 0.57 \text{ s}$$

EXAMPLE

In an Atwood's machine, two masses of 4 kg and 7 kg hang over a 6 kg pulley. The radius of the pulley is 0.4 m. The 7 kg mass starts 2 m above the ground. What is its speed as it hits the ground?

Tutor: How are you going to attack this?
Student: Find the torque, find the moment of inertia, find the angular acceleration, and solve the constant velocity problem.
Tutor: You could do that. The difficulty is that we don't know the tensions, and we'd need to write a Newton's second law equation for each mass. Remember that we did this before, and had two equations. We could do the same thing, but now the tensions aren't the same and we need a third equation for the rotation of the pulley.
Student: That sounds like a lot of work.
Tutor: We could do it. The pulley equation is something like

$$T_2 R - T_1 R = \left(\frac{1}{2}mR^2\right)\underbrace{\left(\frac{a}{R}\right)}_{\alpha}$$

Student: It sounds like you think there's an easier way. After we did forces, we did energy. Energy techniques work well when the acceleration is not constant and when we don't care about the time. We don't care about the time here, so let's try energy.

$$KE_i + PE_i + W = KE_f + PE_f$$

Tutor: What is the initial kinetic energy?
Student: Nothing is moving, so zero.
Tutor: What is the initial potential energy?
Student: I'll take the starting point as height equals zero, so no potential energy.
Tutor: The two masses might not start at the same height, so you have a different $h = 0$ spot for the two masses.
Student: I can handle that. For each mass, when I find the final potential energy, I measure compared to where it started. The 4 kg mass moves up 2 m, and the 7 kg mass moves down 2 m.

$$\cancelto{0}{KE_i} + \cancelto{0}{PE_i} + W = KE_f + (m_4 g(+2\text{ m}) + m_7 g(-2\text{ m}))$$

Tutor: What is the kinetic energy?
Student: Both of the masses are moving, so I need $\frac{1}{2}mv^2$ for each of them.
Tutor: The pulley is also rotating, so you need the kinetic energy of rotation. It takes energy to rotate something, just like it takes energy to move it.
Student: So $\frac{1}{2}mv^2$ becomes $\frac{1}{2}I\omega^2$.

$$W = \left(\frac{1}{2}m_4 v^2 + \frac{1}{2}m_7 v^2 + \frac{1}{2}I\omega^2\right) + (m_4 - m_7)g(2\text{ m})$$

Tutor: When the 7 kg mass hits the ground, are the two masses going the same speed?
Student: Yes. If they weren't, the rope length would change.
Tutor: Good. How fast is the pulley rotating?
Student: Can I use $v = R\omega$?
Tutor: If the rope is not slipping over the pulley, then the speed of a point on the edge of the pulley is the same as the speed of the rope.

$$W = \left(\frac{1}{2}m_4 v^2 + \frac{1}{2}m_7 v^2 + \frac{1}{2}\left(\frac{1}{2}m_p R^2\right)\left(\frac{v}{R}\right)^2\right) + (m_4 - m_7)g(2\text{ m})$$

Student: Look, the radius cancels.

$$W = \left(\frac{1}{2}m_4 v^2 + \frac{1}{2}m_7 v^2 + \frac{1}{4}m_p v^2\right) + (m_4 - m_7)g(2\text{ m})$$

$$W = \frac{1}{2}\left(m_4 + m_7 + \frac{1}{2}m_p\right)v^2 + (m_4 - m_7)g(2\text{ m})$$

Student: Now we need to find the work done by forces other than gravity.
Tutor: There are the tensions in the rope.
Student: Consider the tension in the rope attached to the 7 kg mass. It holds back on the mass, doing negative work, and pulls on the pulley, doing positive work.
Tutor: And since the forces are the same, and the distances are the same, the total work is zero.
Student: There's also a force at the pivot, holding the pulley up, and gravity on the pulley, but the moment arm for these forces is zero.

$$\cancelto{0}{W} = \frac{1}{2}\left(m_4 + m_7 + \frac{1}{2}m_p\right)v^2 + (m_4 - m_7)g(2\text{ m})$$

$$\frac{1}{2}\left(m_4 + m_7 + \frac{1}{2}m_p\right)v^2 = -(m_4 - m_7)g(2\text{ m})$$

$$v = \sqrt{\frac{-2(m_4 - m_7)g(2\text{ m})}{(m_4 + m_7 + \frac{1}{2}m_p)}} = \sqrt{\frac{-2(-3\text{ kg})(9.8\text{ m/s}^2)(2\text{ m})}{(4 + 7 + \frac{1}{2}\cdot 6)\text{ kg}}} = 2.9\text{ m/s}$$

Chapter 11

Rolling, Torque, and Angular Momentum

This chapter is about angular momentum. But first an example of rolling motion.

EXAMPLE

A 7.0 kg bowling ball rolls without slipping down a ramp. The ramp is 1.4 m high and 9.6 m long. How fast is the bowling ball moving when it reaches the bottom of the ramp?

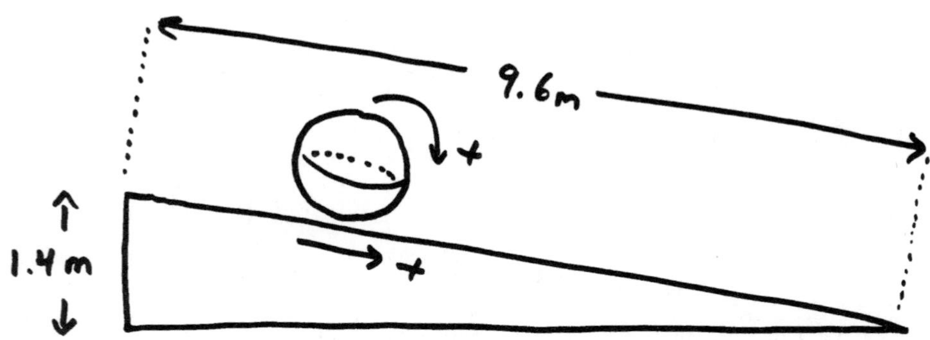

Student: We don't care about the time, so I'll try energy.
Tutor: Energy looks especially promising because we want the speed at the end of the ramp, and speed connects directly to kinetic energy.

$$KE_i + PE_i + W = KE_f + PE_f$$

Student: The ball isn't moving at the top, so the initial kinetic energy is zero. I'll choose the bottom as $h = 0$, so the final potential energy is zero. The initial potential energy is mgh.

$$\cancelto{0}{KE_i} + mgh + W = KE_f + \cancelto{0}{PE_f}$$

Tutor: What is the final kinetic energy?
Student: Ah, we had this in the last chapter. It has $\frac{1}{2}mv^2$ but it also has $\frac{1}{2}I\omega^2$.

$$mgh + W = \frac{1}{2}mv^2 + \frac{1}{2}I\omega^2$$

Tutor: What is the moment of inertia I?
Student: The bowling ball is a solid sphere, so we look in the table and $I = \frac{2}{5}mR^2$.
Tutor: What is the angular velocity ω?
Student: Is this where we use $v = R\omega$? The point on the edge is moving around the center of the bowling ball at the same speed that the bowling ball is moving down the ramp?
Tutor: As long as the bowling ball rolls without slipping, this is true.

$$mgh + W = \frac{1}{2}mv^2 + \frac{1}{2}\left(\frac{2}{5}mR^2\right)\left(\frac{v}{R}\right)^2$$

$$mgh + W = \frac{1}{2}mv^2 + \frac{1}{5}mv^2$$

Student: Now we need to find the work done by forces other than gravity.
Tutor: What other forces are there?
Student: The bowling ball is in contact with the ramp, so there is a normal force from the surface. If there's a normal force, there could be a friction force.
Tutor: If there was no friction force, there would be no torque, and the bowling ball wouldn't roll. Is it static or kinetic friction?
Student: The ball isn't sliding, so it must be static friction.
Tutor: Yes. How large is the friction force?
Student: Is it the limiting case where $F = \mu N$?
Tutor: No, so we don't know how large the friction force is. Does the friction force do any work?
Student: The normal force doesn't do any work because it is perpendicular to the motion. I'm not sure about the friction force.
Tutor: It may seem strange, but because there is no sliding, there is no displacement at the friction force, so the friction force doesn't do any work. Or, does the bowling ball heat up as it rolls, no, only if it slides, so no mechanical energy is being turned into thermal energy by friction.
Student: If the work is zero, then we can solve the equation.

$$\cancel{m}gh = \frac{1}{2}\cancel{m}v^2 + \frac{1}{5}\cancel{m}v^2$$

$$gh = \frac{7}{10}v^2$$

$$v = \sqrt{\frac{10}{7}gh} = \sqrt{\frac{10}{7}(9.8 \text{ m/s}^2)(1.4 \text{ m})} = 4.4 \text{ m/s}$$

Tutor: Did you notice how both m and R cancel from the equation? It doesn't matter what the size or mass of the ball is, it will have the same speed when it gets to the bottom. But if we used a hollow sphere or a tube instead of a solid sphere, the 2/5 would have been something else, and we would have gotten a different result.
Student: So a bowling ball and a Ping-Pong ball have different speeds at the bottom, not because the Ping-Pong ball is smaller or lighter, but because it is hollow?
Tutor: Correct. Of course, air resistance would also play a part, but we've conveniently ignored it.
Student: Again. Could we do this problem using forces?
Tutor: Yes. We would need to find all of the forces. Then we would have to write the Newton's second law equation. Then we would need the Newton's second law equation for rotation ($\tau = I\alpha$), and we'd need to use $v = R\omega$ to connect the two.

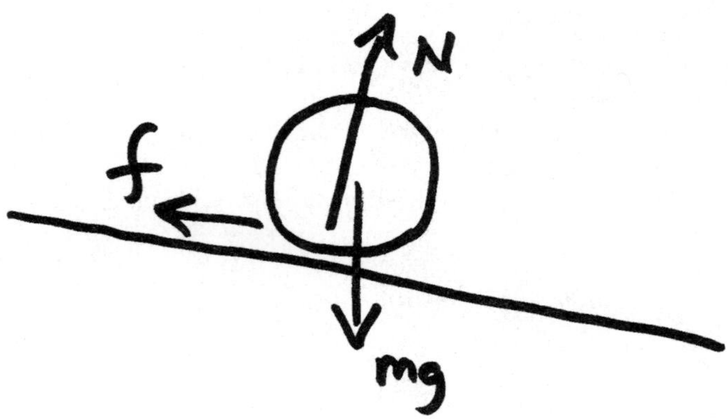

Student: So we'd need to find the normal force and the friction force.
Tutor: Alternatively, we could find the torque about the bottom, where the bowling ball contacts the ramp. Then we could find α and find a. Since the normal force and friction force go through that point, they wouldn't create any torque.
Student: It sounds like energy is easier.
Tutor: If you can use energy, it usually is easier.

Everything we did with momentum works with angular quantities. Angular momentum is traditionally designated L.

$$\begin{aligned} P = mv \quad &\rightarrow \quad L = I\omega = PR\sin\phi \\ \Delta P = Ft \quad &\rightarrow \quad \Delta L = \tau t \\ P_i + \Delta P = P_f \quad &\rightarrow \quad L_i + \Delta L = L_f \end{aligned}$$

For a rotating object, we use $I\omega$, but for a point object moving around a point we use $PR\sin\phi$. Angular impulse is the name given to the change in the angular momentum, which thus has the same units as angular momentum (kg m^2/s).

EXAMPLE

An ice skater doing an axel pulls her arms in to spin faster. In order to triple her rotational speed, what must she do to her moment of inertia?

Student: Shouldn't that be an "axle," as in the thing a wheel turns around?
Tutor: No, it was invented by someone named "Axel."
Student: Well, we just heard about angular momentum, so we must need to use that.
Tutor: Are there any other indications?
Student: We don't know the angular acceleration, but maybe energy would work.
Tutor: It wouldn't, but we'll see why later. What is the angular momentum beforehand?
Student: Angular momentum L equals moment of inertia I times angular velocity ω. We don't know any of them.
Tutor: So we leave them as variables. What is the angular momentum afterward?
Student: Angular momentum L equals moment of inertia I times angular velocity ω. We still don't know any of them.
Tutor: Are they all the same as they were before?
Student: No, so afterward the angular momentum L' equals moment of inertia I' times angular velocity ω'.

$$L_i + \Delta L = L_f$$

$$I\omega + \Delta L = I'\omega'$$

Tutor: What do we know about the angular velocities?
Student: We know that $\omega' = 3 \times \omega$.
Tutor: Correct. What about the angular impulse ΔL?
Student: That's equal to the torque times the time. We need to find all of the torques.
Tutor: What forces are acting on her?
Student: Gravity, the normal force from the ice, and maybe friction from the ice.
Tutor: The axis of rotation is vertically through her center. All of those forces are on that line.
Student: So the moment arms are zero and the torques are all zero.

$$I\omega + \cancel{\Delta L} = I'(3\omega)$$

Student: ω cancels!

$$I\cancel{\omega} = I'(3\cancel{\omega})$$

Tutor: This is a "scaling" problem, so it's not surprising that something cancels.

$$I = 3I'$$

$$I' = \frac{1}{3}I$$

Student: She must reduce her moment of inertia by a factor of three.
Tutor: She does this by pulling her arms in.
Student: Just pulling her arms in does this? Her arms are a small part of her mass.
Tutor: True, but because they are *so* much further out than the rest of her, and distance is squared in the formula for moment of inertia, her arms are most of her rotational inertia.
Student: So why couldn't we use energy for this?
Tutor: Find her energy before and after.
Student: Energy is $\frac{1}{2}I\omega^2$.

$$KE_{\text{before}} = \frac{1}{2}I\omega^2 \qquad \text{and} \qquad KE_{\text{after}} = \frac{1}{2}I'(\omega')^2 = \frac{1}{2}\left(\frac{1}{3}I\right)(3\omega)^2 = \frac{3}{2}I\omega^2$$

Student: They aren't the same!
Tutor: She does work when she pulls her arms in. Unless we know just how much work she does, we can't use energy to solve the problem.

EXAMPLE

A 0.85 kg, 35–cm–long uniform rod falls from vertical to horizontal, pivoting about its end, then strikes a 0.56 kg block of putty. The block sticks to the end of the rod. What is the speed of the putty after it is hit by the rod?

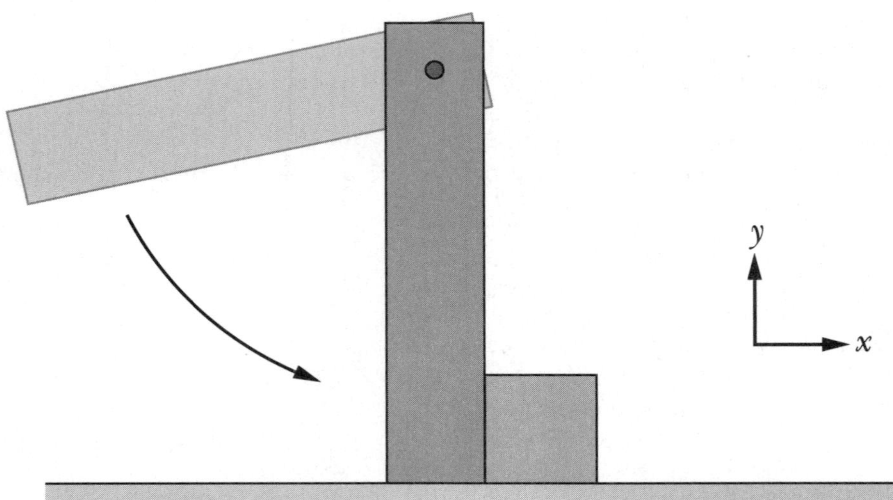

Student: I don't even know where to begin.
Tutor: What you mean by that is that you can't see the end from where you are now. Instead focus on what you can do. We've learned about constant acceleration, forces, energy, momentum, and rotation. What can you do?
Student: I can probably use energy to find the speed of the rod after it falls. We did this for objects falling, sliding, and rolling, so it should work here. The initial kinetic energy is zero, and I'll use the initial position as $h = 0$.

$$\cancelto{0}{KE_i} + \cancelto{0}{PE_i} + W = KE_f + PE_f$$

Tutor: All very good. What is the final kinetic energy?
Student: The rod is rotating, so $\frac{1}{2}I\omega^2$. But the center of mass of the rod is also moving, so do I need $\frac{1}{2}mv^2$ too?
Tutor: Very insightful. If you use the moment of inertia about the center of the rod, then you need the motion of the center of the rod. If you use the moment of inertia about the end of the rod, which is larger by $m(L/2)^2$, then you don't need the $\frac{1}{2}mv^2$.
Student: I'll use the moment of inertia about the end. We found earlier that it's $\frac{1}{3}ML^2$.

$$W = \frac{1}{2}\left(\frac{1}{3}ML^2\right)\omega^2 + PE_f$$

Tutor: What is the final potential energy?
Student: The rod goes down, so it's negative.
Tutor: You need to use the center of mass of the rod. How far down does that go?
Student: The center of mass is at the middle, so it goes down by $L/2$.

$$W = \frac{1}{2}\left(\frac{1}{3}ML^2\right)\omega^2 + Mg(-\frac{1}{2}L)$$

Student: Now we need the work done by forces other than gravity. The pivot exerts a force, but that force is at the pivot, so it doesn't exert any torque.

Tutor: Correct. And the work is $\tau\ \Delta\theta$, so no torque means no work.

$$\frac{1}{2}\left(\frac{1}{3}ML^2\right)\omega^2 = \frac{1}{2}MgL$$

$$\frac{1}{3}L\omega^2 = g$$

$$\omega = \sqrt{\frac{3g}{L}}$$

Student: Now we can plug in numbers and find the rotation rate.
Tutor: There's no hurry. Once you know how fast the rod is rotating, what can you do?
Student: The rod hits the putty. That sounds like a collision, so I need to use momentum.
Tutor: During the collision, the pivot at the top of the rod will exert a force to keep the top of the rod in place. Do you know how big that force will be?
Student: No. Can't I just pretend that it's small, and when I multiply it by the small collision time, the impulse will be really small?
Tutor: The shorter the collision time, the larger the force from the pivot will be. We need to keep that force out of the equation, but that's not the right way. What about angular momentum?
Student: If the force occurs at the pivot point, then it doesn't exert any torque, so no *angular* impulse.
Tutor: Correct. And the forces that the rod and putty put on each other will cancel, just like with momentum.

$$L_i + \cancelto{0}{\Delta L} = L_f$$

$$I_{\text{rod}}\omega_{\text{rod}} + L_{\text{putty}} = I_f\omega_f$$

Student: The angular speed of the rod before the collision is the same as the angular speed of the rod after it falls.

$$\left(\frac{1}{3}M_{\text{rod}}L^2\right)\sqrt{\frac{3g}{L}} + L_{\text{putty}} = I_f\omega_f$$

Tutor: What is the moment of inertia of the combined rod and putty?
Student: The rod is $\frac{1}{3}M_{\text{rod}}L^2$, but the table doesn't have an entry for putty.
Tutor: Just treat it as a single point, and use mr^2.

$$\left(\frac{1}{3}M_{\text{rod}}L^2\right)\sqrt{\frac{3g}{L}} + L_{\text{putty}} = \left(\frac{1}{3}M_{\text{rod}}L^2 + m_{\text{putty}}L^2\right)\omega_f$$

Student: And the angular momentum of the putty beforehand has to be zero, because it isn't moving.

$$\left(\frac{1}{3}M_{\text{rod}}L^2\right)\sqrt{\frac{3g}{L}} = \left(\frac{1}{3}M_{\text{rod}} + m_{\text{putty}}\right)L^2\omega_f$$

Tutor: Right. Now it's just algebra.

$$\frac{1}{3}M_{\text{rod}}\cancel{L^2}\sqrt{\frac{3g}{L}} = \left(\frac{1}{3}M_{\text{rod}} + m_{\text{putty}}\right)\cancel{L^2}\omega_f$$

$$\omega_f = \frac{\frac{1}{3}M_{\text{rod}}}{\left(\frac{1}{3}M_{\text{rod}} + m_{\text{putty}}\right)}\sqrt{\frac{3g}{L}}$$

$$\omega_f = \frac{\frac{1}{3}(0.85\text{ kg})}{\left(\frac{1}{3}(0.85\text{ kg}) + (0.56\text{ kg})\right)}\sqrt{\frac{3(9.8\text{ m/s}^2)}{(0.35\text{ m})}}$$

$$\omega_f = 3.08\ /\text{s}$$

Student: Shouldn't it be in meters per second?
Tutor: We found the rotation rate of the rod and putty...
Student: And I want the speed of the point on the end.

$$v = R\omega = (0.35 \text{ m})(3.08 \text{ /s}) = 1.08 \text{ m/s}$$

EXAMPLE

(a) A small object moves in a circle on the end of a string on a horizontal frictionless table. Initially the object moves at 0.74 m/s at the end of a 17 cm string. The string is then pulled through a hole in the center of the table. When the string remaining is only 5.0 cm, how fast is the object moving? (b) A roller coaster enters a spiral in which the radius of the track decreases from 17 m to 5.0 m while staying horizontal. If the coaster enters the spiral at a speed of 7.4 m/s, what is its speed at the end of the spiral?

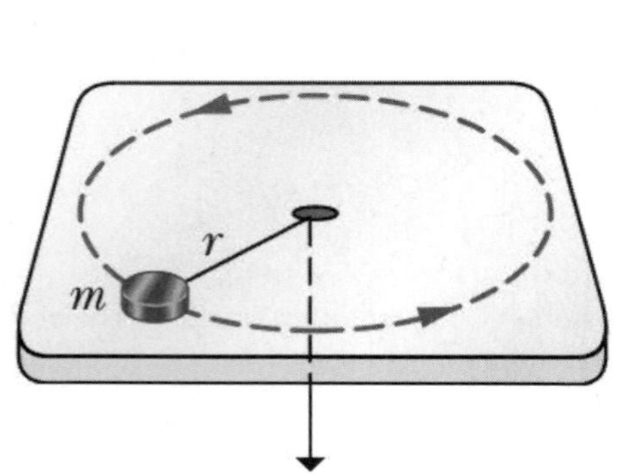

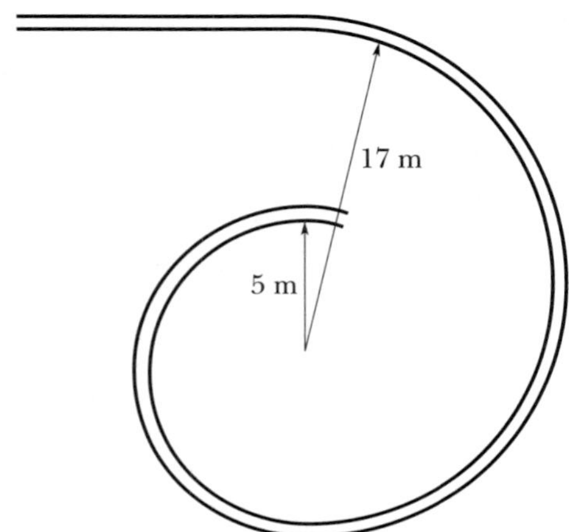

Student: Part *a* sounds a lot like the previous example. We don't know the force of the string, but by using angular moment, the force disappears.

$$L_i + \cancelto{0}{\Delta L} = L_f$$

Tutor: Because it is a single point, we can use $L = PR \sin\phi$ for the angular momentum.

$$PR \sin\phi = P'R' \sin\phi'$$

Student: The angle ϕ is the angle between the momentum and the moment arm, and that's 90°.

$$mvR = mv'R'$$

$$v' = \frac{vR}{R'} = \frac{(0.74 \text{ m/s})(17 \text{ cm})}{(5.0 \text{ cm})} = 2.5 \text{ m/s}$$

Student: Now I can do the same thing to part *b*.
Tutor: Not so fast! What is the direction of the force that the tracks put on the roller coaster?
Student: It's not toward the middle of the circle?
Tutor: No, it's a normal force so it's perpendicular to the track.
Student: So it's perpendicular to the motion of the roller coaster. It's not toward the center, so the torque doesn't cancel. I need the force and direction.
Tutor: Try a different way. If angular momentum won't work, try forces, energy, or momentum.
Student: Well, energy is supposed to be easiest. If the force is perpendicular to the motion, the work is zero. So the speed of the roller coaster doesn't change.

Tutor: The difference between part a and part b is the direction of the force. In part b it was not toward the center, or "pivot point," so the angular momentum could change.
Student: Why couldn't I use momentum? The force is perpendicular to the motion so it doesn't create an impulse.
Tutor: It does create an impulse. Momentum is a vector and the direction of the roller coaster is changing, so the momentum is changing.
Student: So I have to use energy for part b.
Tutor: Yes, but it was easy.
Student: True. Zero equations.

Chapter 12

Equilibrium and Elasticity

The equilibrium point is where something can sit without moving. That means that the total force is zero, so that there is no acceleration. It also means that the total torque is zero, and that there is no angular acceleration.

EXAMPLE

A 2.0–m–long, 6.0 kg board is supported by two scales, one at the left end and the other 1.2 m from the left end. What is the reading on each scale?

Student: Back to forces. I draw the free-body diagram, choose axes, add the forces, and write Newton's second law equations.
Tutor: It is good that you remember. What are the forces on the board?
Student: There is a normal force from the first scale and a normal force from the second scale. The forces might not be equal, so I'll call them N_1 and N_2. There's also gravity acting at the center of mass.

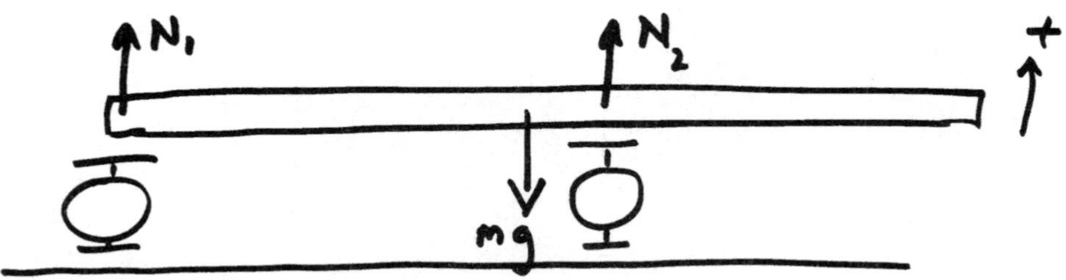

Tutor: Time to choose axes.
Student: I only need one, since all of the forces are colinear. I choose up as positive.

$$+N_1 + N_2 - mg = m\cancel{a_y}^{\,0}$$

Tutor: Can you solve this equation?
Student: No, there are two unknowns and only one equation. I need to do the x forces.
Tutor: There are no x forces, so that doesn't help.
Student: But I need another equation.
Tutor: Indeed you do, but you can't get it from forces. What else has to be zero besides the force?
Student: Ah, the torque needs to be zero. I add the torques about the center of mass and set them equal

to zero.

$$\sum \tau = I\cancelto{0}{\alpha}$$

$$\sum_i F_i R_i \sin\phi_i = 0$$

$$-N_1(1.0 \text{ m})(1) + N_2(0.2 \text{ m}) \underbrace{(1)}_{\sin 90^\circ} + (mg)(0)\sin(?) = 0$$

$$-N_1(1.0) + N_2(0.2) = 0$$

Student: Now I have two equations and two unknowns. I can solve that.
Tutor: Yes you can. You could have used some other point as the pivot point. Try using the left end.

$$\sum_i F_i R_i \sin\phi_i = I\alpha = 0$$

$$N_1(0)\sin(?) + N_2(1.2 \text{ m}) \underbrace{(1)}_{\sin 90^\circ} - (mg)(1.0 \text{ m})(1) = 0$$

$$+N_2(1.2) - (mg)(1.0) = 0$$

Student: Hey, I can solve this one immediately.

$$N_2 = (mg)(1.0)/(1.2) = \frac{1.0}{1.2}(6.0 \text{ kg})(9.8 \text{ m/s}^2) = 49 \text{ N}$$

Tutor: You can always eliminate at least one torque by your choice of pivot point. To make the math easier, it helps to eliminate the torque from an unknown force, like N_1, rather than the torque from a known force like mg.
Student: But does it still work?
Tutor: Yes. The torque has to be zero around *any* point, or the board would start to rotate about that point.
Student: Then I do another torque equation where N_2 is applied so that I can solve for N_1?
Tutor: You could, but you could also go back to your y forces equation to find N_1.
Student: That's easier.

$$N_1 + N_2 - mg = 0$$

$$N_1 + (49 \text{ N}) - (6.0 \text{ kg})(9.8 \text{ m/s}^2) = 0$$

$$N_1 = 10 \text{ N}$$

You can always eliminate at least one torque by your choice of pivot point. If the forces are all colinear, you might not be able to eliminate more than one, but if they are in two dimensions, then you can eliminate two or more by choosing a pivot point where forces "intersect." **To make the math easier, it helps to eliminate the torque from an unknown force** rather than the torque from a known force.

EXAMPLE

A 75 kg diver (165 lb) stands at the end of a diving board. The nearly massless board is 3.7 m long and has two supports 0.90 m apart, with one at the far end from the diver. What is the force from each support on the diving board?

Student: There are two normal forces and gravity.
Tutor: It's a massless board.
Student: The gravity force of the diver.

Tutor: The gravity force on the diver acts on the diver, not the board. Newton's third law says that the equal and opposite force is the diver pulling up on the Earth.
Student: Okay, he pushes down on the diving board with a normal force. Is this force the same as his weight?
Tutor: Add up the forces on the diver.
Student: The forces on the diver are his weight and the normal force from the board. They add to zero, so they must have the same magnitude but be in opposite directions. The normal force that he puts on the board is the same as the force the board puts on him, so they are equal.

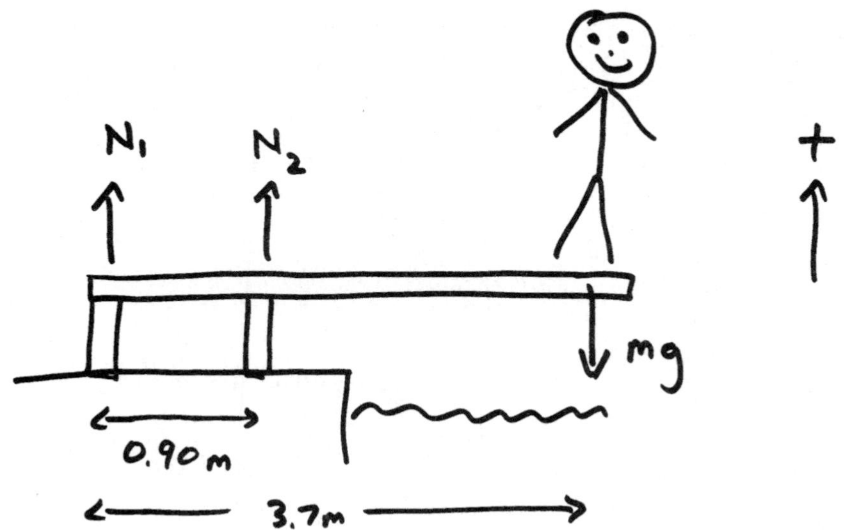

Tutor: Yes, it's our "special case" where there are only two forces and no acceleration. Having said that it's really a normal force acting on the board and not his weight, and having figured out that they're the same, we might just call it his weight sometimes.
Student: Then why did we go through all that?
Tutor: Because sometimes they aren't the same, so we really do need to check.
Student: Okay. I add the vertical forces, using up as positive.

$$\sum F_y = m\cancelto{0}{a_y}$$
$$+N_1 + N_2 - mg = 0$$

Student: I can't solve the equation, so I need a torque equation.
Tutor: What are you going to use as the pivot point?
Student: It wouldn't help to use the center of mass of the board, so I'll use the left end.

$$\sum_i F_i R_i \sin\phi_i = I\cancelto{0}{\alpha}$$
$$N_1(0)\sin(?) - N_2(0.90\text{ m})\underbrace{(1)}_{\sin 90^\circ} + (mg)(3.7\text{ m})(1) = 0$$
$$-N_2(0.90) + (mg)(3.7) = 0$$
$$N_2 = \frac{(mg)(3.7)}{(0.90)} = \frac{3.7}{0.90}(75\text{ kg})(9.8\text{ m/s}^2) = 3020\text{ N}$$

Student: Now I can use the y forces equation to find N_1.

$$N_1 + N_2 - mg = 0$$

$$N_1 + (3020\text{ N}) - (75\text{ kg})(9.8\text{ m/s}^2) = 0$$

$$N_1 = -2290\text{ N}$$

Student: I thought that a normal force couldn't be negative?
Tutor: A normal force has to be out from the surface. The negative means that the force is in the opposite direction that you drew it, so that the left support must pull the left end of the board down.
Student: But a normal force can't do that!
Tutor: Correct. The left end of the board needs to be attached to the support, so that the support pulls it down. Try adding the torques around the center of the board — what do you find?
Student: All of the torques are clockwise.
Tutor: Yes — the board would have to rotate.

EXAMPLE

A 25 kg sign is hanging from a 3.0-m-long, 14 kg beam. The end of the beam is supported by a wire that makes a 39° angle with the beam. What is the tension in the wire?

Student: First we draw the free-body diagram. There is the weight of the beam and the weight of the sign. It's *really* a tension, but it's the same as the weight of the sign. There is also a tension up and a normal force out from the wall.
Tutor: All true. The beam is attached to the wall, like the diving board was attached to the left side support.
Student: So the wall could be pulling the beam in?
Tutor: Yes, and it could also be pushing the end of the beam up or down.
Student: So we don't even know the direction of the force from the wall. How do we handle that?
Tutor: Whatever the force is, it can be divided into components. We just use the components.

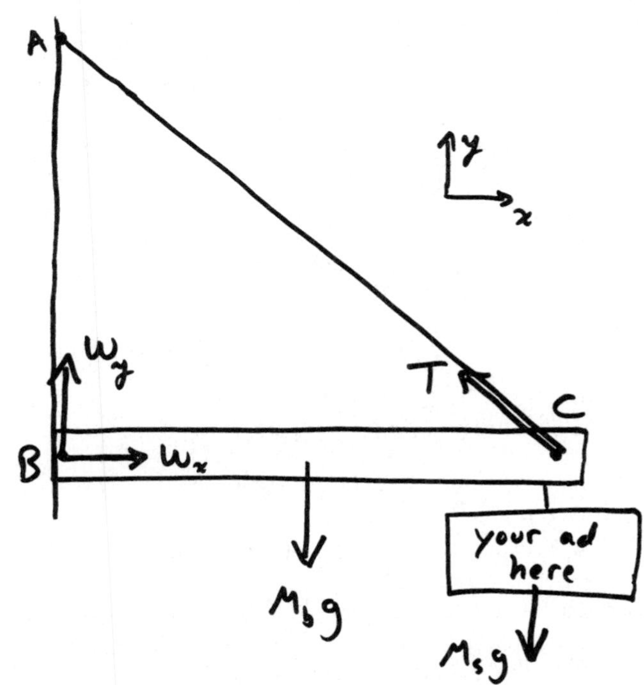

Student: Now I choose axes and add up the x and y forces.

$$F = ma$$
$$+\mathcal{W}_x - T\cos 39^\circ = m_{\text{beam}}\cancelto{0}{a_x}$$
$$+\mathcal{W}_y - M_{\text{beam}} - M_{\text{sign}} + T\sin 39^\circ = m_{\text{beam}}\cancelto{0}{a_y}$$

Tutor: We don't know $\mathcal{W}_x$, $\mathcal{W}_y$, or T, so we can't solve either of these yet. We need to add up the torques.
Student: By choosing the pivot point wisely, I can eliminate two of the torques from the equation. I'll use $\mathcal{B}$ as the pivot point.
Tutor: You could also use $\mathcal{A}$ or $\mathcal{C}$. Using $\mathcal{A}$ eliminates the torques from $\mathcal{W}_y$ and T.
Student: Can I use $\mathcal{A}$ even though it's not in the object?
Tutor: You could use Paris as your pivot point and it would work, but the choice of Paris doesn't make the equations easier to solve.
Student: I'll still use $\mathcal{B}$ as the pivot point.

$$\sum_i F_i R_i \sin\phi_i = I\alpha = 0$$
$$\mathcal{W}_x(0)\sin(?) + \mathcal{W}_y(0)\sin(?) - M_{\text{beam}}g(\frac{1}{2}L)(1) - M_{\text{sign}}g(L)(1) + T(L)\sin 39^\circ = 0$$
$$-\frac{1}{2}M_{\text{beam}}g - M_{\text{sign}}g + T\sin 39^\circ = 0$$
$$T = \left(\frac{1}{2}M_{\text{beam}}g + M_{\text{sign}}g\right)/\sin 39^\circ$$
$$T = \left(\frac{1}{2}(14\text{ kg}) + (25\text{ kg})\right)(9.8\text{ m/s}^2)/\sin 39^\circ$$
$$T = 498\text{ N}$$

Tutor: Using $\mathcal{B}$ is a good choice because the unknown force remaining was the one you wanted to solve for.

The other topic here is elasticity. When an object is in contact with a surface, there can be a normal force. The normal force keeps the object from going into the surface. The normal force is really a spring force, and the surface bends or deforms as it applies a force. For many common surfaces, this bending is so small that we don't see it. This idea applies for all solid materials. **Materials stretch and compress like springs whenever they experience a force**.

The elasticity is a measure of how much a material deforms under force.

$$\frac{F}{A} = E\,\frac{\Delta L}{L}$$

Given that a force F applies to a material with an area A, the Young's modulus E lets us determine how much the material will stretch ΔL. The stretching is expressed as a fraction of the total length $\Delta L/L$.

The strength is a measure of when the material fails completely and breaks.

$$\frac{F}{A} = S_u \text{ or } S_y$$

For rigid materials like steel, glass, or wood, once the stretching has reached about 0.1% of the total length, the material will fail. Either it will snap or it will deform so much that it won't return to its "normal" length.

EXAMPLE

The sign in the previous example is to be supported by a steel wire ($E = 200 \times 10^9$ N/m^2, $S_u = 400 \times 10^6$ N/m^2). If a 0.3–mm–diameter wire is used, how much will the wire stretch under load, and how close will the wire be to its ultimate strength?

Student: To find the amount of stretching, we need to know F, A, E, and L.
Tutor: Good, what is the force F?
Student: We did that in the last example. The force is the tension in the wire, 498 N.
Tutor: What is the area?
Student: The outside area of the wire is the circumference times the length, $(2\pi r)L$.
Tutor: The area we need is the "cross-section" area of the wire. Imagine cutting the wire in half and looking at the circle that you exposed. What is the area of that circle?
Student: That area is π times the radius of the wire squared, or

$$A = \pi r^2 = \pi(0.15 \text{ mm})^2 = \pi(1.5 \times 10^{-4} \text{ m})^2 = 7.07 \times 10^{-8} \text{ m}^2$$

Student: That seems like a very small area.
Tutor: Stretch out your arms to make a big circle, and the area of the circle is less than one square meter. The wire has an area much, much less than that. What is the length of the wire?
Student: We could use the Pythagorean theorem, but we don't know how high above the beam the wire is attached.
Tutor: No, so you need to do some trigonometry. Look at the right triangle formed by the wire.
Student: Okay, the wire is the hypotenuse and the beam is the adjacent side, so

$$\cos 39° = \frac{\text{adj}}{\text{hyp}} = \frac{3 \text{ m}}{L}$$

$$L = \frac{3 \text{ m}}{\cos 39°} = 3.86 \text{ m}$$

Tutor: The Young's modulus E is given, so you can find the stretching ΔL.

$$\frac{F}{A} = E\frac{\Delta L}{L} \quad \longrightarrow \quad \Delta L = \frac{FL}{AE}$$

$$\Delta L = \frac{(498 \text{ N})(3.86 \text{ m})}{(7.07 \times 10^{-8} \text{ m}^2)(200 \times 10^9 \text{ N/m}^2)} = 0.136 \text{ m}$$

Student: The wire stretches 13.6 cm.
Tutor: Is that a lot?
Student: It seems like a lot for a steel wire to stretch. Wouldn't it break?
Tutor: You can check. How does the stress F/A compare to the ultimate strength S_u?

$$\frac{F}{A} = \frac{(498 \text{ N})}{(7.07 \times 10^{-8} \text{ m}^2)} = 7.0 \times 10^9 \text{ N/m}^2 = 7000 \times 10^6 \text{ N/m}^2$$

Student: Wow! That's more than 10 times the stress needed to break the wire.
Tutor: How big would the wire have to be to avoid breaking?
Student: So I need to find the area so that the stress is less than the ultimate strength.
Tutor: Perhaps you would like to be a *little* safer than that. Engineers use a "safety factor," which is how much you could increase the force and it still wouldn't break. What safety factor would you be comfortable with, if you were walking underneath the sign?
Student: How about a million? No, really. So I want the area of the wire so that the stress is one-tenth of the ultimate strength.

$$\frac{F}{\pi r^2} = \frac{1}{10} \times S_u$$

$$d = 2r = 2\sqrt{\frac{10F}{\pi S_u}} = 2\sqrt{\frac{10(498\text{ N})}{\pi(400 \times 10^6\text{ N/m}^2)}} = 0.00398\text{ m} = 3.98\text{ mm}$$

Student: I'd be okay walking under the sign if they used a 4-mm-diameter steel wire.

Chapter 13

Gravitation

Newton said that any two masses pull on each other with a force of

$$F = \frac{GM_1M_2}{r^2}$$

where G is the universal gravitational constant ($G = 6.67 \times 10^{-11}$ N$\cdot$ m^2/kg^2) and r is the distance between the masses. There is a gravitational force between you and this book of about

$$F = \frac{GM_1M_2}{r^2} = \frac{(6.67 \times 10^{-11}\text{ N}\cdot\text{ m}^2/\text{kg}^2)(60\text{ kg})(1\text{ kg})}{(0.3\text{ m})^2} = 4 \times 10^{-8}\text{ N}$$

You pull on the book with a force this big, and the book pulls back on you with the same force. Because this force is so much smaller than the other forces acting on you or the book, we don't bother to include it in our free-body diagrams.

Because the force depends on the distance, does the acceleration of gravity change as we move up and down? Yes. When we move up 3 m (about 10 ft) the gravity force becomes weaker by about one millionth, or 0.0001% weaker. The acceleration of gravity is the force of gravity divided by the mass of the object.

$$a = \frac{F}{m} = \frac{GM_{\text{planet}}m}{r^2} \times \frac{1}{m} = \frac{GM_{\text{planet}}}{r^2}$$

At the height at which airplanes fly, gravity is about 0.3% weaker than on the surface of the Earth. We usually ignore this effect because it is small.

EXAMPLE

The Moon goes around the Earth in a circular path of radius 3.85×10^8 m. The mass of the Earth is 5.98×10^{24} kg and the mass of the Moon is 7.35×10^{22} kg. How long does it take for the Moon to orbit the Earth?

Student: Couldn't we just look this up in the appendix?
Tutor: We could, but the goal is to understand gravity, not to know one number. What are the forces on the Moon?
Student: Gravity from the Earth attracts the Moon. Because the Moon is not at the surface of the Earth, the force is not mg.
Tutor: Good. Are there any other forces on the Moon?
Student: It is not in contact with anything, so just gravity.

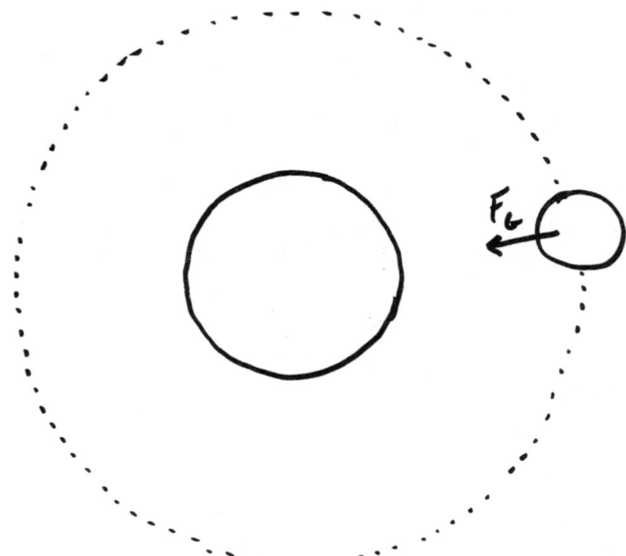

Tutor: Newton says that when you add up all of the forces, all one of them, it's equal to the mass times the acceleration.
Student: The Moon is going in a circle, so the acceleration is v^2/r inward.

$$\frac{GM_{\text{Earth}}M_{\text{Moon}}}{r^2} = M_{\text{Moon}}\frac{v^2}{r}$$

Student: The mass of the Moon cancels.
Tutor: Anything orbiting the Earth at the height of the Moon has the same orbit. What does the variable r represent?
Student: It's the height of the Moon above the Earth.
Tutor: On the left side of the equation, it's the distance between the Earth and the Moon, from center to center. On the right side, it's the radius of the Moon's orbit.
Student: But the Earth is at the center of the orbit, so they're the same.

$$v = \sqrt{\frac{GM_{\text{Earth}}}{r}}$$

Tutor: Now you can use the speed to find the time.
Student: We don't have an equation for the period of an orbit.
Tutor: Not everything needs a new equation. Use $d = vt$.
Student: The circumference of an orbit equals the speed times the period.

$$t = \frac{d}{v} = \frac{2\pi r}{\sqrt{\frac{GM_{\text{Earth}}}{r}}} = 2\pi\sqrt{\frac{r^3}{GM_{\text{Earth}}}}$$

$$t = 2\pi\sqrt{\frac{(3.85\times 10^8\text{ m})^3}{(6.67\times 10^{-11}\text{ N}\cdot\text{ m}^2/\text{kg}^2)(5.98\times 10^{24}\text{ kg})}}$$

$$t = \left(2.4\times 10^6\text{ s}\right)\left(\frac{1\text{ day}}{86,400\text{ s}}\right)$$

$$t = 27.5\text{ days}$$

EXAMPLE

A "binary star" consists of two stars of equal mass rotating about the point midway between them. The mass of each star is 1.7×10^{30} kg and the distance between them is 9×10^{10} m. How long does it take for the stars to make an orbit?

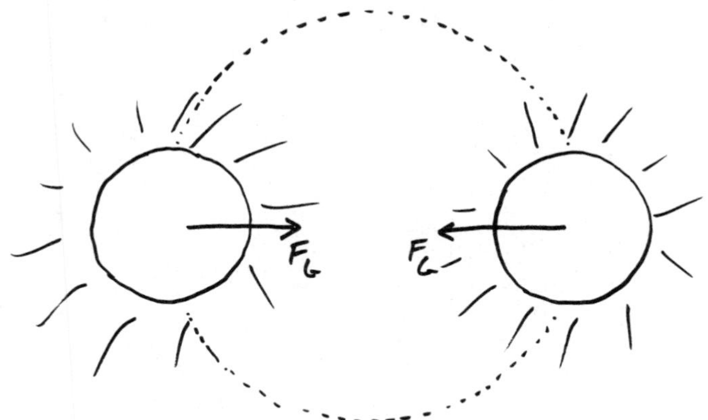

Student: This is just like the last example!

$$\frac{GMM}{r^2} = M\frac{v^2}{r} \quad \textbf{?}$$

Tutor: What does the variable r represent?
Student: It's the distance between the two stars.
Tutor: On the left side of the equation, it's the distance between the two stars, from center to center. On the right side, it's the radius of each star's orbit.
Student: But those aren't equal. How is it that we use r for both when they aren't always equal?
Tutor: That's why it's important to know what each variable means.
Student: Okay. I'll let r be the radius of the orbit. Then the distance between stars is $2r$.

$$\frac{GMM}{(2r)^2} = M\frac{v^2}{r}$$

$$\frac{GM}{4r} = v^2$$

$$v = \sqrt{\frac{GM}{4r}}$$

Student: Now I can repeat the $d = vt$ process like before.

$$t = \frac{d}{v} = \frac{2\pi r}{\sqrt{\frac{GM}{4r}}} = 4\pi\sqrt{\frac{r^3}{GM}}$$

$$t = 4\pi\sqrt{\frac{(4.5 \times 10^{10}\text{ m})^3}{(6.67 \times 10^{-11}\text{ N}\cdot\text{ m}^2/\text{kg}^2)(1.7 \times 10^{30}\text{ kg})}}$$

$$t = \left(1.1 \times 10^7\text{ s}\right) \times \left(\frac{1\text{ day}}{86,400\text{ s}}\right) = 130\text{ days}$$

Earlier, we said that the potential energy of gravity was mgh. This came from using a constant force of gravity mg. Now that we know that the force of gravity is not constant, the potential energy is no longer

mgh. Since the force of gravity weakens as we move further away from the planet, the work we must do to raise something is less than mgh would say.

For simplicity, we choose zero potential energy to be when two masses are an infinite distance from each other. To move something from the surface of a planet to an infinite distance takes work. We must add energy to lift something to energy equals zero, so the potential energy of gravity is always negative. To determine the potential energy, we must find out how much work we need to do. Since the force is not constant, finding this work involves an integral. Leaving the proof to more detailed texts, the potential energy of gravity is

$$U = PE = -\frac{GMm}{r}$$

where G is the same universal gravitational constant, M and m are the two masses, and r is the distance between them.

Remember that to use conservation of energy, we use

$$KE_i + PE_i + W = KE_f + PE_f$$

and that the potential energy appears on both sides of the equation. What matters is the *change* in the potential energy. If the gravity force between the initial and final points is close to constant, then we can still use mgh as an approximation.

EXAMPLE

How much energy does it take to lift an 875 kg communications satellite from Cape Canaveral (latitude 28.6°) into geosynchronous orbit?

Student: We want the energy, so we'll obviously be doing conservation of energy.
Tutor: Very astute.

$$KE_i + PE_i + W = KE_f + PE_f$$

Student: The initial kinetic energy is zero.
Tutor: Not this time. The Earth is rotating, and the rocket moves with the Earth.
Student: That's why they gave us the latitude.
Tutor: Yes. Cape Canaveral sits on a circle of radius $R_E \cos \text{latitude}$.
Student: So the initial velocity is $(2\pi R_E \cos 28.6°)/T$, where $T = 1$ day $= (24$ hr$)$.
Tutor: Almost. The Earth rotates 360° in a "sidereal" day of 23 hours and 56 minutes. This is because the Earth also rotates around the Sun, so that after it has rotated 360°, the same point on Earth is not facing the Sun. The Earth rotates a little more than 360° in a solar day of 24 hours – $360° + 360°/365.25$. We can fudge this. What is the initial potential energy?
Student: The satellite is probably high enough that we can't use mgh. We'll have to use the new formula for the potential energy. The initial distance is equal to $R_E \cos \text{latitude}$.
Tutor: The distance r is the distance from the rocket to the center of the Earth, not the center of the circle in which it rotates.
Student: So r is just R_E.

$$\frac{1}{2}m\left(\frac{2\pi R_E \cos 28.6°}{T}\right)^2 + -\frac{GM_E m}{R_E} + W = KE_f + PE_f$$

Student: m is the mass of the satellite, but where do we get R_E and M_E?
Tutor: We look them up in a reference, but we can do that later. In many textbooks they are in the inside front or back cover of the book. What is the final potential energy?
Student: Isn't that zero?
Tutor: No, we don't have to move the satellite an infinite distance away.

Student: What is geosynchronous orbit?
Tutor: That's when the satellite orbits the Earth in exactly 24 hours. Earth also rotates once in 24 hours. The result is that the satellite is always above the same point on the Earth. It's in synchro with the geo. It only works over the equator.
Student: Cool. That must be where the TV satellites go.
Tutor: Yes, and that's why we want to put a communications satellite there. How far away from Earth is it?
Student: This sounds like the first example in the chapter. Can we reuse the result?
Tutor: Taking an equation out of an example is a good way to screw up. Look what happened in the second example in this chapter. Go through the first example and see how close it is.
Student: Okay. We have a planet and something is orbiting the planet. The only force is gravity, so

$$\frac{GM_E m}{r^2} = m\frac{v^2}{r}$$

$$v = \sqrt{\frac{GM_E}{r}}$$

$$t = \frac{d}{v} = \frac{2\pi r}{v}$$

Student: All of that still works. I don't have r or v, but I have $t = T$. I'll substitute v into the last equation.

$$T = \frac{2\pi r}{\sqrt{\frac{GM_E}{r}}} = 2\pi r\sqrt{\frac{r}{GM_E}} = \sqrt{\frac{4\pi^2 r^3}{GM_E}}$$

Student: Now I can solve for r.
Tutor: Good.

$$r = \sqrt[3]{\frac{GM_E T^2}{4\pi^2}}$$

Student: Whew! Now I can put the numbers in.
Tutor: Let's hold off on that. Symbols are easier to write than numbers.
Student: Okay. The final potential energy is

$$PE_f = -\frac{GM_E m}{r} = -GM_E m\sqrt[3]{\frac{4\pi^2}{GM_E T^2}}$$

$$\frac{1}{2}m\left(\frac{2\pi R_E\cos 28.6^\circ}{T}\right)^2 - \frac{GM_E m}{R_E} + W = KE_f - GM_E m\sqrt[3]{\frac{4\pi^2}{GM_E T^2}}$$

Tutor: What is the final kinetic energy?
Student: Can we get the velocity from the equation above?
Tutor: Sure.

$$KE = \frac{1}{2}mv^2 = \frac{1}{2}m\frac{GM_E}{r} = \frac{1}{2}mGM_E\sqrt[3]{\frac{4\pi^2}{GM_E T^2}}$$

$$\frac{1}{2}m\left(\frac{2\pi R_E\cos 28.6^\circ}{T}\right)^2 - \frac{GM_E m}{R_E} + W = \frac{1}{2}mGM_E\sqrt[3]{\frac{4\pi^2}{GM_E T^2}} - GM_E m\sqrt[3]{\frac{4\pi^2}{GM_E T^2}}$$

$$W = +\frac{GM_E m}{R_E} - \frac{1}{2}GM_E m\sqrt[3]{\frac{4\pi^2}{GM_E T^2}} - \frac{1}{2}m\left(\frac{2\pi R_E\cos 28.6^\circ}{T}\right)^2$$

Student: Do we really have to put in all those numbers?
Tutor: I'm happy to skip it if you are — we've already done the physics. Remember that everything else

is in newtons and seconds, so the time T needs to be in seconds.
Student: Ah, yes. T is 24 hours times 60 minutes times 60 seconds.

Chapter 14

Fluids

Fluids are liquids and gases. Fluids exert forces and have energy. We want to know where and how much, so that we can use fluids.

Fluids exert a pressure,

$$p = \frac{F}{A}$$

A fluid exerts its pressure in every direction. Air at ground level has a pressure of 15 pounds per square inch, and by being in air, this pressure is exerted on every square inch of you. The same pressure, and force, is exerted on each side of you, so that the air doesn't push you over (wind pushing you over is different pressures on your two sides).

The pressure in a fluid increases with depth. This is because the lower fluid has to hold the upper fluid up, so the change in pressure is

$$\Delta p = \rho g \, \Delta y$$

where ρ is the density of the fluid.

$$\rho = \frac{m}{V} = \frac{\text{mass}}{\text{volume}}$$

When something is in the fluid, then the fluid above it pushes down and the fluid beneath it pushes up. The fluid beneath it has a greater pressure and exerts a greater force. The difference in the forces is the buoyancy force or buoyant force

$$F_{\mathrm{B}} = m_f g$$

where m_f is the mass of fluid displaced, or mass of the fluid that would be where the object is if the object wasn't there. The air exerts a buoyant force on you that is equal to your volume times the density of air times g.

When the pressure of air increases, often its density does as well. At high altitude, the density of air is less, and airplanes need to be pressurized so the occupants can breathe. We usually ignore this complication, just as we ignored air resistance, because the calculations would be difficult and we can get most of the important information without it. We treat fluids as being incompressible, so that their densities do not change. One result of this is that the volume flow rate R_V is a constant as the fluid moves.

$$R_V = Av = \text{constant}$$

What the constant is we can't say now, but as a fluid moves the volume flow rate will stay the same. This is equivalent to saying that fluid can't build up anywhere.

As a fluid moves, its pressure changes because of its motion. This is Bernoulli's equation,

$$p + \frac{1}{2}\rho v^2 + \rho g y = \text{a constant}$$

Again, we can't say what the constant is, but for two points in the same fluid, the constant is the same, so that we can calculate one missing piece, like conservation of energy.

Some Useful Numbers	
density of air	1.21 kg/m^3
density of water	998 kg/m^3
density of ice	917 kg/m^3
pressure at sea level	1.0×10^5 N/m^2
	all at 20°C and 1 atm

EXAMPLE

A submarine is approximately a cylinder 10 m in diameter and 88 m long, with a mass of 6.0×10^6 kg. How much water (in m^3) must the submarine take on as ballast so that it keeps a constant depth while submerged?

Student: This looks complicated.
Tutor: Not really. What is the weight of the submarine?
Student: The weight is mg, or

$$mg = (6.0 \times 10^6 \text{ kg})(9.8 \text{ m/s}^2) = 5.9 \times 10^7 \text{ N}$$

Tutor: What is the buoyant force on the submarine?
Student: The buoyant force is $F_\text{B} = m_f g$.
Tutor: That's true, but it's not what I meant. Come at the buoyant force from the other direction. What is the total force on the submarine?
Student: It doesn't say, so it must be zero.
Tutor: An interesting conclusion. If the total force was upward, then the submarine would accelerate up.
Student: And if it was downward then it would accelerate down. To keep a constant depth, the force must be zero.
Tutor: Good.
Student: So the buoyant force is equal to the weight.
Tutor: Yes, and it's also equal to $F_\text{B} = m_f g$.

$$mg = m_f g$$

Student: g cancels. What does that mean?
Tutor: It means that the same submarine, in an ocean on a different planet with a different gravity, would have to take on the same amount of water.
Student: Interesting. What's m_f?
Tutor: The mass of the fluid that isn't there because the submarine is there.
Student: So it's the mass of water that would fit in the volume of the submarine.
Tutor: Yes. Find the volume of the submarine and multiply by the density of water to find the mass of the fluid.

$$m_f = \rho V = \rho \pi r^2 L = (998 \text{ kg/m}^3)\pi(5 \text{ m})^2(88 \text{ m}) = 6.9 \times 10^6 \text{ kg}$$

Student: That's the mass of the fluid displaced, and it has to be equal to the mass of the submarine. It isn't. Something is wrong.

Tutor: The buoyancy force on the submarine is greater than its weight, so the submarine would normally float to the surface. To stay submerged under water, the submarine takes on water in ballast tanks, increasing the mass of the submarine.
Student: So I need to find out how much they increase the mass by taking on water. That's the difference in the two masses.

$$m_{\text{water}} = m_f - m_{\text{sub}} = 6.9 \times 10^6 \text{ kg} - 6.0 \times 10^6 \text{ kg} = 9 \times 10^5 \text{ kg}$$

Student: That's a lot of water.
Tutor: The question was to find the volume of the water.
Student: The connection between volume and mass is density.

$$\rho = \frac{m}{V}$$

$$V = \frac{m}{\rho} = \frac{9 \times 10^5 \text{ kg}}{998 \text{ kg/m}^3} = 900 \text{ m}^3$$

EXAMPLE

A 6 cm cube of wood with density 650 kg/m^3 is placed into a pail of oil and water. The 1–cm–thick layer of oil has a density of 850 kg/m^3. How much of the wood cube sticks out above the oil?

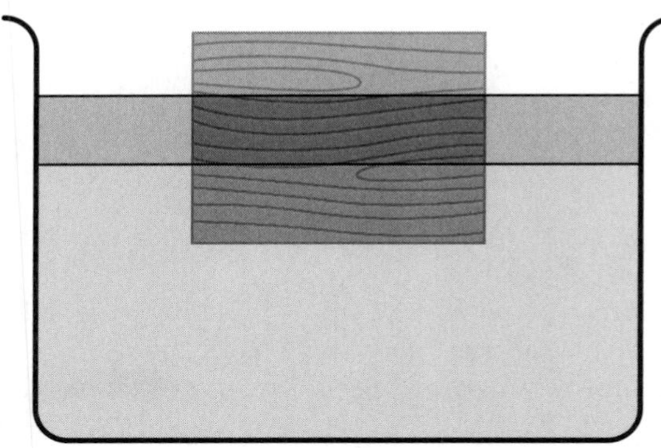

Student: The buoyancy force is equal to the weight.
Tutor: How do you know?
Student: It was last time.
Tutor: All problems are different. That's why it is dangerous to take any equation out of an example, because it applies to the example and not necessarily to any other situation. But we reuse techniques.
Student: The wood cube is not accelerating, so the total force on it is zero. The forces are the buoyancy force and the weight, so they have to cancel; they must be equal.
Tutor: Much better.
Student: The buoyancy force is equal to the weight of the liquid displaced.

$$F_{\text{B}} = m_f g$$

Student: But the densities of oil and water are different. We can't just multiply the volume by the density.
Tutor: No, we have to do that separately for the oil and water. Could the oil by itself hold up the wood cube?

Student: The most oil that could be displaced would be 6 cm by 6 cm by 1 cm. The mass of that much oil is

$$m_{\text{oil}} = \rho V = \left(850 \text{ kg/m}^3\right)\left((0.06 \text{ m})(0.06 \text{ m})(0.01 \text{ m})\right) = 30.6 \text{ g}$$

Student: The mass of the block is

$$m_{\text{wood}} = \rho V = \left(650 \text{ kg/m}^3\right)\left((0.06 \text{ m})(0.06 \text{ m})(0.06 \text{ m})\right) = 140.4 \text{ g}$$

Student: No, the oil by itself could not hold up the wood cube.
Tutor: Correct. If there was 1 cm of oil and nothing else, the wood would rest on the bottom. The wood cube extends down into the water.
Student: How do you know that the oil is on top?
Tutor: Because it has a lower density. The air is above the oil because the density of air is less than the density of oil.
Student: Okay. So the difference must be the mass of the water displaced.

$$m_{\text{water}} = m_{\text{wood}} - m_{\text{oil}} = (140.4 \text{ g}) - (30.6 \text{ g}) = 109.8 \text{ g}$$

Student: Is there a reason for keeping more significant figures than usual?
Tutor: When we subtract two numbers, we often lose significant figures. If we think we might subtract later, then we keep more digits.
Student: Now that I have the mass of the water, I can find the volume.
Tutor: And that volume depends on how much of the wood cube is submerged.
Student: I'll let x be the amount of the cube in the water.

$$m_{\text{water}} = \rho V$$

$$109.8 \text{ g} = \left(998 \text{ kg/m}^3\right)\left((0.06 \text{ m})(0.06 \text{ m})(x)\right)$$

$$x = 0.0306 \text{ m} = 3.06 \text{ cm}$$

Student: The height of wood sticking out is 6 cm − 1 cm − 3.06 cm = 1.94 cm.
Tutor: There's another way to do the problem, using pressure.
Student: Is it easier?
Tutor: I'll let you decide. What is the force that the liquid has to exert on the wood cube?
Student: A force equal to the weight of the cube.
Tutor: And if we divide that force by the area of the bottom of the wood ...
Student: ... we get the pressure at the bottom of the wood cube.
Tutor: Correct. So if we can find the pressure, then we can determine the depth at which that pressure occurs.
Student: Okay. The weight is

$$m_{\text{wood}} g = (0.1404 \text{ kg})(9.8 \text{ m/s}^2) = 1.376 \text{ N}$$

Student: Divide by the area of the bottom to find the pressure.

$$p = \frac{F}{A} = \frac{1.376 \text{ N}}{(0.06 \text{ m})^2} = 382.2 \text{ N/m}^2$$

Tutor: A newton per meter squared (N/m^2) is sometimes called a pascal.
Student: But this pressure is less than the pressure of the air. The air would push it down.
Tutor: You are confusing absolute pressure and gauge pressure. Gauge pressure is the difference between the pressure and atmospheric pressure, because gauges can record only the difference. The absolute pressure of the air at the surface is 1.0×10^5 N/m^2, but the gauge pressure is zero. Your 382.2 N/m^2 is the increase in pressure due to the liquid. The water pushes up that much harder than the air pushes down, which just balances the weight of the wood cube.
Student: Where does the oil come into this?
Tutor: It causes a pressure difference between the top and bottom of the oil. The pressure change due to

the oil is

$$\Delta p = \rho g\ \Delta y = (850\ \text{kg/m}^3)(9.8\ \text{m/s}^2)(0.01\ \text{m}) = 83.3\ \text{N/m}^2$$

Student: So if I take the pressure at the surface to be zero, using gauge pressure, then the pressure at the bottom of the oil is 83.3 N/m^2?
Tutor: Yes.
Student: But I need a pressure of 382.2 N/m^2.
Tutor: The rest of the increase must be due to the water.
Student: So I need a pressure change from water of

$$\Delta p = \rho g\ \Delta y$$

$$382.2\ \text{N/m}^2 - 83.3\ \text{N/m}^2 = (998\ \text{kg/m}^3)(9.8\ \text{m/s}^2)(x)$$

$$x = 0.0306\ \text{m} = 3.06\ \text{cm}$$

Student: It's the same thing.
Tutor: As it should be.
Student: But the top of the wood cube is above the surface of the oil, so it's at a lower pressure. Can pressure be negative?
Tutor: An absolute pressure cannot be negative. A gauge pressure *can* be negative, which means a lower pressure than atmospheric pressure. Since you're using gauge pressure, the pressure at the top of the wood is negative. But the density of air is so much lower than the density of water or oil that we typically ignore it.
Student: As you would say, let's check.
Tutor: Okay. The pressure at the top of the wood is

$$p = -\rho g\ \Delta y = -(1.21\ \text{kg/m}^3)(9.8\ \text{m/s}^2)(0.0194\ \text{m}) = -0.23\ \text{N/m}^2$$

Tutor: We should have subtracted that, or something close to that, from the 382.2 N/m^2 pressure of water, so the error we made by ignoring it was

$$\%\ \text{error} = \frac{\text{error}}{\text{true}} \times (100\%) = \frac{0.23}{382} \times (100\%) = 0.06\%$$

Student: I can live with that.

EXAMPLE

One way to create "downforce" on a race car is with a narrowing channel on the underside. Assume that the car is traveling at 90 m/s and is 0.6 m high. Calculate the downforce.

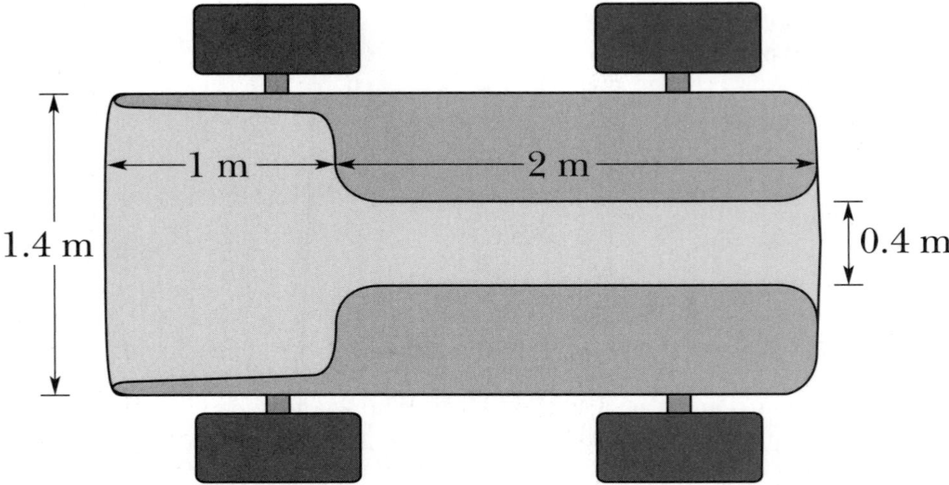

Tutor: Downforce is created when the pressure under the car is less than the pressure above it. The net force from pressure is downward.
Student: How can the pressure be less below the car than above it? Pressure increases as you go down.
Tutor: True, but in the last example we saw that the change in air is fairly small. Pressure also changes as the speed of the fluid changes. Pretend that the car is stationary and the air is moving across it at 90 m/s. What is the force on the top of the car?
Student: The force is

$$F_t = p_t A$$

Tutor: What's the pressure in area 1 of the bottom?
Student: The bottom is lower than the top, so the pressure is greater.
Tutor: Also, the air in area 1 is going the same speed as the air on the top.
Student: And speed changes the pressure. So we're using Bernoulli's equation.

$$p_1 + \frac{1}{2}\rho v_1^2 + \rho g y_1 = \text{a constant}$$

Student: What is the constant?
Tutor: The same as it was on top, whatever that was.

$$p_1 + \frac{1}{2}\rho v_1^2 + \rho g y_1 = p_t + \frac{1}{2}\rho v_t^2 + \rho g y_t$$

Student: The speeds are the same. What about the ρ's?
Tutor: We treat the air as an incompressible fluid, so ρ doesn't change.

$$p_1 + \rho g y_1 = p_t + \rho g y_t$$

$$p_1 = p_t + \rho g\ \Delta y = p_t + (1.21\ \text{kg/m}^3)(9.8\ \text{m/s}^2)(0.6\ \text{m}) = p_t + 7\ \text{N/m}^2$$

Tutor: Now do area 2.
Student: So area 2 is the same as area 1, except the air is moving faster?
Tutor: Yes.

Student: How much faster?
Tutor: Use the volume-rate equation, that the rate of flow is the same.
Student: So the area of area 1 times the speed is the same as the area of volume 2 times the speed there.
Tutor: Careful about what you mean by area. The area in that equation is the cross-sectional area, which is the width times the height. The height of the cross-sectional area is into the page.
Student: So we can't see it.
Tutor: Assume it's the same along the whole length of the car.
Student: So

$$\cancel{h_1} w_1 v_1 = \cancel{h_2} w_2 v_2$$

$$v_2 = \frac{w_1 v_1}{w_2} = \frac{(1.4\text{ m})(90\text{ m/s})}{(0.4\text{ m})} = 315\text{ m/s}$$

Student: The air in the channel is moving at 315 m/s.
Tutor: Now you can find p_2.

$$p_1 + \frac{1}{2}\rho v_1^2 + \cancel{\rho g y_1} = p_2 + \frac{1}{2}\rho v_2^2 + \cancel{\rho g y_2}$$

$$p_2 = (p_t + 7\text{ N/m}^2) + \frac{1}{2}\rho\left(v_1^2 - v_2^2\right)$$

$$p_2 = p_t + 7\text{ N/m}^2 + \frac{1}{2}(1.21\text{ kg/m}^3)\left((90\text{ m/s})^2 - (315\text{ m/s})^2\right)$$

$$p_2 = p_t + 7\text{ N/m}^2 - 55131\text{ N/m}^2 = p_t - 55124\text{ N/m}^2$$

Student: That's a big change.
Tutor: Yes, it's about half of atmospheric pressure.
Student: What's the pressure p_3?
Tutor: There is air there, and it's also moving at 90 m/s, so it's the same as p_1. Really it's a couple of inches lower, so the pressure is maybe 1 kg/m^3 higher.
Student: So now I'm ready to find the downforce.

$$F = (p_t)(1.4\text{ m})(3\text{ m}) - (p_1)\left((1.4\text{ m})(3\text{ m}) - (0.4\text{ m})(2\text{ m})\right) - (p_2)(0.4\text{ m})(2\text{ m})$$

$$F = (p_t)(4.2\text{ m}^2) - (p_t + 7\text{ N/m}^2)(3.4\text{ m}^2) - (p_t - 55124\text{ N/m}^2)(0.8\text{ m}^2)$$

$$F = -(7\text{ N/m}^2)(3.4\text{ m}^2) + (55124\text{ N/m}^2)(0.8\text{ m}^2)$$

$$F = 4.4 \times 10^4\text{ N/m}^2$$

Student: Is that a lot?
Tutor: It's about 9000 pounds, more than the weight of the race car.
Student: So a race car could drive upside down?
Tutor: Some of them, but only as long as they were driving fast. This isn't the same as a roller coaster doing a loop. There, the coaster accelerated down, but the acceleration was what was needed to go in a circle. Here, the race car could drive in a straight line upside down, but would have to get going fast before turning upside down.

Chapter 15

Oscillations

Imagine a box attached to a spring and sliding on a frictionless horizontal surface. The vertical forces are the weight and the normal force, and they add to zero since there is no vertical acceleration. The only horizontal force is the spring.

$$F = -kx = ma$$

When the spring is neither stretched nor compressed the force is zero, and we call this the equilibrium point. When the box is to the left of the equilibrium point the force is toward the right, and vice versa.

If the box starts on the left then it accelerates to the right. It gains speed until it reaches equilibrium. At equilibrium the force is zero, but that doesn't mean that the box stops — the acceleration is zero (not the velocity). It coasts through equilibrium. Once the box reaches the right side the spring slows it, stops it, then pulls it back to the left. This process continues until the cows come home.

The acceleration is the change in the velocity, which is the change in the displacement from equilibrium.

$$-kx = m\frac{d^2}{dt^2}x \qquad \longrightarrow \qquad \frac{d^2}{dt^2}x = -\frac{k}{m}x$$

To find out how the box behaves we need to solve this differential equation. We do this using the standard method for solving all differential equations: guess. We want something that oscillates back and forth like a sine wave, so let's try a sine or cosine function. When we do this we find that the solution is

$$x(t) = x_m \cos(\omega t + \phi)$$

where A, ω, and ϕ are constants. We could also use sine instead of cosine. Anything that behaves in this fashion, where the acceleration is minus a constant times the displacement, is called a simple harmonic oscillator.

x_m and ϕ can be anything, but the angular frequency ω must be equal to

$$\omega = \sqrt{\frac{k}{m}}$$

Since the biggest that cosine gets is 1, the biggest that xt gets is x_m, and it is called the amplitude.

Once we know the displacement as a function of time, we can find the velocity and acceleration as well. The velocity is the derivative of the displacement, or $v(t) = -x_m\omega \sin(\omega t + \phi)$. The acceleration is the derivative of the velocity, or $a(t) = -x_m\omega^2 \cos(\omega t + \phi)$. Earlier we said that $\omega = \sqrt{\frac{k}{m}}$, so we find that the acceleration is $a(t) = -(k/m)x(t)$, as it had to be.

The motion of a simple harmonic oscillator is characterized by its frequency, amplitude, and phase constant ϕ. It is often useful to convert between the angular frequency ω (rad/s), the frequency f (cycles/s), and the period T (s).

$$\frac{\omega}{2\pi} = f = \frac{1}{T}$$

Also, while the maximum displacement of an oscillator is x_m, the maximum speed is

$$v_m = x_m\omega$$

This comes from the equation for the velocity, where the greatest that sine can be is 1, so the velocity can only be as large as $x_m\omega$.

The maximum speed does not occur at the same time as the maximum displacement. Instead, the maximum speed occurs when the displacement is zero, as the oscillator coasts through equilibrium. When the oscillator is at maximum displacement, it is momentarily stationary, and all of its energy is potential energy. When the oscillator is at equilibrium, the potential energy is zero and all of its energy is kinetic energy.

$$E = \frac{1}{2}kx^2 + \frac{1}{2}mv^2$$

EXAMPLE

A 56 g mass attached to a spring oscillates as shown. Write an equation for the displacement of the mass as a function of time.

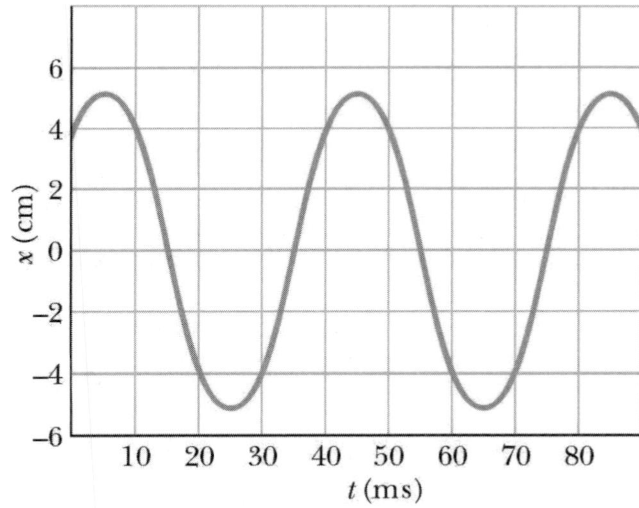

Student: The equation is $x(t) = x_m \cos(\omega t + \phi)$. It says so above.
Tutor: But what are the values for x_m, ω, and ϕ?
Student: Oh, that's what they mean. Is x_m equal to 4 because the displacement at $t = 0$ is 4?
Tutor: x_m is the biggest that the displacement ever gets. The displacement is not always biggest at $t = 0$.
Student: So $x_m = 5$ cm. Or is it -5 cm, or 10 cm?
Tutor: Cosine goes from $+1$ to -1, so the displacement goes from as high as $+x_m$ to as low as $-x_m$.
Student: Then $x_m = 5$ cm. The frequency is 40, because the motion repeats after 40.
Tutor: The period is the time before the motion repeats. So the period $T = 40$ ms.
Student: The displacement is zero at $t = 15$ ms, and again at $t = 35$ ms. Should the period be 20 ms?
Tutor: The displacement is the same but the velocity is not in the same direction, so that isn't " repeating

the motion." The displacement at $t = 10$ ms is the same as at $t = 0$, but the period is not 10 ms either.
Student: From the period I can find the angular frequency ω.

$$\frac{\omega}{2\pi} = f = \frac{1}{T}$$

$$\omega = \frac{2\pi}{T} = \frac{2\pi}{40\text{ ms}} = 157\text{ /sec}$$

Student: So the equation is $x(t) = (5\text{ cm})\cos\left((157\text{ /sec})t + \phi\right)$. How do I determine t?
Tutor: t is an independent variable. Any equation for a function includes at least one independent variable. When you are done with your equation, I can come along and pick any time t, and you can use your equation to tell me what the displacement is.
Student: So I don't need a value for t.
Tutor: You need to *not* have a value for t. The displacement is a function of the time. If your equation did not have t in it, then it would be a constant displacement, not changing with time.
Student: But I do need a value for ϕ. What does ϕ do?
Tutor: The phase constant ϕ determines the starting point. That is, where the oscillator is at $t = 0$.
Student: The oscillator is at 4 when $t = 0$. I can set $t = 0$ in my equation, and $x = 4$ cm, and solve for ϕ.

$$\cancel{4\text{ cm}} = (\cancel{5\text{ cm}})\cos\Big((157\text{ /sec})(0) + \phi\Big)$$

$$0.8 = \cos(\phi)$$

$$\phi = \arccos 0.8 = 37^\circ = 0.64\text{ rad} \quad \textbf{?}$$

Tutor: You could, but the problem is that there are two angles for which $\cos\phi = 0.8$. You need to be sure to get the correct one.
Student: What's the difference?
Tutor: What's the difference between $t = 0$ and $t = 10$ ms?
Student: The displacement is the same, but the velocity is in opposite directions.
Tutor: That's the difference between the two angles.
Student: So I have to find both values for ϕ and put them in the velocity equation to see which gives a positive result?
Tutor: That is one way to do it. Here's another: what is the cosine of zero?
Student: One.
Tutor: So when the phase, or everything inside the parentheses, is zero, the displacement is a positive maximum.
Student: What's the difference between phase and phase constant?
Tutor: The phase constant is ϕ. The phase is $\omega t + \phi$, or the whole thing that we take the sine or cosine of. The phase at $t = 0$ is equal to the phase constant.
Student: So the *phase* is zero at $t = 5$ ms?
Tutor: Yes.

$$\Big((157\text{ /sec})(5\text{ ms}) + \phi\Big) = 0$$

$$\phi = -(157\text{ /sec})(0.005\text{ s}) = -0.78$$

Student: We didn't get the same value for ϕ. Something must be wrong.
Tutor: It's hard to read a graph exactly. The difference is about 0.14 out of 2π, or 8° out of 360°, or 2% of a cycle.
Student: And the final equation is

$$x(t) = (5\text{ cm})\cos\Big((157\text{ /sec})t - 0.78\Big)$$

EXAMPLE

A 0.24 kg mass is attached to a spring with spring constant 50 N/m. At $t = 0$, the displacement of the mass is $x(0) = 0.19$ m and the velocity is $v(0) = -4.1$ m/s. Find the frequency, maximum speed, and acceleration one period later.

Student: So I need to come up with an equation for the displacement as a function of time, and find the first and second derivatives.
Tutor: You could do that. But there are easier ways. Energy is usually an easier way, when it works.
Student: It must work here or you wouldn't suggest it. Why didn't it work in the last example?
Tutor: Since the energy typically doesn't depend on the time, we can't use energy to find the time.
Student: So I can use energy to find the maximum speed.
Tutor: Yes.
Student: I'll start with the frequency.

$$\omega = \sqrt{\frac{k}{m}} = \sqrt{\frac{50 \text{ N/m}}{0.24 \text{ kg}}} = 14.4 \text{ /s}$$

$$f = \frac{\omega}{2\pi} = \frac{14.4 \text{ /s}}{2\pi} = 2.3 \text{ Hz}$$

Student: If hertz is really just one over seconds, and radians per second is one over seconds, then why can't we write the angular frequency in Hz?
Tutor: Radians per second and cycles per second are not the same, even though neither radians nor cycles is a proper unit, so that both are really one over seconds. f and ω aren't interchangeable, so we often use Hz as a reminder that we're using f rather than ω.
Student: Okay. Now I'll use energy as you suggest.

$$E = \frac{1}{2}kx^2 + \frac{1}{2}mv^2 = \frac{1}{2}(50 \text{ N/m})(0.19 \text{ m})^2 + \frac{1}{2}(0.24 \text{ kg})(-4.1 \text{ m/s})^2 = 2.92 \text{ J}$$

Student: Okay, I have the energy, what do I do with it?
Tutor: Using the energy meant using $E_i + W = E_f$. We're not really interested in what the energy is, but in comparing the energy at two times or places.
Student: So if I want the maximum speed, I need to pick the other time to be when the mass has the maximum speed.
Tutor: Yes. When is that?
Student: When it coasts through equilibrium.
Tutor: Good.

$$\left(\frac{1}{2}kx_0^2 + \frac{1}{2}mv_0^2\right) + W = \left(\frac{1}{2}kx_e^2 + \frac{1}{2}mv_e^2\right)$$

Student: Is the work W equal to zero?
Tutor: Are there any other forces acting?
Student: The weight?
Tutor: If the motion is horizontal, then the weight doesn't do any work. If the motion is vertical, then the spring stretches until it balances the weight at the equilibrium point. As the mass moves from there, the total of the weight and spring force is $-kx$, so it's the same as before, and we can do all of the same stuff as long as we measure x from the equilibrium point.
Student: So I don't need the potential energy of gravity, even if the weight is moving vertically.

$$\left(\frac{1}{2}kx_0^2 + \frac{1}{2}mv_0^2\right) + \cancel{W} = \left(\frac{1}{2}kx_e^2 + \frac{1}{2}mv_e^2\right)$$

$$kx_0^2 + mv_0^2 = kx_e^2 + mv_e^2$$

$$(50 \text{ N/m})(0.19 \text{ m})^2 + (0.24 \text{ kg})(-4.1 \text{ m/s})^2 = (50 \text{ N/m})(0)^2 + (0.24 \text{ kg})(v_m)^2$$

$$v_m = 4.9 \text{ m/s}$$

Student: Is it positive or negative?
Tutor: It's a speed. Can a speed be negative?
Student: No, because it's a magnitude.
Tutor: What is the acceleration after one period?
Student: Now I have to find an equation for the oscillator.
Tutor: Not necessarily. If you knew the acceleration at $t = 0$, could you find the acceleration one period later?
Student: It would be the same, wouldn't it? After one period, the oscillator repeats its motion.
Tutor: Correct. What would you have to know to find the acceleration?
Student: The maximum acceleration is $x_m\omega^2$.
Tutor: True. Where does maximum acceleration occur?
Student: At maximum displacement. The farther the spring is stretched, the greater the force and the greater the acceleration.
Tutor: Is the displacement a maximum at $t = 0$?
Student: No.
Tutor: But if you knew the force, could you find the acceleration?
Student: I have the mass, so yes. Ah, the force of the spring is $-kx$, so

$$a = \frac{F}{m} = \frac{-kx}{m} = \frac{-(50 \text{ N/m})(0.19 \text{ m})}{(0.24 \text{ kg})} = -39.6 \text{ m/s}^2$$

Student: Is the negative important? When we did springs before, the minus sign was a reminder that the force was in the opposite direction as the displacement x.
Tutor: True. If we care about the direction of the acceleration, what should it be?
Student: The displacement is positive, so the acceleration should be negative. It is important.

We have three more topics is this chapter: the pendulum, damping, and resonance.

For a pendulum, it is possible to show that

$$\frac{d^2}{dt^2}\theta = -\frac{mgh}{I}\sin\theta$$

where θ is the angular displacement, h is the distance between the center of mass and the pivot point, and I is the moment of inertia of the pendulum. This is not quite the same equation that we had before. If the angle θ is small, then

$$\theta(\text{in radians}) \approx \sin\theta \approx \tan\theta \qquad \text{when } \theta \text{ is small}$$

This approximation is the same as using the first term of the Taylor series (don't panic yet — wait for the appropriate time to panic). So if we restrict the pendulum to small angles (about 15° or less), then it is approximately the same equation with the same solutions. The angular frequency is

$$\omega = \sqrt{\frac{mgh}{I}}$$

or

$$\omega = \sqrt{\frac{g}{l}}$$

for a simple pendulum (a point mass on the end of a massless string) of length l.

Friction or air resistance in an oscillator will do negative work, decreasing the amplitude over time. If we include this effect, then the displacement as a function of time is

$$x(t) = x_m e^{-bt/2m}\cos(\omega' t + \phi)$$

where the frequency is

$$\omega' = \sqrt{\frac{k}{m} - \frac{b^2}{4m^2}}$$

EXAMPLE

An experimenter tries to build a clock from a mass on a spring. He chooses the mass and spring constant so that the period will be one second. When he uses the clock, he notices that damping reduces the amplitude by one-sixth in 73 minutes. How much of the original amplitude will remain after 24 hours? How much does the damping change the rate of the clock?

Student: We need to determine the damping constant b. How do we find b?
Tutor: We may not need to. What does the information about 73 minutes tell you?
Student: That

$$\frac{1}{6}x_m = x_m e^{-b(73 \text{ min})/2m} \cos\left(\omega'(73 \text{ min}) + \phi\right) \quad ?$$

Tutor: A good start, but we need to fix some things. The cosine part is the oscillation, and the other stuff is the amplitude. It is not clear that the oscillation is at maximum displacement at $t = 73$ min. Also, the amplitude decreased *by* a sixth, not *to* a sixth.
Student: Okay, so five-sixths is left.

$$\frac{5}{6}x_m = x_m e^{-b(73 \text{ min})/2m}$$

Tutor: Much better.
Student: The amplitude x_m cancels.
Tutor: The original amplitude cancels, because the later amplitude is 5/6 of the initial amplitude, regardless of the initial amplitude.

$$\frac{5}{6} = e^{-b(73 \text{ min})/2m}$$

Student: I still can't solve for b.
Tutor: No, but you can solve for the combination b/m.
Student: Is that enough?
Tutor: It's enough to answer the first part of the question, anyway.

$$\ln\left(\frac{5}{6}\right) = -\frac{b(73 \text{ min})}{2m}$$

Tutor: Are you bothered by the minus sign? Will b be negative?
Student: No, because the log of a number less than one is negative, so b will be positive.

$$-\frac{b}{2m} = \frac{\ln(\frac{5}{6})}{(73 \text{ min})}$$

Tutor: You can use this to find the amplitude after 24 hours.
Student: Does x_m still cancel?
Tutor: Because the question is what fraction of the original amplitude remains, rather than what is the amplitude remaining, it cancels.

$$\text{fraction} = e^{-\frac{b}{2m}(24 \text{ hr})} = e^{\frac{\ln(5/6)}{(73 \text{ min})}(24 \text{ hr})}$$

Tutor: Remember to get your units straight.
Student: Do the units always have to cancel?
Tutor: We can take the exponent or log only of a number, so we have to get rid of units.

Student: Okay. 24 hours is (24 hr)(60 min/hr) = 1440 min.

$$\text{fraction} = e^{\left(\ln\frac{5}{6}\right)\frac{(1440\ \cancel{\text{min}})}{(73\ \cancel{\text{min}})}} = e^{(-0.182)(19.7)} = 0.0274 \quad \text{or} \quad 2.74\%$$

Student: Isn't there an easier way?
Tutor: Well, we could say that $1440/73 = 19.7$ 73 minute periods pass, and during each one the amplitude is decreased by a factor of (5/6), so that after 24 hours the amplitude is

$$\left(\frac{5}{6}\right)^{19.7} = 0.0274$$

Tutor: But the way we did it is more general.
Student: Now we can find the *real* frequency.

$$\omega' = \sqrt{\frac{k}{m} - \frac{b^2}{4m^2}} = \sqrt{\omega^2 - \left(\frac{b}{2m}\right)^2}$$

Student: The period is 1 second, so

$$\omega = \frac{2\pi}{T} = 2\pi\ /\text{s}$$

Tutor: And we know the combination $b/2m$.

$$\frac{b}{2m} = -\frac{\ln(\frac{5}{6})}{(73\ \text{min})} = \frac{\ln(\frac{6}{5})}{(4380\ \text{s})}$$

Student: So

$$\omega' = \sqrt{(2\pi/\text{s})^2 - \left(\frac{\ln(\frac{6}{5})}{(4380\ \text{s})}\right)^2}$$

Tutor: Excellent.
Student: And we don't even need to know b or k or m. Cool.
Tutor: And what is it equal to?
Student: 6.28/s.
Tutor: Well, we knew that. The change in the period is really small, but it's that difference that we care about. Take your number and subtract 2π to find the difference.
Student: Zero.
Tutor: But you know it's not zero. The equation is clearly not equal to 2π.
Student: But it's really close.
Tutor: Yes it is. To find the difference, we need to use a Taylor series.
Student: What! We did those in calculus class. I hated them.
Tutor: As do many calculus students, because in calculus class they make you derive them. In physics, we look them up in a reference book.

$$\sqrt{1+x} = 1 + \frac{1}{2}x - \frac{1}{8}x^2 + \frac{1}{16}x^3 + \cdots \qquad \text{for } x << 1$$

Student: And I just look this up?
Tutor: Yes.
Student: But I don't have $1 + x$.
Tutor: Make it $1 + x$ by taking the 2π outside the square root.

$$\omega' = \sqrt{(2\pi/\text{s})^2 - \left(\frac{\ln(\frac{6}{5})}{(4380\ \text{s})}\right)^2} = (2\pi/\text{s})\ \sqrt{1 - \left(\frac{\ln(\frac{6}{5})}{2\pi(4380)}\right)^2}$$

Tutor: The right-hand part is much smaller than one, so it becomes x.

$$\omega' \approx (2\pi/\text{s})\left(1 - \frac{1}{2}\left(\frac{\ln(\frac{6}{5})}{2\pi(4380)}\right)^2\right)$$

Student: You included only the first two terms.
Tutor: Yes. That's the other reason that physicists and engineers like Taylor series. We only include the first nontrivial term.
Student: The 1 is "trivial?"
Tutor: The 1 says that the real frequency ω' is very nearly the same as ω. The second term is the difference between them.

$$\frac{\omega' - \omega}{\omega} \approx \frac{1}{2}\left(\frac{\ln(\frac{6}{5})}{2\pi(4380)}\right)^2 = 2.2 \times 10^{-11}$$

Student: That looks small.
Tutor: And that is why your calculator had trouble telling the difference between 2π and 2π minus that. The difference shows up in the twelfth significant digit.
Student: So why do we care?
Tutor: The difference tells us how much the damping has changed the frequency. There are about $\pi \times 10^7$ seconds in a year, and if we multiply by the error, we find that the damping causes the clock to lose 0.7 milliseconds per year.
Student: That's not much.
Tutor: No, but then the damping is very weak.

The last topic in this chapter is resonance. The calculations for resonance are typically saved for a differential equations or more advanced physics class. The behavior of resonance is worth knowing now.

It is possible to apply an oscillating force to an oscillator. That is, as it oscillates, we apply another force $F_d \cos(\omega_d t + \phi_d)$, called a driving force. We can do this without matching our frequency ω_d to the natural frequency of the oscillator $\omega = \sqrt{k/m}$.

When this happens, the oscillator can oscillate at two frequencies, the natural frequency ω and the driving frequency ω_d. Damping will cause the natural frequency to disappear (sometimes quicker than others), but the driving force will cause the oscillator to continue to oscillate at the driving frequency. The result is that **any driven oscillator will oscillate at the driving frequency rather than the natural frequency**. The closer these two frequencies are to each other, the greater the amplitude of the oscillation. When the two frequencies are equal, the oscillation amplitude is the greatest and we call this resonance. For example, when you hear sound, your ear vibrates at the frequency at which the sound pushes on it, rather than the natural frequency of your ear. If this was not true, then any sound would cause your ear to vibrate at the same frequency, and you could only hear one frequency — you could tell whether there was sound or not, but that is all.

Chapter 16

Waves — I

Imagine a group of oscillators, connected one after the other. We take the one on the near end and shake it. It puts a force on the next one, which drives the next one, and so on. Remember that driven oscillators oscillate at the driving frequency. So the oscillation that we start will be passed down the line at the same frequency.

The individual oscillators move up and down, but do not travel along the line of oscillators. Each oscillator stays in its place and vibrates. Similarly, when fans in a stadium do the "wave," each fan moves up and down as the wave passes, but the fans don't move to the next seat. What does not happen is that one section of fans stands up and then all of the fans scootch around the stadium. The *wave* that moves is the point at which the displacement is a maximum. That is, an oscillator along the wave moves up and down, and we find one that is up; an instant later it is no longer up but moving down. The next one is now up. An instant after that, the next one is up. This is the motion of the wave — the point that is at maximum displacement moves, and we call this the movement of the wave.

An important question then is how fast does the wave move. This is different from asking how fast do the individual oscillators move (still $v_m = \omega x_m$). An important principle of waves is that **all waves travel through a medium at the same speed** regardless of the frequency or amplitude of the wave. (A "medium" is whatever waves travel through.) Shaking our end of the train of oscillators will not cause the wave to move down the line any faster.

For any medium, it is possible to derive a "wave equation" of the form

$$\frac{d^2}{dx^2}y(x,t) = \frac{1}{v^2}\frac{d^2}{dt^2}y(x,t)$$

where v is a constant, and y is the displacement of the medium at a spot x and at time t. The solution to this equation is a wave that travels with speed v. Solutions come in two forms:

$$y(x,t) = y_m \sin(kx \pm \omega t + \phi_0)$$

$$y(x,t) = y_m \sin(kx)\cos(\omega t)$$

The first form is a traveling wave and the second form is a standing wave. (Like oscillators, the sines could be cosines and the cosines could be sines.)

In a traveling wave, the phase ϕ is the entire contents inside the parentheses (ϕ_0 is the phase constant). When the phase is $\pi/2$, $5\pi/2$, or $9\pi/2$, then the sine is 1 and the displacement is a maximum. The distance along x between successive maxima (from $\phi = \pi/2$ to $\phi = 5\pi/2$) is a wavelength λ.

As time increases, the phase increases or decreases, depending on the sign of the $\pm\omega t$ term. To follow a maximum, we need to keep the phase the same, so x must decrease or increase. The ratio of k and ω determines how much x must change as t increases, and thus the speed of the wave.

$$v = \frac{\omega}{k} = f\lambda = \frac{f}{T}$$

The second form $v = f\lambda$ is the most commonly used, and can be understood as: the speed of the wave is the rate at which oscillations occur f times the distance moved per oscillation λ.

The wavenumber k performs the same task for distance that the angular velocity ω does for time. If the time increases by one period T, then there is one oscillation and the phase changes by 2π. ω is then the radians per second by which the phase changes. If the position x increases by one wavelength λ, then there is one oscillation and the phase changes by 2π. k is then the radians per meter by which the phase changes.

$$\omega = \frac{2\pi}{T} \qquad k = \frac{2\pi}{\lambda}$$

The signs in the phase determine the direction of the wave. If the signs of kx and ωt are the same, then as time increases, x must decrease to keep the phase constant, and the wave moves in the $-x$ direction. If the signs of kx and ωt are opposite, then as time increases, x must increase to keep the phase constant, and the wave moves in the $+x$ direction.

Remember that the speed of the wave depends on the medium. If we increase the frequency at which we shake the end of the medium, even though $v = f\lambda$ the speed of the wave does not increase. Instead the wavelength decreases. For any given medium we could measure or calculate the speed of waves. One common medium is a tightly stretched string, for which the speed is

$$v_{\text{string}} = \sqrt{\frac{\tau}{\mu}}$$

Here τ is the tension in the string, and we use τ instead of T because we are using T for the period of the oscillation. μ is the mass density, or mass per length of the string ($mu = m/L$).

EXAMPLE

The figure shows a "picture" of a wave traveling along a steel wire at time $t = 2$ s. The wave travels at 52 m/s to the right and the tension in the wire is 60 newtons. Find the wavelength and frequency of the wave, the mass density of the wire, and write an equation for the wave.

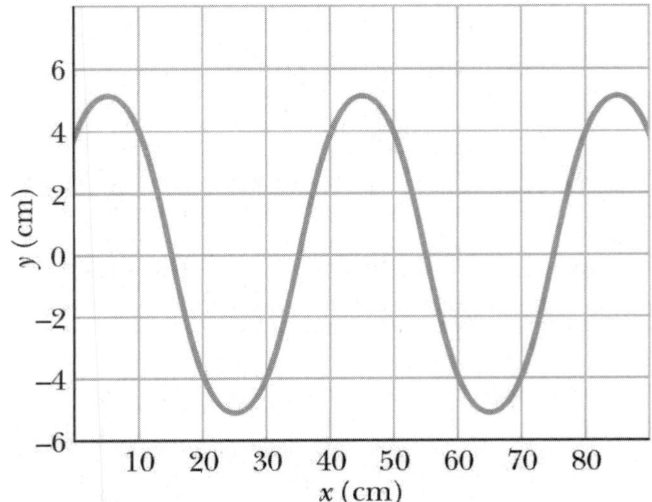

Student: This picture looks a lot like the one from the last chapter.
Tutor: Yes, but the axes have changed. x is no longer the displacement, but the position along the wire, and y is the displacement.
Student: Why couldn't we use x for the displacement?
Tutor: We could. $x(y,t) = x_m \sin(ky - \omega t)$ is a wave moving in the $+y$ direction, with the oscillations happening in the x direction.
Student: Okay. The amplitude y_m is 5 cm, because that's the greatest that the displacement gets from zero.
Tutor: Correct. Can you identify the wavelength?
Student: There are peaks at $x = 5$ cm and $x = 45$ cm so the wavelength λ is 40 cm.
Tutor: What is the velocity of the wave?
Student: That's the $v = f\lambda$ one, rather than the $v_m = \omega x_m$, right?
Tutor: Correct. $v_m = \omega y_m$ is the maximum speed of each piece of the medium. Imagine tying a ribbon to part of the wire. Then $v_m = \omega y_m$ is the fastest that the ribbon moves up and down.
Student: And it's y_m rather than x_m because the motion of the ribbon is in the y direction.
Tutor: Yes.
Student: But I want the wave speed $v = f\lambda$.

$$v = f\lambda \quad \rightarrow \quad f = \frac{v}{\lambda} = \frac{52 \text{ m/s}}{40 \text{ cm}} \frac{100 \text{ cm}}{1 \text{ m}} = 130 \text{ /s}$$

Tutor: Good.
Student: Now I can use $v_{\text{string}} = \sqrt{\frac{\tau}{\mu}}$ to find the mass density μ.

$$v_{\text{string}} = \sqrt{\frac{\tau}{\mu}} \quad \rightarrow \quad \mu = \frac{\tau}{v^2} = \frac{60 \text{ N}}{(52 \text{ m/s})^2} = 0.0222 \text{ kg/m} = 22.2 \text{ g/m}$$

Student: Now I need to write the equation for the wave. It should look like $y(x,t) = y_m \sin(kx \pm \omega t + \phi_0)$. I know the amplitude $y_m = 5$ cm. The wavenumber $k = \frac{2\pi}{\lambda} = \frac{2\pi}{0.40 \text{ m}} = 15.7$ /m. Radians are a virtual unit, not a real unit, and I can write $k = 15.7$ rad/m.
Tutor: I don't think I've heard the phrase "virtual unit" before, but it fits.
Student: I can get the angular velocity ω from the frequency f, $\omega = 2\pi f = 2\pi(130 \text{ /s}) = 817$ /s.

$$y(x,t) = (5 \text{ cm}) \sin\left[(15.7 \text{ /m})x \pm (817 \text{ /s})t + \phi_0\right]$$

Student: The x and t are the independent variables, and don't have specific values.
Tutor: All very good. All that's left is to choose between + and −, and to determine ϕ_0.
Student: If the signs of kx and ωt are opposite, then the wave moves in the $+x$ direction, so they must be opposite. Does that mean I could write $(kx - \omega t)$ or $(-kx + \omega t)$?
Tutor: That's what it means. You might get different values of ϕ_0 depending on which you choose.
Student: I choose $(kx - \omega t)$.

$$y(x,t) = (5 \text{ cm}) \sin\left[(15.7 \text{ /m})x - (817 \text{ /s})t + \phi_0\right]$$

Student: Now I look at the peak at $x = 5$ cm, and that corresponds to zero phase, like in the last chapter.
Tutor: Not so fast. You have the right idea, but we're using sine instead of cosine.
Student: Why did we change?
Tutor: We use sine here for the same reason that we used cosine with oscillators — it improves the chances that the phase constant is zero. But that doesn't mean that the phase constant here is zero. Because we're using sine, a maximum occurs when the phase is $\pi/2$ rather than at zero.
Student: So at $x = 5$ cm and $t = 2$ s the phase is $\pi/2$?
Tutor: Yes, or $5\pi/2$, or $9\pi/2$, or so on. It doesn't matter which you use.
Student: Then I'll use $\pi/2$.

$$\left[(15.7 \text{ /m})(0.05 \text{ m}) - (817 \text{ /s})(2 \text{ s}) + \phi_0\right] = \frac{\pi}{2}$$

Tutor: There's a quicker way to an answer.

$$y(x,t) = (5 \text{ cm}) \sin\left[(15.7\ /\text{m})(x - 0.05 \text{ m}) - (817\ /\text{s})(t - 2 \text{ s}) + \frac{\pi}{2}\right]$$

Student: It seems simple, I guess. How do you know that it works?
Tutor: It's clear that if $x = 5$ cm and $t = 2$ s the phase is $\pi/2$, yes?
Student: Sure, because the first two terms of the phase are zero.
Tutor: As we move away from $x = 5$ cm, the phase changes by 15.7 rad/m, the wavenumber. And as time changes, the phase changes by 817 rad/s, the angular velocity.
Student: Yes, those must both be true, and the signs for x and t are opposite, so the direction is correct.

If you take two solutions to the wave equation and add them, you get another solution. This is how we get waves that are not shaped like sine or cosine, by adding many sine or cosine waves together. All of the waves travel the same speed because they are traveling in the same medium.

There are three important cases of adding waves that appear many times, and are worth learning about:

- two (or more) waves traveling the same direction, identical except perhaps for a phase difference,
- two waves traveling the same direction, identical except for a slight difference in frequency, and
- two waves, identical except that they travel in opposite directions.

When two identical waves travel the same direction, they add together to form a wave that is twice as big, having twice the amplitude. When the waves are exactly "out of phase," or one is half a wavelength behind the other, they are exact opposites and they add to zero. We call these constructive interference and destructive interference, respectively. If they are somewhere in between, then using a trig identity

$$\sin\alpha + \sin\beta = 2\sin\left(\frac{\alpha+\beta}{2}\right)\cos\left(\frac{\alpha-\beta}{2}\right)$$

$$y_m \sin(kx - \omega t + \phi) + y_m \sin(kx - \omega t) = 2\cos\left(\frac{1}{2}\phi\right) y_m \sin\left(kx - \omega t + \frac{1}{2}\phi\right)$$

so they add to form a new wave, identical to the first two, but with an amplitude of $2\cos\left(\frac{1}{2}\phi\right)$ times the original amplitudes. Sure enough, for $\phi = 0$ this is 2 and for $\phi = \pi$ this is 0.

When two waves have slightly different frequencies,

$$\sin\alpha + \sin\beta = 2\sin\left(\frac{\alpha+\beta}{2}\right)\cos\left(\frac{\alpha-\beta}{2}\right)$$

$$y_m \sin(kx - \omega_1 t) + y_m \sin(kx - \omega_2 t) = 2y_m \cos\left((\omega_2 - \omega_1)\,t\right) \sin\left(kx - \frac{\omega_1 + \omega_2}{2}t\right)$$

The result is a wave with the average frequency, with an amplitude that oscillates at the difference in the two frequencies. We call this phenomena beats, and the difference frequency is the beat frequency.

EXAMPLE

What phase difference must there be between two otherwise identical waves so that their sum has half the amplitude of each of them?

Student: This seems pretty straightforward.
Tutor: Okay. How are you going to do it?

Student: With the formula above.

$$2\cos\left(\frac{1}{2}\phi\right) y_m = \frac{1}{2} \quad ?$$

Tutor: The units on the left are length, but the right side doesn't have units. Something must be wrong.
Student: Oh, I need the amplitude that the waves start with.

$$2\cos\left(\frac{1}{2}\phi\right) y_m = \frac{1}{2} y_m$$

Student: But the original amplitude cancels.
Tutor: Yes, because in this problem we are comparing to the original amplitude, instead of to meters or centimeters.

$$\cos\left(\frac{1}{2}\phi\right) = \frac{1}{4}$$

$$\frac{1}{2}\phi = \arccos\left(\frac{1}{4}\right)$$

Student: Does it matter whether I use radians or degrees?
Tutor: No. Like any unit, you need to write the units and use them consistently.

$$\phi = 2\arccos\left(\frac{1}{4}\right) = 151^\circ = 2.64 \text{ rad}$$

Student: There, I did both for good measure. But what if the waves don't have the same amplitude? Do we have a formula for that?
Tutor: We have a technique that we call phasors.
Student: Isn't that some kind of stun gun?
Tutor: Only in science fiction. In physics, it's a way to add waves by adding vectors. Imagine a vector that initially points in the $+x$ direction, but then rotates counterclockwise. What would the x component be?
Student: It would be $\cos\theta$.
Tutor: As long as we measure θ from the x axis, yes. As the wave passes a point, let the vector rotate at angular velocity ω, and then the x component is the displacement. Now imagine that we have another wave, similar to the first but $\pi/2$ behind and with half the amplitude. We can draw another vector, in the $-y$ direction and half as long. Because the waves have the same frequency, these rotate at the same rate. To add the waves together, we add the vectors together.

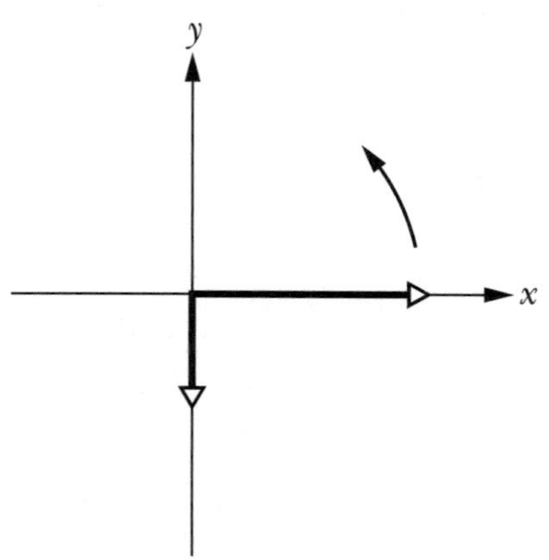

Student: So the sum is a wave with amplitude $\sqrt{(1)^2 + (0.5)^2} = 1.12$?
Tutor: Yes, and the angle of the resultant vector tells us the phase of the resulting wave.
Student: That's kind of cool. And we can add more than one vector this way?
Tutor: As many as you want.
Student: So how can you explain beats with phasors?
Tutor: Because the two waves have slightly different frequencies, the vectors rotate at slightly different frequencies. When they point in the same direction, the amplitude is large, and when they point in opposite directions, they add to zero and the amplitude is small.
Student: But they don't rotate together.
Tutor: No, but they rotate much faster than the difference frequency, which is how the difference in the vectors changes.

The last important case is two waves traveling in opposite directions.

$$\sin\alpha + \sin\beta = 2\sin\left(\frac{\alpha+\beta}{2}\right)\cos\left(\frac{\alpha-\beta}{2}\right)$$

$$y_m\sin(kx+\omega t) + y_m\sin(kx-\omega t) = 2y_m\cos(\omega t)\sin(kx)$$

Each piece of the medium is an oscillator, but instead of moving one after the other, they all move in sync. **Some places have an amplitude of $2y_m$ and others have an amplitude of zero**. Because these waves don't move left or right, but oscillate in place, we call them standing waves. We call the spots where the amplitude is zero nodes, and the spots where the amplitude is a maximum antinodes.

How can you get two identical waves but going in opposite directions? You tie down one end of your string. **It is a general principle of waves that when a wave encounters a change of medium, in which the wave would go a different speed, some of the wave is reflected and some is transmitted**. If the boundary is especially hard, so that none of the wave is transmitted, then the entire wave is reflected and we have an identical wave going the opposite direction. If the medium off of which the wave reflects is one in which it would go slower, then the wave is flipped upon reflection, so that the incoming and reflected waves add to zero, producing a node at the endpoint.

If both ends of a string are bound then the wavelength is constrained. Nodes are half a wavelength apart, so if both ends must be nodes, then the length of the medium must be an integer times half a wavelength.

$$L = n\frac{\lambda}{2} \qquad \text{or} \qquad \lambda = \frac{2L}{n}$$

where n is any positive integer. This means that there are many acceptable wavelengths, but they are a discrete set — only those wavelengths will work. This is the basis of musical instruments. In a guitar, for example, the wavelength must be twice the length of the string, or an integer times that.

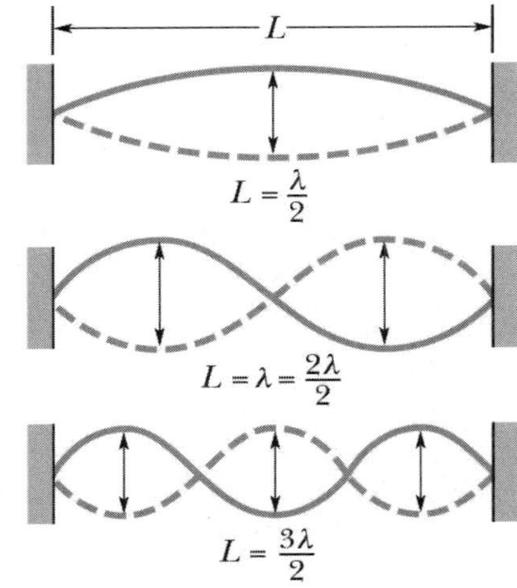

We can then use $v = f\lambda$ to determine the frequency.

$$f = \frac{v}{\lambda} = nf_1$$

The important point is that **there is a fundamental frequency f_1, and the other allowed frequencies are integers times the fundamental, called harmonics**. The integer n is the same integer for how many half-wavelengths fit in the medium, and is the number of antinodes present. Sometimes these higher frequencies are called overtones. The difference is that, if a medium is more than one dimensional, there may be allowed frequencies that are not integer multiples of the fundamental, and the word overtones covers these.

I strongly recommend *against* memorizing an equation for the frequencies of a standing wave. One of the most common mistakes made with standing waves is to pull out a formula for the frequencies and use it improperly, or where it doesn't apply. Instead, picture the waveform, apply $L = n\frac{\lambda}{2}$ or $\lambda = \frac{2L}{n}$, use $v = f\lambda$, and find the frequencies. It is always better to learn to use a few equations well than to memorize more equations, and especially here.

EXAMPLE

A 74 cm string clamped at both ends with a mass density of 0.310 g/m has a fundamental frequency of 440 Hz (the note A). How far from the end would you have to hold the string down so that the third harmonic is the note F (1398 Hz)?

Student: I don't see how to get from here to there.
Tutor: So start with what you would need.
Student: If the third harmonic is 1398 Hz, then the fundamental is

$$f_3 = 3f_1 \quad \rightarrow \quad f_1 = \frac{f_3}{3} = \frac{1398\text{ Hz}}{3} = 466\text{ Hz}$$

Student: Does that mean that the first harmonic is equal to the fundamental?
Tutor: Indeed it does. The first harmonic is one times the fundamental, or equal to the fundamental. The first overtone is the first frequency above the fundamental, or the second harmonic. What is the wavelength of the 466 Hz fundamental?
Student: How can the fundamental be both 440 Hz and 466 Hz?
Tutor: It isn't both, so much as first one and then the other. Initially the fundamental is 440 Hz, but then we hold the string down partway along its length. When someone plays a guitar, they hold the string down on a "fret" along the guitar's neck. That shortens the string, so the wavelength will be shorter and the frequency higher.
Student: Because the medium hasn't changed and the velocity is the same, right? But we've changed the string, so the medium has changed.
Tutor: The tension is the same and the mass density is the same, so the speed of the wave is the same.
Student: So it's really two different strings.
Tutor: Yes, but with the same speed.
Student: So I can write

$$f_{\text{before}}\lambda_{\text{before}} = v_{\text{before}} = v_{\text{after}} = f_{\text{after}}\lambda_{\text{after}}$$

Tutor: Very good.
Student: Do I even need to find the wave speed?
Tutor: Nope.
Student: I know the frequency before and after holding down the string.

$$(440\text{ Hz})\lambda_{\text{before}} = (1398\text{ Hz})\lambda_{\text{after}}$$

Student: How does the fundamental afterward come into it.
Tutor: You could use it. I'll show you in a minute. But since you've opted for the third harmonic, we can

use that. What is the wavelength of the fundamental before?
Student: It's the length of the string.
Tutor: Think about what the waveform looks like. Is that one wavelength?
Student: No, it's only half a wavelength. The fundamental wavelength (can we say that?) is twice the length of the string, or 148 cm.

$$(440 \text{ Hz})(148 \text{ cm}) = (1398 \text{ Hz})\lambda_{\text{after}}$$

Student: Now I can solve.

$$\lambda_{\text{after}} = \frac{(440 \text{ Hz})(148 \text{ cm})}{(1398 \text{ Hz})} = 46.6 \text{ cm}$$

Tutor: Good. How can you use the wavelength of the third harmonic?
Student: That's right, it's the third harmonic, not the fundamental. So the wavelength is $2L/3$.

$$\frac{2L}{3} = \lambda_{\text{after}} = 46.6 \text{ cm}$$

$$L = 69.9 \text{ cm}$$

Student: The guitar player has to hold the string down 69.9 cm from one end.
Tutor: Or $74 - 69.9 = 4.1$ cm from the other end, as he probably thinks about it. If you had used the fundamental frequency afterward instead of the third harmonic, then you would have gotten a wavelength of about 140 cm, but then the wavelength would have been $2L$ instead of $2L/3$. It didn't matter which you used, so long as you matched wavelength and frequency.
Student: Is that why grand pianos are curved, because the strings are different lengths?
Tutor: Yes. You could change the length, or you could change the tension or mass density and thus change the speed. Remember that changing the speed will change the frequency because the wavelength is fixed by the length of the string. (You tune a guitar or piano by changing the tension.) The high notes are short strings and they get longer for lower notes. Eventually the strings would become too long, and we can't really put more tension in the strings, so thicker strings are used for the lowest notes. Guitar strings are of different thicknesses so that they can have the same length and tension but play different notes.
Student: Can you explain standing waves using phasors?
Tutor: If you like. The two vectors now rotate in opposite directions. If the start both vertical, then they will always add to a vertical vector with no x component, and we get a node — no displacement ever. If they start both to the right ($+x$), then they will add to a horizontal vector that varies from $+2$ to -2, and we get an antinode.

The fundamental (pardon the pun) equation for all waves is

$$v = f\lambda$$

Here's how to keep straight which of the three variables is determined.

- Any two waves in the same medium have the same speed.
- A wave keeps the same frequency as it crosses from one medium into another.
- Standing waves, with nodes at fixed points, have certain allowed wavelengths.

We've seen examples of the first and last. For the second, consider a piano. The strings vibrate at a frequency f, and strike the air molecules, causing them to vibrate at the same frequency f.

Chapter 17

Waves — II

Air may seem empty, but that is merely relative. Air contains about 10^{19} molecules per cubic meter. Liquids and solids, for comparison, contain about 10^{29} molecules per cubic meter.

It can be helpful to think of sound as the molecules in the air vibrating, but this is not strictly true. Because of the lower density, the molecules in air move around, travelling a short distance (centimeter-like) before colliding with another air molecule. Instead the pressure of the air oscillates, changing by a small amount of maybe $\frac{1}{1000}$ of atmospheric pressure, and this region of slightly higher pressure is the wave that moves.

The speed of sound depends on the molecules in the air and the conditions, so that it changes with temperature, for example. At standard conditions (atmospheric pressure, room temperature) the speed of sound is 343 m/s. This is much slower than light, so that when a spaceship in a science fiction movie blows up, you would see the explosion before hearing it — like lightning on Earth. Really, because there are so few molecules in outer space, you wouldn't hear the explosion at all.

Having described string instruments in the last chapter, we examine wind instruments here. Imagine a pipe with one open end and one closed end. At the closed end, the air can't move very far, because it would leave a vacuum in the end. At the open end, the air molecules are free to move. But the open end is literally out in the open, so the pressure there remains atmospheric pressure, while the pressure in the closed end can change. So an open end is a pressure node but a displacement (of the air molecules, in and out of the pipe) antinode. A closed end is a displacement node but a pressure antinode.

Confusing? Remember that **the two ends of an open pipe have to be the same, but the two ends of a closed pipe have to be different**. So the allowed wavelengths of an open pipe (open at both ends) are similar to those of a string. But for a closed pipe (closed at one end; nothing would happen if both ends were closed) the length must be $\frac{1}{4}\lambda$, $\frac{3}{4}\lambda$, $\frac{5}{4}\lambda$, and so on.

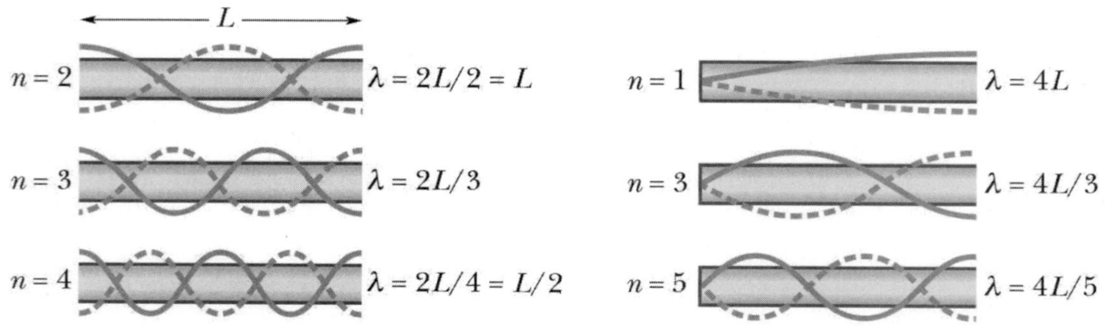

L

$n = 2$ $\lambda = 2L/2 = L$ $n = 1$ $\lambda = 4L$

$n = 3$ $\lambda = 2L/3$ $n = 3$ $\lambda = 4L/3$

$n = 4$ $\lambda = 2L/4 = L/2$ $n = 5$ $\lambda = 4L/5$

EXAMPLE

A closed pipe in standard conditions has a length of 32.7 cm. When an open pipe is played at the same time, the beat note between the fundamental of the open pipe and the third harmonic of the closed pipe is 3 Hz. How much longer must the open pipe be so that the two frequencies match?

Student: This looks complex! Where can I start?
Tutor: Can you find the frequency of the closed pipe?
Student: Which, the fundamental or the third harmonic?
Tutor: Either. Of course we'll need the third harmonic eventually, but let's start with the fundamental.
Student: Well, because it's open at one end and closed at the other, one is a node and the other is an antinode. Does it matter which is which?
Tutor: Usually not. It would depend on whether you're looking at air pressure or at air molecule displacement. You do have the important thing, which is that they are opposites.
Student: In a standing wave, nodes and antinodes are a quarter of a wavelength apart. So for the fundamental, the length of the pipe is a quarter wavelength ($L = \frac{1}{4}\lambda$).
Tutor: Very good.
Student: The second harmonic is when the length is $\frac{3}{4}\lambda$ and the third is when it is $\frac{5}{4}\lambda$.
Tutor: You have the right idea, but incorrect terminology. The first overtone is indeed when the length is $\frac{3}{4}\lambda$ and the second overtone is when it is $\frac{5}{4}\lambda$. But when $L = \frac{3}{4}\lambda$, the wavelength is $\frac{1}{3}$ what it was, and the frequency is three times higher, so we call it the third harmonic.
Student: So what's the second harmonic?
Tutor: There is no second harmonic for a closed pipe. A medium with a node and an antinode at the two ends doesn't have even harmonics, only odd ones.
Student: So if $L = \frac{5}{4}\lambda$ it's the fifth harmonic?
Tutor: Yes, because the frequency is five times the fundamental frequency.
Student: Okay. For the third harmonic

$$L = \frac{3}{4}\lambda \quad \rightarrow \quad \lambda = \frac{4}{3}L = \frac{4}{3}(32.7\text{ cm}) = 43.6\text{ cm}$$

$$f = \frac{v}{\lambda} = \frac{343\text{ m/s}}{0.436\text{ m}} = 787\text{ Hz}$$

Tutor: Okay. When the two pipes are played at the same time, we hear a beat frequency of 3 Hz.
Student: So the frequency from the open pipe is 784 Hz.
Tutor: Or 790 Hz. You can't tell from the beat frequency which of the two is the higher frequency.
Student: So I'm stuck? The open pipe frequency is 784 or 790 Hz and I can't tell which?
Tutor: The problem does say that to make the two pipes match, we need to make the open pipe longer. Does the frequency increase or decrease as the pipe gets longer?
Student: If the pipe gets longer, then the wavelength gets longer too. If λ gets bigger, then f gets smaller.
Tutor: So to make them match, we're going to make f smaller.
Student: So the frequency of the open pipe must be 790 Hz. Do we have an equation for the change in frequency?
Tutor: No, but we could determine the length, and then the needed length, and subtract.
Student: I was hoping for an easier way.
Tutor: You mean a single step?
Student: Yes.
Tutor: Having fewer equations and putting them together *is* easier than having an equation for every need. We'd have nearly infinite equations.
Student: Okay. The length of the open pipe is half the fundamental wavelength.

$$L = \frac{1}{2}\lambda \quad \& \quad v = f\lambda \quad \rightarrow \quad L = \frac{1}{2}\frac{v}{f}$$

$$L_1 = \frac{v}{2f_1} = \frac{(343\text{ m/s})}{2(790\text{ Hz})} = 21.71\text{ cm}$$

$$L_2 = \frac{v}{2f_2} = \frac{(343 \text{ m/s})}{2(787 \text{ Hz})} = 21.79 \text{ cm}$$

$$L_2 - L_1 = 21.79 \text{ cm} - 21.71 \text{ cm} = 0.08 \text{ cm} = 0.8 \text{ mm}$$

Student: That's not very much.
Tutor: To tune a wind instrument, like a clarinet, you take two of the pieces and pull them just a little apart. Pulling them 1 mm apart can change the frequency by a couple of hertz.

In general, when two or more waves add we call it interference. There are many examples of interference, such as standing waves when you add two waves traveling in opposite directions.

We now consider adding two waves coming from arbitrary directions, but with the same frequency. If peaks arrive at the same time, then we have constructive interference where the waves meet. It doesn't matter that the waves came from different directions. If a peak from one wave arrives at the same time as a dip (maximum negative displacement) from the other, then we get destructive interference.

Usually the two sources are coherent, or "in-phase," meaning that they produce peaks at the same time. If the wave from one has to travel for an extra period, then a peak from that wave arrives just as the next peak from the closer source arrives. The same is true if the further wave travels for an extra two, three, or any integer number of periods. Because a wave travels one wavelength in one period, **constructive interference occurs when one source is an integer number of wavelengths further away than the other**. If the wave from one has to travel for an extra half period, then a peak from that wave arrives just as the next dip from the closer sources arrives. The same is true if the further wave travels for an extra one and a half, two and a half, or any half-integer (integer plus a half) number of periods. Therefore, **destructive interference occurs when one source is a half-integer number of wavelengths further away than the other**.

$$\Delta L = |L_2 - L_1| = \begin{cases} m\lambda & \text{constructive interference} \\ (m + \frac{1}{2})\lambda & \text{destructive interference} \end{cases}$$

EXAMPLE

A person stands near two in-phase speakers as shown. Find the three lowest frequencies at which destructive interference occurs.

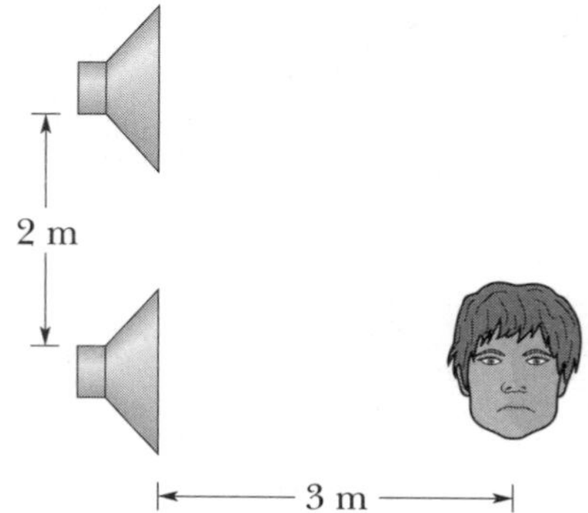

Student: I want destructive interference, so I need $\Delta L = (m + \frac{1}{2})\lambda$.
Tutor: Good. What is ΔL?
Student: The difference between the two distances. L_2 is 3 meters, or is that L_1?
Tutor: It doesn't matter, and we don't care whether the difference is positive or negative.

Student: So I can just say that one distance is 3 m and the other is $\sqrt{2^2+3^2} = 3.60$ m?
Tutor: Yes.
Student: Then the difference is $3.60 - 3.00 = 0.60$ m. But the frequency hasn't shown up in any of the equations.
Tutor: You can still use $v = f\lambda$.
Student: Ah yes. If I can find the wavelength, then I can find the frequency. ΔL is half a wavelength, or one and a half wavelengths, and so on.

$$(m + \tfrac{1}{2})\lambda = \Delta L = 0.60 \text{ m}$$

$$f = \frac{v}{\lambda} = \frac{(m+\frac{1}{2})v}{\Delta L} = (m+\tfrac{1}{2})\frac{343 \text{ m/s}}{0.60 \text{ m}} = (m+\tfrac{1}{2})(572 \text{ Hz})$$

Tutor: Very good.
Student: And m can be any integer, so I put in 1, 2, ...
Tutor: The lowest frequency occurs when $m = 0$.
Student: Oh yes, because then ΔL is $\frac{1}{2}\lambda$.

$$\begin{aligned} m = 0 \quad &\rightarrow \quad f = (\tfrac{1}{2})(572 \text{ Hz}) = 286 \text{ Hz} \\ m = 1 \quad &\rightarrow \quad f = (\tfrac{3}{2})(572 \text{ Hz}) = 858 \text{ Hz} \\ m = 2 \quad &\rightarrow \quad f = (\tfrac{5}{2})(572 \text{ Hz}) = 1430 \text{ Hz} \end{aligned}$$

Student: How far can we go?
Tutor: You could keep going, but the human ear doesn't hear well above about 20,000 Hz.
Student: Couldn't we just combine what we've done and get a formula for the destructive frequencies?
Tutor: We could do that for any problem. The question is whether we should. If an equation is used over and over, is difficult to do mathematically, and not prone to misuse, then it makes sense. This was true for elastic collisions, where we had two equations and one of them had v^2 in it. Is this something that we'll need many times?
Student: It seems like it could come up more than once.
Tutor: It could. Is it hard to do the math?
Student: No, it wasn't.
Tutor: But experience teaching this topic has taught me that it is prone to misuse. Two out of three says to stick to two simple equations and not try to memorize something else.
Student: Fair enough.

Waves carry energy. Light from the Sun brings or brought us most of the energy we use (brought for fossil fuels). We describe this energy by its power and intensity. The power is the energy per time, as before. The intensity is the power per area.

Imagine turning on a flashlight. The flashlight bulb emits a certain amount of power in the form of light. As the light leaves the flashlight it spreads out. The further away from the flashlight, the lower the intensity, as the light spreads out over a larger area, but the power is the same. The brightness or **intensity of any wave is proportional to the amplitude squared**.

In many cases we use a point source that radiates uniformly in all directions. The math is easier, and there are many sources that come close to this. The intensity is the power divided by the area of a sphere over which the power is spread.

$$I = \frac{P}{4\pi r^2}$$

This is true for any type of wave emanating from a point source.

We often describe not the intensity of sound waves in W/m^2 but the sound level in decibels

$$\beta = (10 \text{ dB}) \log \frac{I}{I_0}$$

where $I_0 = 10^{-12}$ W/m^2 is approximately the threshold of human hearing. Decibels are "factors of ten," so 10 dB is 10 times greater than 0 dB (0 dB is not zero sound), and 20 dB is 10 times greater than 10 dB, or 100 times greater than 0 dB. Think of 20 dB as "2 factors of ten above 0 dB." 3 dB is approximately a factor of 2, so 23 dB is two times greater than 20 dB, or 200 times greater than 0 dB. Normal conversation is about 60 dB, or about a million times more intense than the threshold of hearing (six factors of ten). Decibels can be used in a relative or absolute sense, so one signal can be +23 dB relative to another (200 times as large), or 23 dB (200 times I_0, or threshold). (We use decibels because the ear is logarithmic.)

EXAMPLE

A man hears a jackhammer at a sound level of 95 dB. If we moves twice as far away, what will the sound level be? What if he moves 20 times further away?

Student: So I have to use the decibel equation to find the power of the jackhammer, then double r and find the new intensity, then ... wait a minute, it doesn't say how far away from the jackhammer he starts. How can I find the power?
Tutor: You can't. But see where it says "twice as far away"?
Student: It's a scaling problem. I write the equation before and after.

$$I_{\text{before}} = \frac{P_{\text{before}}}{4\pi r_{\text{before}}^2} \qquad \text{and} \qquad I_{\text{after}} = \frac{P_{\text{after}}}{4\pi r_{\text{after}}^2}$$

Tutor: Yes. What is the same between the two?
Student: Well, π before is the same as π afterward.
Tutor: True but not enough.
Student: Okay, the power is the same before and after, but the radius has changed.

$$\frac{I_{\text{after}}}{I_{\text{before}}} = \frac{\frac{\cancel{P_{\text{after}}}}{4\pi r_{\text{after}}^2}}{\frac{\cancel{P_{\text{before}}}}{4\pi r_{\text{before}}^2}} = \frac{r_{\text{before}}^2}{r_{\text{after}}^2} = \left(\frac{r_{\text{before}}}{r_{\text{after}}}\right)^2$$

Student: Now $r_{\text{after}} = 2r_{\text{before}}$, so $\frac{I_{\text{after}}}{I_{\text{before}}} = \frac{1}{4}$.
Tutor: Good.
Student: We're using dB in a relative sense, rather than compared to I_0.

$$\beta = (10 \text{ dB}) \log \frac{I_{\text{after}}}{I_{\text{before}}} = (10 \text{ dB}) \log \frac{1}{4} = -6.02 \text{ dB}$$

Tutor: True. You could also say that 4 is two factors of two, so −3 dB for the first factor of two, then −3 dB more for the second factor of two.
Student: And you don't multiply?
Tutor: You never multiply dB, because dB is "factors of ten."
Student: Can you always do this tricky stuff? What's −17 dB?
Tutor: If you go down a factor of ten (−10 dB), then down another factor of ten (−10 dB), then up a factor of 2 (+3 dB), you're at −17 dB, so −17 dB is down a factor of 50.
Student: And if he stands three times further away, the intensity is 9 times less.
Tutor: −10 dB is 10 times less, and $-9 = (-3) + (-3) + (-3)$ dB is three factors of 2 or 8 times less, so it's between −9 and −10 dB.
Student: Seems simple.
Tutor: You need to remember that **you add dB to multiply the intensity**. By the way, the problem didn't ask for the relative sound level, it asked for the sound level.
Student: Ah, he starts at 95 dB, and the change is −6 dB, so the result is $95/6 = 16$ dB. No, you add dB. It's $95 - 6 = 89$ dB. That doesn't seem like much of a change.
Tutor: It isn't. The ear perceives a factor of 10 in intensity to be "one notch" in sound, so he hasn't really reduced the sound much. What if he stands 20 times further away?

Student: Then the intensity is $20^2 = 400$ times lower. 400 is $10 \times 10 \times 2 \times 2$, so $(-10)+(-10)+(-3)+(-3) = -26$ dB. 26 dB below 95 dB is 69 dB.
Tutor: Excellent. 69B is three factors of 2 or 8 times louder than normal conversation.
Student: Even though he's moved 20 times further away?!
Tutor: That's why you should wear ear protection when operating a jackhammer, and it's hard to have a conversation near one.

Our last sound phenomenon is the Doppler effect. If we move, or the source of the sound moves, then the frequency of the sound changes.

Peaks of the sound, spots of slightly higher air pressure, are a wavelength apart as they move toward us. Waves travel one wavelength in one period, so we receive the peaks at intervals of a period. If we move toward the source of the sound, then it takes less time than a period (as measured by the source) for the next peak to get to us. We perceive a shorter period than the source does, and a higher frequency. If the source moves toward us, then one period later, as it emits the next peak, the previous one is not a full wavelength away — it is a wavelength minus the distance the source moved in a period.

The frequency that a "detector" hears is

$$f' = f\left(\frac{v \pm v_d}{v \mp v_s}\right)$$

where f is the frequency of the source, v_d and v_s are the velocities of the detector and source, respectively, and v is the speed of sound. Which of + and − do you use? One way is to remember that motion of the source or detector toward the other increases the frequency, and use the sign to produce the desired result. Written the way I have it here, use the top sign for motion toward the other and the bottom sign for motion away from the other. If the detector moves toward the source, then use the + sign in the numerator. If the source moves away from the detector, use the bottom sign or + sign in the denominator.

EXAMPLE

A ship is chasing a submarine. To detect the submarine, the ship uses sonar, sending out a sound wave and detecting the reflected sound. The submarine is moving at 8 m/s and the ship chases it at 20 m/s. If the ship sends out a 700 Hz sound wave, what frequency do they hear for the return wave? The speed of sound in water is 1500 m/s.

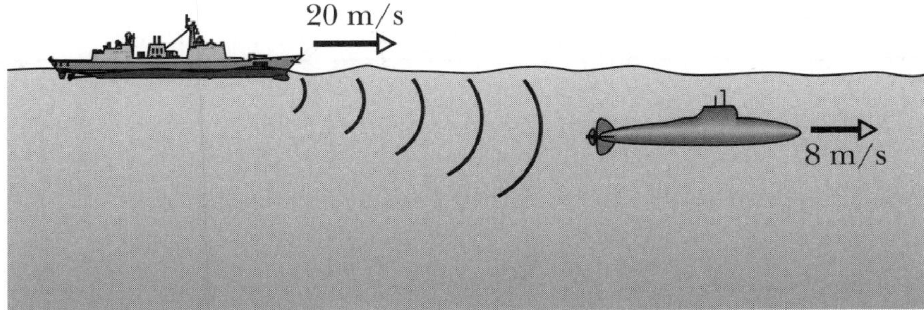

Student: The ship is both source and detector, so there's no change.
Tutor: It's not quite that simple. The ship is the source and the sub is the detector. The sub hears a different frequency than the ship sends.
Student: The detector moves away from the source at 8 m/s, and because it's away we use the bottom sign or − in the numerator. The source moves toward the detector, and because it's toward we use the top

sign or − in the denominator.

$$f_{\text{sub}} = f\left(\frac{v \pm v_d}{v \mp v_s}\right) = (700\text{ Hz})\left(\frac{(1500\text{ m/s}) - (8\text{ m/s})}{(1500\text{ m/s}) - (20\text{ m/s})}\right) = (700\text{ Hz})\left(\frac{1492\text{ m/s}}{1480\text{ m/s}}\right) = 705.7\text{ Hz}$$

Student: The sub is moving away from the ship. Shouldn't they hear a lower frequency?
Tutor: The ship is moving toward the sub faster than the sub is moving away from it.
Student: Okay. Couldn't we have used the relative velocity, saying that the sub was stationary and the ship moving 12 m/s?
Tutor: We can't because the sound travels through the water. If the sub moves at the speed of sound, then the peaks never reach the sub, even though the ship is moving faster than the sub.
Student: But the ship would be moving faster than sound. Can that happen?
Tutor: It is possible to move faster than sound. Supersonic airplanes are ones that move faster than sound in air. If the source moves at the speed of sound, then each peak is on top of the previous one, because the source is at the same place as the previous peak as it releases the next one.
Student: That's how you get a sonic boom.
Tutor: More or less. And the equation would say that the received frequency was infinite, reflecting that all of the peaks would arrive at the same time. Here, if we used relative velocity as you suggest, we would get a very small error in our answer. The error is small because the speeds of the vessels are so much slower than the speed of sound.
Student: Okay. But we have the frequency that the submarine hears. We want the frequency that the ship hears.
Tutor: True. The waves reflect off of the submarine.
Student: Is that because the steel of the submarine is a different medium than the water?
Tutor: Very good. Whenever a wave encounters a change in medium, some of the wave is reflected. Most of the sound reflects off of the air inside the submarine, rather than the steel of the submarine. Now the submarine is the source of the reflected wave, and that reflected wave starts with a frequency of 705.7 Hz.
Student: So the ship is the detector and hears

$$f_{\text{ship}} = f_{\text{sub}}\left(\frac{v \pm v_d}{v \mp v_s}\right) = (705.7\text{ Hz})\left(\frac{(1500\text{ m/s}) + (20\text{ m/s})}{1500\text{ m/s}) + (8\text{ m/s})}\right) = (705.7\text{ Hz})\left(\frac{(1520\text{ m/s}}{1508\text{ m/s}}\right) = 711.3\text{ Hz}$$

Tutor: Yes, and you used the + sign in the numerator because the ship was moving toward the source, and the + sign in the denominator because the sub was moving away from the detector.
Student: And the ship can use this to determine where the sub is?
Tutor: They can use the shift in frequency to determine the speed of the sub. They measure the time that it takes for the reflected wave to arrive to determine how far away the sub is.
Student: Can't they just look?
Tutor: Light does not travel far through seawater, so the submarine isn't visible. Sound travels quite well through seawater.

Chapter 18

Temperature, Heat, and the First Law of Thermodynamics

What happens when we cook food? If there is no force exerted over a distance, then how can we move energy into the food, and where does the energy go?

There is a second way to move energy into an object, other than by doing work on it. Heat is the energy we move into an object other than by work. **No object has heat, because heat is the transfer of energy.**

Every object has internal energy E_{int}. **Temperature is a measure of the density of the internal energy**, in energy per atom rather than per volume. Two objects at the same temperature do not necessarily have the same amount of internal energy. A large bucket of water has more internal energy than a small bucket of water even though they have the same temperature. When placed in contact, no heat flows between the two because they have the same "energy density." The small bucket could have the same internal energy as the large one, but would need to have a higher energy density or temperature. Then when placed in contact, heat would flow from the hotter, higher temperature bucket of water to the colder, lower temperature bucket.

The first law of thermodynamics is conservation of energy, so

$$\Delta E_{\text{int}} = \Delta Q + \Delta W_{\text{done on}} = \Delta Q - \Delta W_{\text{done by}}$$

where ΔQ is the heat transferred into the object. We can increase the internal energy by transferring heat into or by doing work on the object. **Throughout these chapters, ΔW is the work done *by* the object, so a positive ΔW decreases the internal energy.**

There are many scales for measuring temperature. In the United States, the most familiar is Fahrenheit (65°F is a pleasant spring day). In chemistry, the Celsius scale is often used,

$$^{\circ}\text{F} = \frac{9}{5}\,(^{\circ}\text{C}) + 32^{\circ}$$

where 100°C is the boiling point of water and 0°C is the freezing point of water. In physics, we often use Kelvin.

$$\text{K} = {}^{\circ}\text{C} + 273.15^{\circ}$$

So the boiling point of water is 373 K. We say "100 degrees Celsius" but "373 kelvin," not "373 degrees kelvin."

Because the difference between Celsius and kelvin is only adding 273, an increase of one degree Celsius is the same as an increase of one kelvin. Therefore, when using a temperature difference, Celsius and kelvin are the same. But zero kelvin corresponds to zero internal energy, which is an important point in thermodynamics. Therefore **use Celsius *only* when finding a temperature difference; all temperatures must be in kelvin**. Temperature differences can also be in kelvin, of course.

Now we need to determine how to find the heat and the work and what they do. The work done by an object is

$$W = \int p dV$$

where p is the pressure the object (often a fluid) exerts. Because the pressure of a gas or liquid is exerted outward, an increase in volume means that the motion was in the same direction as the force and the work done is positive. When the volume decreases, then the motion is in the opposite direction as the force and the work done by the object is negative.

EXAMPLE

Find the work done by the gas undergoing the process shown. The curve A→B is pV = constant.

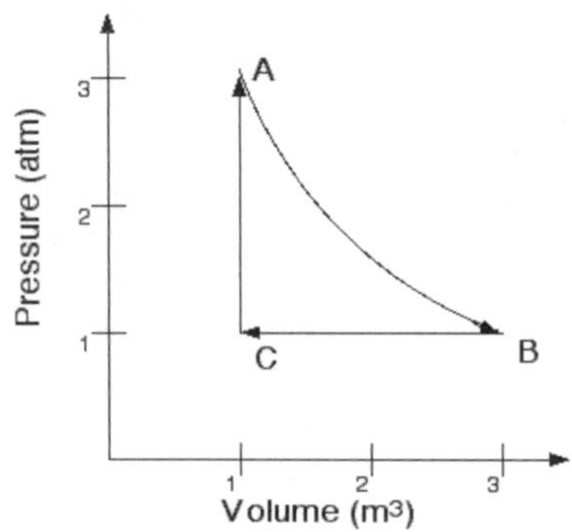

Student: So I have to do an integral $W = \int p dV$.
Tutor: Don't you like doing integrals?
Student: NO!
Tutor: Then we should look for ways to avoid doing any more than we have to.
Student: I'm all for that.
Tutor: Look at the path from B to C and tell me about the pressure.
Student: It's a constant. That means that the integral $W = \int p dV = p\Delta V$.
Tutor: Very good. The work is the area under the curve.
Student: The pressure times the change in volume. I get it.
Tutor: But the change in volume from B to C is negative.
Student: So the work is negative?
Tutor: Yes. We're pushing in on the gas harder than it pushes out, so it is compressed. The work that the gas does is the pressure it exerts times the change in volume.
Student: Shouldn't that be the force, or pressure times area, times dx?
Tutor: But dx times the area is the change in volume.
Student: If we're looking for the work, don't we push harder on the gas, so shouldn't we use our pressure?
Tutor: If the change happens slowly, as most changes in this chapter do, then our pressure is the same as the gas's. We push a little harder to get it compressing...
Student: ...and then a little less hard at the end, but with the same pressure during the compression. Just like work.
Tutor: It is the same as work. What about from C to A?
Student: There is no change in the volume, so no work.
Tutor: Correct, and there is no area under the curve.

Student: But from A to B we have to do an integral.
Tutor: Yes. Along the curve the pressure times the volume is a constant, so we can write $pV = (3\text{ atm})(1\text{ m}^3) = 3\text{ atm·m}^3$.
Student: Then we can set up the integral.

$$W_{\text{AB}} = \int p dV = \int_{1\text{ m}^3}^{3\text{ m}^3} \frac{3\text{ atm·m}^3}{V} dV = (3\text{ atm·m}^3)\,[\ln V]_{1\text{ m}^3}^{3\text{ m}^3}$$

$$= (3\text{ atm·m}^3)\left[\ln(3\text{ m}^3) - \ln(1\text{ m}^3)\right] = (3\text{ atm·m}^3)\ln\frac{(3\text{ m}^3)}{(1\text{ m}^3)} = (3\text{ atm·m}^3)\ln 3 = 3.30\text{ atm·m}^3$$

Student: Shouldn't the answer be in joules?
Tutor: 1 atmosphere (atm) of pressure is 1.013×10^5 N/m^2.
Student: So I can convert...
Tutor: Before you do that, let's finish the problem in the units we have. Add the work from the three steps.
Student: The work from B to C is -2 atm·m^3, and from C to A is zero, so the total work done is 1.30 atm·m^3.
Tutor: Yes.
Student: So the internal energy of the gas decreases by that amount?
Tutor: No, the internal energy is determined by p and V. Because the internal energy is the same after a complete cycle, there must have been heat ΔQ into the gas somewhere during the process.
Student: Okay. Now I can see whether the work is really joules.
Tutor: There are many units we could use for work or energy. Any one of them should convert to any other of them. If we can't turn atm m^3 into joules, then we did something wrong.

$$(1.30\,\cancel{\text{atm}}\,\text{m}^3) \times \frac{(1.013 \times 10^5\text{ N/m}^2)}{(1\,\cancel{\text{atm}})} = 1.32 \times 10^5\text{ N·m} = 1.32 \times 10^5\text{ J}$$

Student: Is that a lot?
Tutor: About enough to run an LCD computer monitor for an hour.
Student: How did you come up with that?
Tutor: The power is the energy over the time, so 1.32×10^5 J/3600 s = 36 W.
Student: All of the energy and work and power stuff still applies?
Tutor: Would you rather you had to learn new energy and work and power stuff?
Student: No.

When heat or work is added to an object, it increases the internal energy. We use temperature to measure this internal energy (energy density, really), so the temperature increases.

The amount of heat needed to increase the temperature is

$$Q = C\Delta T = cm\Delta T$$

where C is the heat capacity and c is the specific heat. **An object has a heat capacity, and a material has specific heat**. The more of the material, the greater the heat capacity. The specific heat of water is 1.00 cal/g·K, or 4180 J/kg·K.

It is also possible for the fluid to change form, to another state of matter. An example of this is ice turning into water, or water turning into steam. The heat needed to transform an object is

$$Q = Lm$$

where L is the heat of transformation. The heat of fusion is the heat needed to go from a solid to a liquid, and the heat of vaporization is the heat needed to go from a liquid to a solid. The heat of fusion of water is 333 kJ/kg, and the heat of vaporization of water is 2256 kJ/kg.

EXAMPLE

How much energy does it take to turn 15 kg of ice at −10°C into steam?

Student: What's the temperature of steam?
Tutor: Water has to be at least 100°C to become steam. Steam can get hotter than that.
Student: So

$$Q = cm\Delta T = (4180\ \text{J/kg·K})(15\ \text{kg})(100°C - (-10°C)) = 6.90 \times 10^6\ \text{J} = 6.90\ \text{MJ} \quad ?$$

Tutor: Partly correct. Your use of specific heat is good, and using the temperature difference is good, and it's true that degrees Celsius and kelvin are the same because it's a temperature *difference* that we need. But you've forgotten the heat of transformation.
Student: The $Q = Lm$ part?
Tutor: Yes. It takes 0.63 MJ to heat the ice up to 0°C. Then we need to melt it into water.
Student: Okay. The energy needed to melt the ice is

$$Q = Lm = (333\ \text{kJ/kg})(15\ \text{kg}) = 5000\ \text{kJ} = 5.00\ \text{MJ}$$

Student: The table in the book has a heat of fusion of hydrogen. You can have frozen hydrogen?
Tutor: Yes. You cool hydrogen down to 20 K, then remove the heat of vaporization to turn it into liquid hydrogen. Then you cool the liquid down to the melting point and remove the heat of fusion and, presto, solid hydrogen.
Student: You make it sound easy.
Tutor: It requires appropriate equipment, of course, but it's not difficult. Liquid nitrogen sells for a couple dollars per liter and is commonly used in physics experiments and demonstrations.
Student: Like the one where they freeze a banana and use it as a hammer?
Tutor: Yes. The heat you need to remove from a banana to freeze it is sufficient to boil only a small amount of liquid nitrogen, so with more liquid nitrogen than that, you can freeze a banana.
Student: Does the heat of fusion work the same each way?
Tutor: Yes, to freeze something you remove the same amount of heat that you added to melt it.
Student: Okay. Then we add 6.30 MJ of energy to raise the temperature to 100°C, and add the heat of vaporization.

$$Q = Lm = (2256\ \text{kJ/kg})(15\ \text{kg}) = 33,840\ \text{kJ} = 33.84\ \text{MJ}$$

Student: The total heat is

$$(0.63 + 5.00 + 6.30 + 33.84) = 45.77\ \text{MJ}$$

Tutor: That's about 13 kilowatt-hours (kWh), or 13 kilowatts for an hour. A typical house uses 1-2 kilowatts, so that would power a house for 6-8 hours.
Student: Or turn 15 kg of ice into steam. It takes a lot of energy.
Tutor: Where did most of it go?
Student: Into the last step, turning the hot water into steam.
Tutor: Yes. If you start with hot water, say from the hot-water tap, then it takes 10 times the energy to vaporize the water as it does to heat it. That's why it takes only a few minutes to boil water on a stove, but a long time to boil it all away.
Student: And the large heat of fusion is why coolers work, right?
Tutor: Yes. The energy to melt ice is the same to cool ten times more water by 8°C. That makes using ice in a cooler effective.

There are three ways to transfer heat. Heat transfers by conduction, convection, and radiation.

Conduction is when heat moves between two objects that are in contact. The speed of the heat transfer is measured by the power, or energy per time,

$$P = \frac{Q}{t} = kA\frac{\Delta T}{L}$$

where k depends on the material through which the heat transfers, A is the cross-sectional area through which it transfers, L is the length or thickness through which it transfers, and ΔT is the temperature difference. This is why thicker insulation keeps houses warmer, because L is larger so the heat transfer is slower. A layer of air can be used to insulate an object, because the thermal conductivity k of air is small at 0.026 W/m·K, less than fiberglass insulation.

Convection is the nonrandom motion of molecules in a gas or liquid. In air, for example, the oxygen and nitrogen molecules race around at speeds close to the speed of sound. They go only a short distance before colliding with another molecule and bouncing off in a random direction. But when air gases get warm they expand, and as the density decreases they tend to rise. As a fire in a fireplace warms the air, this warm air moves up the smokestack because of its lower density, rather than moving randomly either into the room or the chimney. This nonrandom motion is called convection. Calculating the heat transfer from convection is very difficult, and we will not try it here.

Radiation is heat transferred by electromagnetic waves, such as light or infrared radiation. All objects give off thermal radiation. The power or heat transfer rate from radiation is

$$P = \sigma\epsilon AT^4$$

where σ is the Stefan-Boltzmann constant ($\sigma = 5.6704 \times 10^{-8}$ W/m^2·K^4), ϵ is the emissivity of the material, A is the area of the object, and T is the temperature.

Emissivity is a measure of how much radiation an object both absorbs and emits (same number for both), and is one for a completely black object and zero for a perfectly reflective object (neither of which can actually be made). The SR-71 "blackbird" airplane was designed to fly 3 times the speed of sound. The air resistance was expected to heat the plane considerably, and the designers painted it black so that it would radiate heat more quickly.

EXAMPLE

The intensity of sunlight at the Earth's surface is about 1000 W/m^2. The average temperature of the Earth is about 58°F. What must the emissivity of the Earth be?

Student: Don't we need to know the temperature of the Sun?
Tutor: That would help us find the energy that the Sun gives to the Earth, but we already have that.
Student: Okay. The area of the Earth is $4\pi R^2$, times the intensity is the power into the Earth.
Tutor: Except that the Sun is not above every spot on the Earth. If you stand back from a globe and look at it, how much of the Earth do you see?
Student: Half of it.
Tutor: But even that half you don't all see at normal incidence. What is the area you perceive?
Student: Ah. It looks like a circle of area πR^2, right?
Tutor: Yes. Now what's the power out from the Earth?
Student: That's radiation, so it's $P = \sigma\epsilon AT^4$.
Tutor: True, but not what I had in mind. If the Earth is to stay at the same temperature, with the same internal energy, then...
Student: ...the power out has to equal the power in.
Tutor: Yes, what we call "steady-state."

$$(\pi R^2)I = \sigma\epsilon(4\pi R^2)T^4$$

Student: Do I need to use kelvin here?
Tutor: Is it a temperature *difference*?
Student: No. So it's an absolute temperature, and I need to use kelvin.

$$58°\text{F} = \frac{9}{5}(°\text{C}) + 32° \quad \rightarrow \quad 14.4°\text{C}$$

$$\text{K} = 14.4°\text{C} + 273.15° = 287.6\ \text{K}$$

$$\epsilon = \frac{I}{4\sigma T^4} = \frac{(1000\ \text{W/m}^2)}{4(5.6704 \times 10^{-8}\ \text{W/m}^2{\cdot}\text{K}^4)(287.6\ \text{K})^4} = 0.64$$

Tutor: The emissivity of the Earth varies with the surface, with different values for water, land, pavement, even clouds.
Student: Clouds change the emissivity? So the emissivity of the Earth isn't constant.
Tutor: That's one of the things that makes climate modelling difficult.
Student: Can't we take an average value?
Tutor: Because the temperature goes to the fourth power, we need to calculate the power emitted from each area and average the powers, rather than averaging the emissivity.

When the temperature of a solid or liquid increases, the object expands. The expansion is

$$\Delta L = L\alpha\Delta T \qquad \text{or} \qquad \frac{\Delta L}{L} = \alpha\Delta T$$

where α is the coefficient of thermal expansion. Because 2-D objects expand in two dimensions and 3-D objects in three,

$$\Delta A = A(2\alpha)\Delta T \qquad \Delta V = V(3\alpha)\Delta T$$

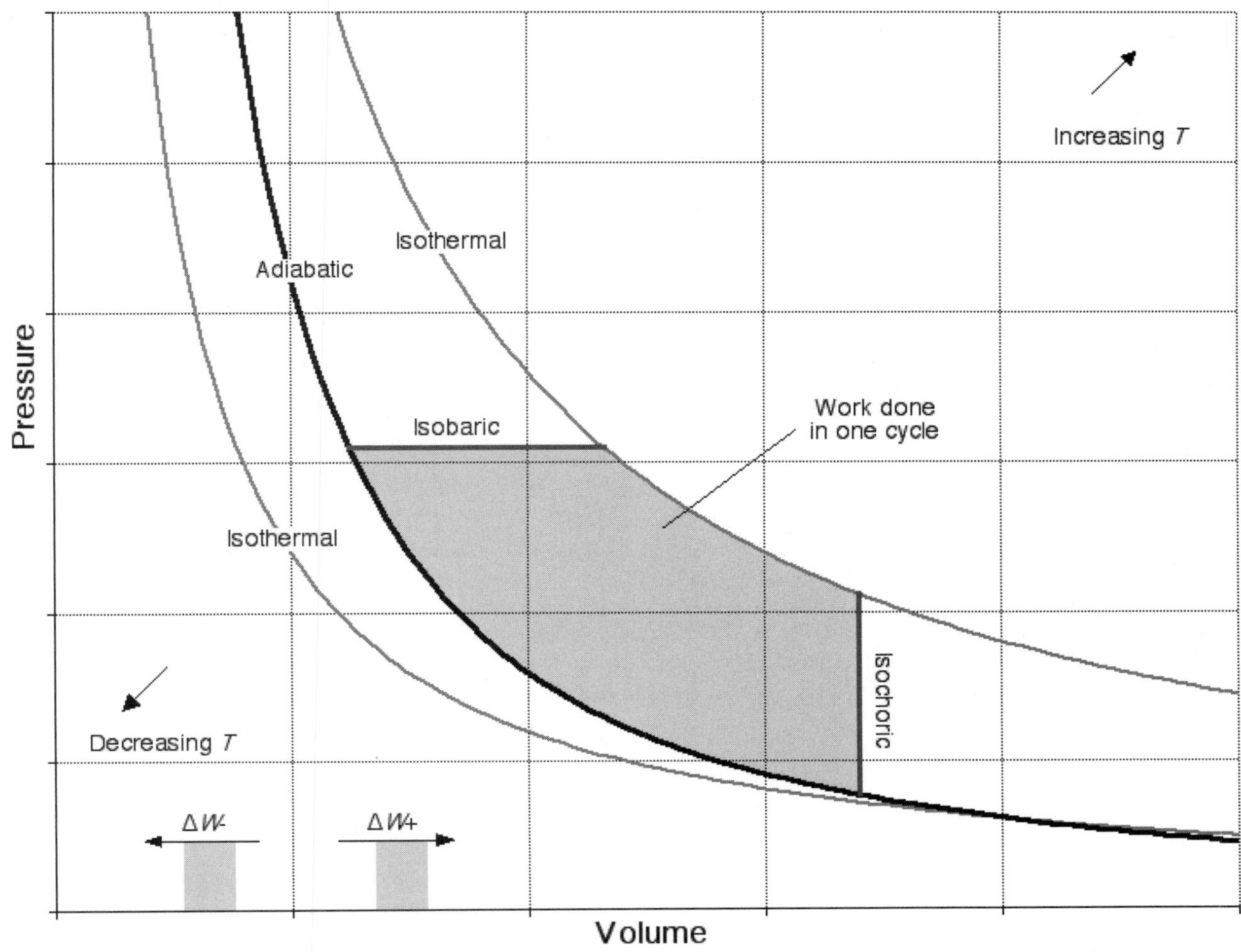

adiabatic	isolated	$Q = 0$ so $\Delta E_{\text{int}} = -W$, $pV^{\gamma} = \text{const}$	$\Delta S = 0$
isothermal	constant T	$\Delta E_{\text{int}} = 0$ so $Q = W = nRT\ln(V_f/V_i)$	$\Delta S = Q/T$
isobaric	constant p	$W = p\Delta V$, $Q = nC_p\Delta T$	
isochoric	constant V	$W = 0$, $Q = \Delta E_{\text{int}} = nC_V\Delta T$	

Chapter 19

The Kinetic Theory of Gases

Gases consist of many molecules, often 10^{20} or more. Not all of these are travelling in the same direction, or even at the same speed. How can we deal with so many objects? When the numbers are that large, statistics works very, very well.

Whatever a single molecule does, it typically travels only a short distance (micrometers at 25°C and 1 atm) before colliding with another molecule, after which it has a new velocity. The average kinetic energy of an atom or molecule in a gas is:

$$E_{\text{ave}} = \frac{3}{2}kT$$

The total energy of a large number of atoms is:

$$E_{\text{int}} = N\left(\frac{3}{2}kT\right) = n\left(\frac{3}{2}RT\right)$$

The difference between the two forms is that the first form uses the number of atoms (N) and the second uses the number of moles of atoms n. A mole is Avogadro's number, $N_A = 6.022 \times 10^{23}$ atoms, so

$$n = N/N_A \quad \text{and} \quad R = N_a k$$

A mole is significant because one mole of atoms has a mass in grams equal to the mass number of the atom, so one mole of nitrogen, with mass number 14, has a mass of 14 grams.

We can evaluate the sum of all atoms or molecules in a gas using the ideal gas law.

$$pV = nRT = NkT$$

Again it comes in two forms, one for atoms (or molecules) and one for moles of atoms. $k = 1.3807 \times 10^{-23}$ J/K is the Boltzmann constant, and $R = 8.3145$ J/mol · K is the gas constant. Because T is a temperature and not a temperature difference, it must be in kelvin.

Our goal is to calculate the change in energy ΔE, the heat flow Q, and the work done W. There are many specific processes that appear often, and that can aid us in our calculations. The chart on page 149 shows these important processes.

An isothermal process is one that occurs at a constant temperature. Because the temperature doesn't change, neither does the internal energy. Conservation of energy still holds, so the heat in Q must equal the work done by the gas W. Finding the work requires an integral, and the result is $W = nRT\ln(V_f/V_i)$.

An isobaric process is one that occurs at a constant pressure. The work $\int p dV = p\Delta V$ is easy to calculate. The heat in $Q = nC_p\Delta T$, where C_p is the molar specific heat at constant pressure.

An isochoric process is one that occurs at a constant volume. The work $\int p dV = 0$ is again easy to calculate. The heat in $Q = nC_V \Delta T$, where C_V is the molar specific heat at constant volume.

An adiabatic process is one that occurs with no heat transfer Q. Because $Q = 0$, the change in internal energy is the opposite of the work done by $\Delta E_{\text{int}} = -W$. In an adiabatic process, the quantity pV^γ is a constant.

What are C_p, C_V, and γ? It depends on the number of "degrees of freedom" of the atoms or molecules. An atom has three: up-down, left-right, in-out. So for an atom, $E_{\text{ave}} = \frac{3}{2}kT$ and $C_V = \frac{3}{2}R$. A diatomic molecule, two atoms together like O_2, has two additional degrees of freedom: spinning around each of the two axes perpendicular to the line joining the atoms. So for a diatomic molecule, $E_{\text{ave}} = \frac{5}{2}kT$ and $C_V = \frac{5}{2}R$. γ is the ratio $\gamma = C_p/C_V$, which is 5/3 for an atom and 7/5 for a diatomic molecule.

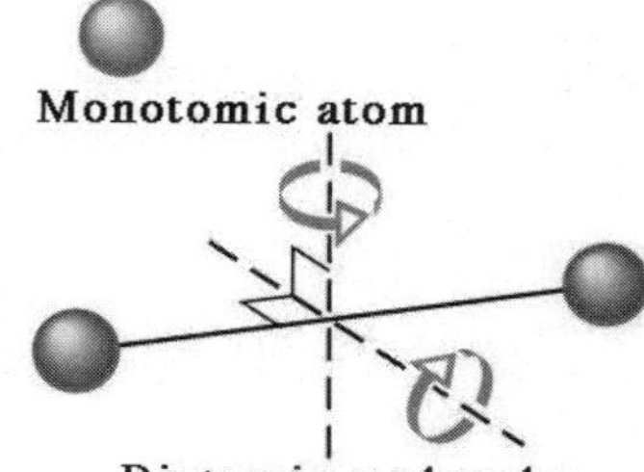

EXAMPLE

Find the heat Q, the work W, and the change in energy ΔE for each step of the process of the diatomic gas. The process goes clockwise around the enclosed area.

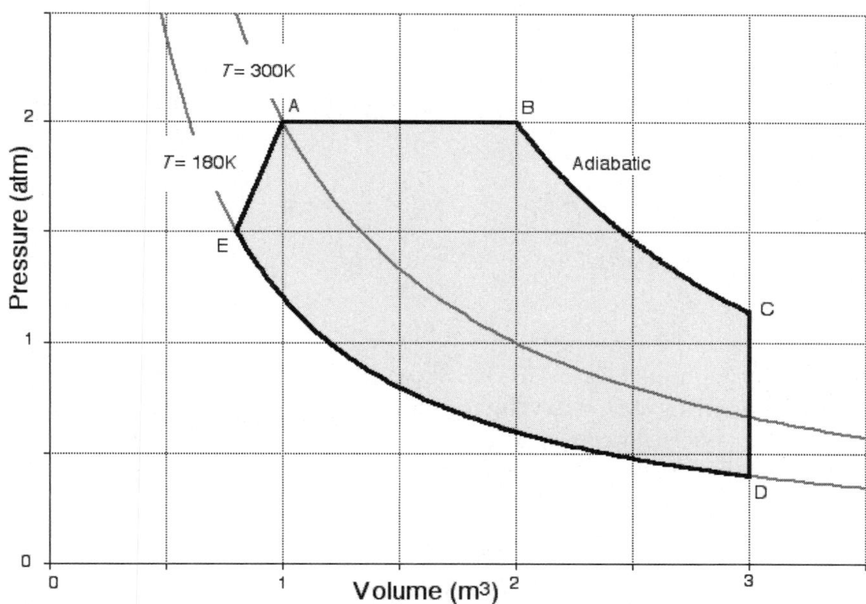

Student: This looks complicated. We don't even have all of the data. What's the pressure at C?
Tutor: Well, start with what you do know. Do you know the volume, pressure, or temperature at A?
Student: We know all three, right? The volume is 1 m^3, the pressure is 2 atm, and the temperature is 300 K.
Tutor: Yep, room temperature but twice the pressure. What about B?
Student: The pressure is still 2 atm, but the volume has increased to 2 m^3 and we don't know the temperature.
Tutor: But we might be able to calculate the temperature using the ideal gas law.
Student: Does that apply both at A and at B?
Tutor: Yes, it applies anywhere on the graph. Better yet, the number of moles n doesn't change, so pV/T is constant anywhere on the graph.

Student: Okay. At C we know that the volume is 3 m^3.
Tutor: Is the temperature at C the same as the temperature at B?
Student: They don't have to be, do they?
Tutor: No. From B to C is an adiabatic path, and any adiabatic change will change the temperature.
Student: So we don't know the temperature at B or C, but they're not the same. At D the temperature is 180 K.
Tutor: Yes, very cold. Below −100°F.
Student: The path from D to E is an isothermal, so the temperature is also 180 K at E. We know the volume at D and the pressure at E, but not the other way around.

	p (atm)	V (m^3)	T (K)
A	2	1	300
B	2	2	?
C	?	3	?
D	?	3	180
E	1.5	?	180
A	2	1	300

	all in atm·m^3		
	Q	W	ΔE
A→B	?	?	?
B→C	?	?	?
C→D	?	?	?
D→E	?	?	?
E→A	?	?	?
loop	$\Sigma=?$	$\Sigma=?$	$\Sigma=?$

Student: Now I can use the ideal gas law to find the temperature at B. I know p, V, T, and of course R, so I can find the number of moles n.
Tutor: We don't really need to know n. What we need is the combination nR, because they are always together.

$$pV = nRT \quad \rightarrow \quad nR = \frac{p_\mathrm{A} V_\mathrm{A}}{T_\mathrm{A}} = \frac{p_\mathrm{B} V_\mathrm{B}}{T_\mathrm{B}}$$

$$\frac{(2\text{ atm})(2\text{ m}^3)}{T_\mathrm{B}} = nR = \frac{(2\text{ atm})(1\text{ m}^3)}{(300\text{ K})} = \frac{1}{150}\text{ atm m}^3/\text{K}$$

$$T_\mathrm{B} = 600\text{ K}$$

Student: Can we keep using units of atm m^3/K? Shouldn't we convert to something?
Tutor: Our pressures are all in atm, our volumes in m^3, and our temperatures in K, so these will work fine. We can convert the energies to joules at the end.
Student: Like we did in the last chapter. I can find p_D and V_E the same way.

$$\frac{p_\mathrm{D}(3\text{ m}^3)}{180\text{ K}} = \frac{(1.5\text{ atm})V_\mathrm{E}}{180\text{ K}} = nR = \frac{1}{150}\text{ atm m}^3/\text{K}$$

$$p_\mathrm{D} = 0.4\text{ atm} \quad \text{and} \quad V_\mathrm{E} = 0.8\text{ m}^3$$

Student: But I can't do point C that way. The temperature isn't the same as either B or D, and I don't think I can read the pressure off of the graph.
Tutor: True. But B→C is an adiabatic, so pV^γ is a constant along the curve.
Student: This is where it matters that our gas is diatomic. $\gamma = \frac{7}{5} = 1.4$.

$$p_\mathrm{B} V_\mathrm{B}^{1.4} = p_\mathrm{C} V_\mathrm{C}^{1.4} \quad \rightarrow \quad p_\mathrm{C} = p_\mathrm{B}\left(\frac{V_\mathrm{B}}{V_\mathrm{C}}\right)^{1.4} = (2\text{ atm})\left(\frac{2\text{ m}^3}{3\text{ m}^3}\right)^{1.4} = 1.13\text{ atm}$$

Tutor: Very good.
Student: Back to the ideal gas law.

$$\frac{(1.13\text{ atm})(3\text{ m}^3)}{T_\mathrm{C}} = nR = \frac{1}{150}\text{ atm m}^3/\text{K} \quad \rightarrow \quad T_\mathrm{C} = 510\text{ K}$$

	p (atm)	V (m^3)	T (K)
A	2	1	300
B	2	2	600
C	1.13	3	510
D	0.4	3	180
E	1.5	0.8	180
A	2	1	300

	all in atm·m^3		
	Q	W	ΔE
A→B	?	?	?
B→C	?	?	?
C→D	?	?	?
D→E	?	?	?
E→A	?	?	?
loop	Σ =?	Σ =?	Σ =?

Tutor: Well done. Now we can start on the right side of the chart.
Student: How do I find the heat transferred to the gas during A→B?
Tutor: The path from A→B is an isobar (constant pressure), so $Q = nC_p\Delta T$.
Student: What's the difference between isobar and isobaric?
Tutor: Isobar is a noun, and refers to the line on the chart. Isobaric is an adjective, and refers to the process. An isobaric process is one that travels along an isobar on the chart.
Student: So an isothermal process is one that travels along an isotherm?
Tutor: Yes.
Student: What do you call the line for an adiabatic process?
Tutor: An adiabat.
Student: Okay. $C_p = C_V + R$, and the gas is diatomic so $C_V = \frac{5}{2}R$. $C_p = \frac{7}{2}R$.

$$Q = nC_p\Delta T = \frac{7}{2}nR\Delta T$$

Tutor: There's the combination nR again.

$$Q = \frac{7}{2}nR\Delta T = \frac{7}{2}\left(\frac{1}{150}\text{ atm m}^3/\text{K}\right)((600\text{ K}) - (300\text{ K})) = 7\text{ atm·m}^3$$

Student: The work is easy.

$$W = \int p dV = p\Delta V = (2\text{ atm})(1\text{ m}^3) = 2\text{ atm·m}^3$$

Student: I can use $\Delta E = Q - W$ to find the change in energy, right?
Tutor: Yes, but you can also calculate it directly. The energy is $E = n\left(\frac{5}{2}RT\right)$.
Student: The $\frac{5}{2}$ is because it's diatomic, yes?
Tutor: Yes.

$$E_{\text{A}} = \frac{5}{2}nRT_{\text{A}} = \frac{5}{2}\left(\frac{1}{150}\text{ atm m}^3/\text{K}\right)(300\text{ K}) = 5\text{ atm·m}^3$$
$$E_{\text{B}} = \frac{5}{2}nRT_{\text{B}} = \frac{5}{2}\left(\frac{1}{150}\text{ atm m}^3/\text{K}\right)(600\text{ K}) = 10\text{ atm·m}^3$$
$$\Delta E = E_{\text{B}} - E_{\text{A}} = 10\text{ atm·m}^3 - 5\text{ atm·m}^3 = 5\text{ atm·m}^3$$

Student: That's the same as $\Delta E = Q - W$!
Tutor: It's nice to know that energy is still being conserved.
Student: Stop being sarcastic. From B→C is adiabatic, so $Q = 0$. Finding the work is going to take an integral.
Tutor: There's an easier way.
Student: Oh, conservation of energy. $Q = 0$, so if I find ΔE then $W = -\Delta E$.
Tutor: Yes. Let's do the integral first.

$$W = \int p dV = \int_2^3 p_{\text{B}}\left(\frac{V_{\text{B}}}{V}\right)^{1.4} dV = p_{\text{B}}V_{\text{B}}\left[\left(-\frac{1}{0.4}\right)\left(\frac{V_{\text{B}}}{V}\right)^{0.4}\right]_2^3$$

$$= -\frac{(2)(2)}{0.4}\left[\left(\frac{(2)}{3}\right)^{0.4} - \left(\frac{(2)}{2}\right)^{0.4}\right]_2^3 = +1.50 \text{ atm·m}^3$$

Student: Now I'll do it the easy way.

$$E_\text{C} = \frac{5}{2}nRT_\text{B} = \frac{5}{2}\left(\frac{1}{150} \text{ atm m}^3/\text{K}\right)(510 \text{ K}) = 8.5 \text{ atm·m}^3$$

$$\Delta E = E_\text{C} - E_\text{B} = 8.5 \text{ atm·m}^3 - 10 \text{ atm·m}^3 = -1.5 \text{ atm·m}^3$$

$$\Delta E = \overset{0}{\cancel{Q}} - W \quad \rightarrow \quad W = -\Delta E = +1.5 \text{ atm·m}^3$$

Student: From C→D is an isochoric, with no volume change, so no work. I can find the change in energy and that's equal to the heat.

$$E_\text{D} = \frac{5}{2}nRT_\text{B} = \frac{5}{2}\left(\frac{1}{150} \text{ atm m}^3/\text{K}\right)(180 \text{ K}) = 3 \text{ atm·m}^3$$

$$\Delta E = E_\text{D} - E_\text{C} = 3 \text{ atm·m}^3 - 8.5 \text{ atm·m}^3 = -5.5 \text{ atm·m}^3$$

Student: What does it mean that Q is negative?
Tutor: We lowered the temperature by removing heat from the system. With the volume held constant, the pressure decreased. Each molecule had less energy, so less speed, and exerted less impulse on the walls as they bounced off. Can you find the heat directly?
Student: You mean using $Q = nC_V\Delta T$?

$$Q = nC_V\Delta T = \frac{5}{2}nR\Delta T = \frac{5}{2}\left(\frac{1}{150} \text{ atm m}^3/\text{K}\right)((180 \text{ K}) - (510 \text{ K})) = -5.5 \text{ atm·m}^3$$

Student: From D→E is an isotherm. The temperature doesn't change, so the energy doesn't change. It looks like we need to do an integral to find the work. I'll cheat by using $W = nRT\ln(V_f/V_i)$.
Tutor: It's not cheating. We did the integral in the last chapter, and there's no reason to do it every time.
Student: That's right. $pV =$ constant is an isotherm.
Tutor: We didn't call it that then because we didn't have the ideal gas law yet.

$$W = nRT\ln(V_f/V_i) = \left(\frac{1}{150} \text{ atm m}^3/\text{K}\right)(180 \text{ K})\ln\left(\frac{0.8 \text{ m}^3}{3 \text{ m}^3}\right) = -1.59 \text{ atm·m}^3$$

Student: ΔE is zero, so Q is also -1.59 atm·m^3.
Tutor: Last one.
Student: From E→A is not any of our special curves.
Tutor: But you could find the work W.
Student: Doing another integral?
Tutor: It's the area under the curve.
Student: Do you mean below the E→A line but above the D→E isothermal curve?
Tutor: The work from E→A is the area under the E→A line and above the axis. To find the total work done in a cycle, we'll add all of the works, and the D→E work was negative because we were going left on the graph.
Student: Okay, it's just a trapezoid.

$$W = \frac{1}{2}\Big(p_\text{E} + p_\text{A}\Big)\Delta V = \frac{1}{2}\Big((1.5 \text{ atm}) + (2 \text{ atm})\Big)\left(0.2 \text{ m}^3\right) = 0.35 \text{ atm·m}^3$$

Student: And we can find the energy like we did before.

$$E_\text{E} = E_\text{D} = 3 \text{ atm·m}^3$$

$$\Delta E = E_\text{A} - E_\text{E} = 5 \text{ atm·m}^3 - 3 \text{ atm·m}^3 = 2 \text{ atm·m}^3$$

	p (atm)	V (m^3)	T (K)
A	2	1	300
B	2	2	600
C	1.13	3	510
D	0.4	3	180
E	1.5	0.8	180
A	2	1	300

		all in atm·m^3	
	Q	W	ΔE
A→B	+7	+2	+5
B→C	0	+1.5	−1.5
C→D	−5.5	0	−5.5
D→E	−1.59	−1.59	0
E→A	2.35	0.35	2
loop	$\Sigma=?$	$\Sigma=?$	$\Sigma=?$

Tutor: If we add all of the ΔE's, we get zero. Does that make sense?
Student: The energy depends on the temperature, and if we're back to the same temperature, then it should be the same energy. What about the work, though? If we add all of the works, then we get +2.26 atm·m^3, or +229 kJ. How do we get work out if the energy is the same?
Tutor: Try adding the Q's.
Student: Oh, they add up to +2.26 atm·m^3 also.
Tutor: We get work out, and it's the same as the net amount of heat that we put in. For a complete cycle this has to be true.
Student: What is "net heat?"
Tutor: In some of the steps we put heat in, and in some we got heat out. Net heat is the difference. In the next chapter, we find that we don't recover the heat out, and this limits our efficiency.

EXAMPLE

An ideal monatomic gas at 600 K has a pressure of 2.5×10^5 Pa and a volume of 3 m^3. It goes through an adiabatic process and then an isothermal process and finishes with a pressure of 1.5×10^5 Pa and a temperature of 390 K. Find the volume of the gas after the adiabatic process but before the isothermal process.

Student: That's a lot to keep track of.
Tutor: Make a table.

$p_0 = 2.5 \times 10^5$ Pa	$V_0 = 3$ m^3	$T_0 = 600$ K
$p_1 = ?$	$V_1 = ?$	$T_1 = ?$
$p_2 = 1.5 \times 10^5$ Pa	$V_2 = ?$	$T_2 = 390$ K

Student: I can use the ideal gas law to find V_2:

$$\frac{p_0 V_0}{T_0} = nR = \frac{p_2 V_2}{T_2}$$

$$V_2 = \frac{p_0 V_0 T_2}{p_2 T_0} = \frac{(2.5 \times 10^5 \text{ Pa})(3 \text{ m}^3)(390 \text{ K})}{(1.5 \times 10^5 \text{ Pa})(600 \text{ K})} = 3.25 \text{ m}^3$$

Student: But I can't find V_1 because I don't know the pressure or temperature.
Tutor: The second step is an isothermal process. What is held constant during an isothermal process?
Student: The temperature. Ah, $T_1 = T_2 = 390$ K.

$p_0 = 2.5 \times 10^5$ Pa	$V_0 = 3\ \text{m}^3$	$T_0 = 600$ K
$p_1 = ?$	$V_1 = ?$	$T_1 = 390$ K
$p_2 = 1.5 \times 10^5$ Pa	$V_2 = 3.25\ \text{m}^3$	$T_2 = 390$ K

Tutor: What has to be true along an adiabat?
Student: The pressure and volume have to match $pV^\gamma =$ constant, but the temperature can be anything it wants.
Tutor: Good. Now write that down, and write down the ideal gas law.
Student: Okay. The gas is monatomic, so $\gamma = 5/3$.

$$p_1 V_1^{5/3} = p_0 V_0^{5/3} \qquad \text{and} \qquad \frac{p_1 V_1}{T_1} = nR = \frac{p_0 V_0}{T_0}$$

Student: I know T_0 and T_1, so it's two equations and two unknowns.

$$p_1 = \frac{p_0 V_0 T_1}{V_1 T_0}$$

$$\frac{\cancel{p_0} V_0 T_1}{V_1 T_0} V_1^{5/3} = \cancel{p_0} V_0^{5/3}$$

$$\frac{T_1}{T_0} V_1^{2/3} = V_0^{2/3}$$

$$V_1 = \left(\frac{T_0}{T_1}\right)^{3/2} V_0 = \left(\frac{600}{390}\right)^{3/2} (3\ \text{m}^3) = 5.7\ \text{m}^3$$

Student: Why would anyone want to do such a thing?
Tutor: In the next chapter we'll want to calculate entropy. Entropy is easy to calculate for adiabatic and isothermal processes, so if we can get from point A to point B using an adiabat and an isotherm, we can calculate entropy.

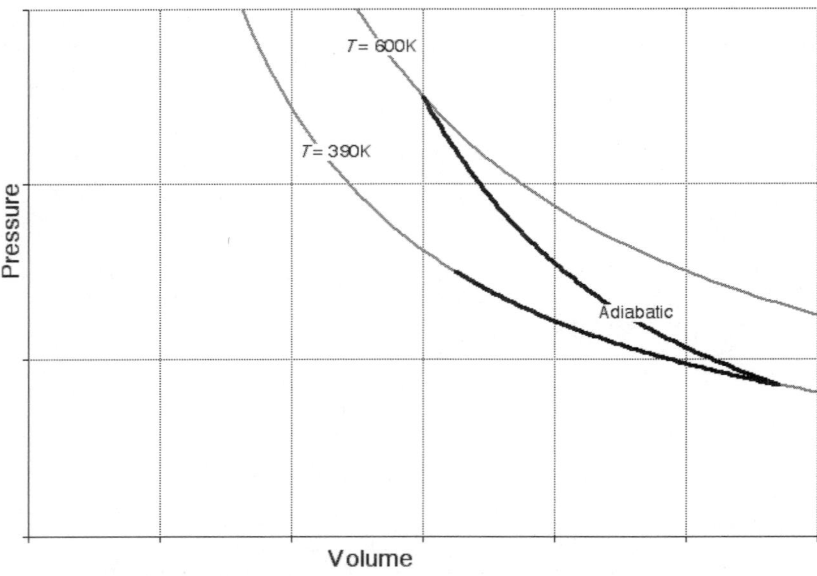

Chapter 20

Entropy and the Second Law of Thermodynamics

I'd like to explain entropy in a clear, intuitive manner. Unfortunately I don't have such an explanation. Robert Jones, in his book *Common-Sense Thermodynamics*, writes:

> No simple "intuitive reason" can be given to explain why a *state function* results when a quantity of heat ... is divided by the temperature. And indeed Clausius [one of the pioneers of thermodynamics] himself took 15 years in developing a concept of entropy that satisfied his own doubts.

So what can we say about entropy? Entropy is a state function, which means that for any combination of p, V, and T there is a well-defined value for the entropy S. Being a state function also means that if we can find even a single way to calculate the entropy, then any other way to calculate the entropy must give the same result.

We are typically interested only in changes in the entropy ΔS. While it is possible to calculate the entropy, usually the change is both easier to calculate and more meaningful.

To decrease the entropy requires work or energy from outside the system. For a closed system, the entropy can never decrease. A reversible process is one where the entropy change is zero, so that a positive entropy change means that the process can't be reversed, at least without some outside source of energy.

Reversible processes happen infinitely slowly (and are therefore no fun to watch). We don't have a good way to calculate ΔS for an irreversible process, so we find a reversible path that goes from the starting configuration to the final one, then we can calculate the change in entropy for the reversible process.

The change in entropy is

$$\Delta S = \int_i^f \frac{dQ}{T}$$

Two important special cases are adiabatic and isothermal processes. They are important because they are easy to calculate. For an adiabatic process, $\Delta Q = 0$, so $\Delta S = 0$. For an isothermal process, T is a constant so $\Delta S = \Delta Q/T$. Lengthier calculations include an ideal gas

$$\Delta S_{\text{ideal gas}} = nR\ln\frac{V_f}{V_i} + nC_V\ln\frac{T_f}{T_i}$$

and for a solid or liquid, with no volume change

$$\Delta S_{\text{solid,liquid}} = \int\frac{dQ}{T} = \int\frac{mc\Delta T}{T} = mc\ln\frac{T_f}{T_i}$$

Remember that C_V is energy per kelvin per mole, and c is energy per kelvin per kg, so these two equations are very similar.

EXAMPLE

A 30 g ice cube at −5°C is placed in 508 g of 23°C water. Find the change in entropy ΔS of the water and ice as the ice melts and the system comes to equilibrium.

Student: What's equilibrium?
Tutor: After the ice has melted, it will be 0°C. The melted ice will then warm as the cooled water further cools, until they are all the same temperature.
Student: What is that temperature?
Tutor: We don't know. We have to figure it out. How much energy does it take to warm the ice?
Student: The mass times the specific heat times the temperature change.

$$Q = mc\Delta T = (0.030\text{ kg})(4180\text{ J/kg·K})(5\text{ K}) = 627\text{ J}$$

Tutor: Good. You saw that a change of 5°C is the same as 5 K.
Student: Yes. Now I need to find the energy to melt the ice.

$$Q = Lm = (333\text{ kJ/kg})(0.030\text{ kg}) = 9.99\text{ J} \quad ?$$

Tutor: Uh, that's kilojoules.
Student: Oops.

$$Q = Lm = (333\text{ kJ/kg})(0.030\text{ kg}) = 9.99\text{ kJ} = 9990\text{ J}$$

Student: I was wondering why it took so little heat to melt the ice.
Tutor: It's good to see whether your answer makes sense. How much will the water cool while the ice melts?

$$Q = mc\Delta T \quad \rightarrow \quad (627 + 9990)\text{ J} = (0.508\text{ kg})(4180\text{ J/kg·K})\Delta T \quad \rightarrow \quad \Delta T = 5\text{ K}$$

Student: The water is cooled to 18°C as the ice melts.
Tutor: Good. Now find the equilibrium temperature. The heat that the melted ice gains is equal to the heat that the cooled water loses, until the temperatures are equal.
Student: But I don't know the temperature.
Tutor: Then make up a variable.

$$m_1 \not{c}\Delta T_1 = Q_1 = Q_2 = m_2 \not{c}\Delta T_2$$
$$(0.030\text{ kg})(T - 0°) = (0.508\text{ kg})(18° - T)$$
$$(T)(0.030\text{ kg} + 0.508\text{ kg}) = (0.508\text{ kg})(18°)$$
$$(T) = \frac{(0.508\text{ kg})(18°)}{(0.030\text{ kg} + 0.508\text{ kg})} = 17°$$

Student: And we still haven't done the entropy.
Tutor: Now we can. What is the change in entropy of the ice as it warms to 0°?
Student: Can we use $\Delta S_{\text{solid,liquid}} = mc\ln\frac{T_f}{T_i}$?
Tutor: Yes.

$$\Delta S = (0.030\text{ kg})(4180\text{ J/kg·K})\ln\frac{0°}{-5°} \quad ?$$

Tutor: How are you going to take the logarithm of zero?
Student: I must have made a mistake, yes?
Tutor: Yes. When you found the heat to warm or melt the ice, it was a temperature difference...

Student: ...so I could use Celsius, but when I divide temperatures I need to use kelvin.

$$\Delta S = (0.030\text{ kg})(4180\text{ J/kg·K})\ln\frac{0+273}{-5+273} = 2.32\text{ J/K}$$

Student: What kind of unit is a joule per kelvin?
Tutor: I don't have a good explanation for that, just like I don't have a good one for entropy.
Student: Okay. Then the ice melts, and that happens at a constant temperature.

$$\Delta S = \frac{Q}{T} = \frac{9990\text{ J}}{273\text{ K}} = 36.59\text{ J/K}$$

Student: And then the melted ice warms.

$$\Delta S = (0.030\text{ kg})(4180\text{ J/kg·K})\ln\frac{17+273}{0+273} = 7.58\text{ J/K}$$

Student: And the water cools.
Tutor: Remember to include the cooling that happens as the ice melts.
Student: Oh yes, the initial temperature of the water was 23°.

$$\Delta S = (0.508\text{ kg})(4180\text{ J/kg·K})\ln\frac{17+273}{23+273} = -43.48\text{ J/K}$$

Student: Then the total change in entropy is:

$$\Delta S = (2.32\text{ J/K}) + (36.59\text{ J/K}) + (7.58\text{ J/K}) + (-43.48\text{ J/K}) = 3.01\text{ J/K}$$

Tutor: Good.
Student: The change in entropy is positive. That means that this process is irreversible.
Tutor: Correct.
Student: But I could put 30 g of the water in the freezer and turn it into ice again, and 23°C is room temperature, so if I leave it sitting out it will warm again. I'll be back where I started, so the entropy will be the same as it was. That's what being a "state function" means, right?
Tutor: All true. The change in entropy of the water as you move it back to the initial conditions will be −3.01 J/K. But the freezer will use electrical energy, and the water will absorb energy from the room, and if we include that in our system the change in entropy will be positive. You can reduce the entropy only by bringing work or energy in from outside the system.
Student: So if I take the water back to the initial conditions, the change in entropy of the freezer and the room must be positive and greater than 3.01 J/K.
Tutor: Yes.

The effect of entropy is to limit the efficiency of thermodynamic processes. In the last chapter, we said that we could not recover the heat out. Now we tell why.

First we need to know about reservoirs. A reservoir is an external body at a particular temperature. The reservoir is large enough that we can add heat or take heat out and not change the temperature of the reservoir. Though not perfectly true, it works well for doing calculations.

To make a heat engine — an engine that turns heat into work — requires two reservoirs. Find a liquid that boils at a temperature that is lower than the hotter of the two reservoirs. When the "working fluid" is in contact with the hotter or higher temperature reservoir, it boils into gas or steam. When the working fluid is in contact with the colder or lower temperature reservoir, it either condenses to liquid or is still gas, but lower temperature gas with a lower pressure. If we put a turbine or propeller or fan between the two reservoirs, the higher pressure from the hotter side will spin the turbine. We can connect this turbine to a generator and get electricity, or use the torque directly. If we had only one reservoir, then both sides of the turbine would be at the same pressure and the turbine wouldn't spin or do work.

The process must lose heat to the lower temperature reservoir. When the working fluid is in contact with the high-temperature reservoir, it absorbs heat from the reservoir. After going through the turbine, the fluid comes in contact with the low-temperature reservoir, and heat comes from the working fluid to the reservoir. Not all of the heat taken from the high-temperature reservoir can get turned into work.

The efficiency is the ratio of the work out to energy in. The highest possible efficiency is

$$\epsilon \equiv \frac{|W|}{|Q_{\mathrm{H}}|} \leq 1 - \frac{T_{\mathrm{L}}}{T_{\mathrm{H}}} = \frac{T_{\mathrm{H}} - T_{\mathrm{L}}}{T_{\mathrm{H}}}$$

which is obtained through the "Carnot cycle," consisting of adiabats and isotherms.

A heat pump takes heat from the low-temperature reservoir and moves it to the high-temperature reservoir. Heat does not normally flow this way, so we need to put energy in to the heat pump in the form of work. The coefficient of performance for a heat pump is the ratio of heat extracted to work supplied, and is also limited by the laws of thermodynamics:

$$K \equiv \frac{|Q_{\mathrm{L}}|}{|W|} \leq \frac{T_{\mathrm{L}}}{T_{\mathrm{H}} - T_{\mathrm{L}}}$$

Air conditioners and refrigerators are heat pumps. A refrigerator takes heat out of the colder inside and puts it into the warmer room, typically through coils on the back. An air conditioner moves heat from the cooler room to the hotter outdoors. To do so takes work, and we use electricity to power an electric motor to do this work.

How can work move heat from a cold reservoir to a hot one? Imagine that you have a box of gas, designed so that you can push in one side to change the volume, and so that heat can flow through the sides of the box. Standing inside, you pull out the side of the box, expanding the gas. As it expands, the gas cools. It is now colder than the room, and heat flows from the room to the box. You now walk outside while pushing in on the box, compressing it. As the gas in the box compresses, it becomes hotter, so that it is hotter than the outside air. Heat now flows from the box to the air. As you walk back inside, you expand the box again, repeating until the inside air is cool enough.

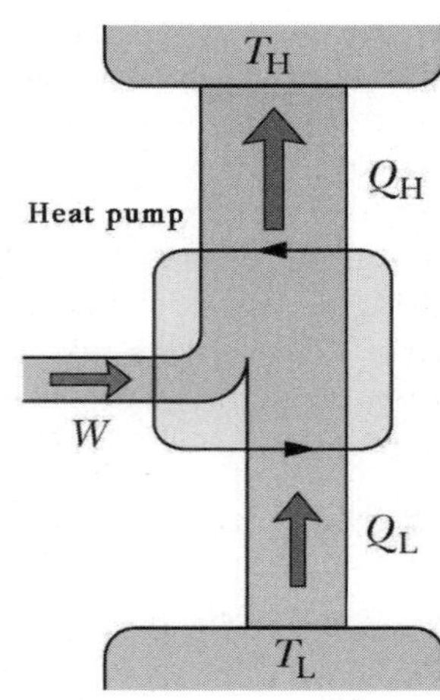

The second law of thermodynamics can be expressed in many ways. You can't turn all of the heat into work, heat doesn't go from cold to hot on its own, and $\Delta S_{\text{closed-system}} \geq 0$ are just some expressions of the second law of thermodynamics.

EXAMPLE

A *solar pond* uses salt to trap the heated water at the bottom of the pond, so that the energy does not escape to the atmosphere. Such a pond can achieve 80°C temperature water in this bottom, heated layer. This water is used to make steam out of a working fluid with a suitably low boiling point. Use 450 W/m^2 for the intensity of solar radiation, with the Sun shining only 6 hours per day. How large a solar pond would be needed to make a 1 GW power plant?

Student: One gigawatt seems like a lot of power.
Tutor: About enough for an American city of half a million people.
Student: Why is the Sun shining for only 6 hours per day?
Tutor: The Sun is not always directly overhead. Take a piece of paper (or this book) and turn it so that

you are looking somewhat along the page, rather than normal to the page. The area you see is less than the area of the page. When the Sun is not directly overhead, we need to include a factor of $\cos\theta$, where θ is the angle between the Sun's rays and the normal to the surface. It averages out to about 6 hours a day, 7 or 8 in Arizona, 5 in Ohio.

Student: Okay. If I need 1 GW of power, then I need 4 GW for 6 hours a day, because I get nothing for the other 18 hours a day.

Tutor: You haven't taken the efficiency into account. You can't collect all of the energy that reaches the pond. That comes from the second law of thermodynamics.

Student: So I need to multiply by the efficiency.

Tutor: Multiply the power reaching the pond by the efficiency to see how much power you can get.

$$P_{\text{in}} \times \epsilon = P_{\text{out}} = 4 \text{ GW}$$

Student: To calculate the efficiency, I need the high and low temperatures. The high temperature is 80°C, but what's the low temperature?

Tutor: Take something reasonable.

Student: Okay. The average temperature of the Earth is 15°C. I'll use that.

Tutor: Good choice.

Student: And I need to use kelvin because it's not a temperature difference.

Tutor: Really you only need to use kelvin in the denominator, but it doesn't hurt.

$$\epsilon = \frac{T_{\text{H}} - T_{\text{L}}}{T_{\text{H}}} = \frac{(80+273)-(15+273)}{(80+273)} = 0.18 = 18\%$$

Student: I can get only 18% of the power out?

Tutor: Probably not even that much.

$$P_{\text{in}} \times (0.18) = P_{\text{out}} = 4 \text{ GW}$$

$$P_{\text{in}} = \frac{4 \text{ GW}}{0.18} = 22 \text{ GW}$$

$$I = \frac{P}{A} \quad \rightarrow \quad A = \frac{P}{I} = \frac{22 \times 10^9 \text{ W}}{450 \text{ W/m}^2} = 4.9 \times 10^7 \text{ m}^2$$

Tutor: About 19 square miles.

Student: So one of these "solar ponds" could power a city with an area of 4 miles × 5 miles?

Tutor: Theoretically. There are issues involved with doing so. For example, you're collecting 5.5 GW, on average, but generating 1 GW, so you need a reservoir where you can dump 4.5 GW of waste heat.

Student: Do coal and nuclear plants have the same problem?

Tutor: Yes, but they operate at higher temperatures and are about 33% efficient, so they only have 2 GW of waste heat to dump. Most power plants are located by rivers or lakes so that they can dump the waste heat into the water. Remember that water has a high specific heat and can store a lot of energy.

Chapter 21

Electric Charge

So far, every force on an object has been exerted by something else in contact with it, except for gravity. There is a second type of force that can act without being in contact, and that is electricity. Eventually there will be a third such force — magnetism.

Electric forces occur between two objects that each have a special character about them. For lack of anything better, we call this special trait charge. Charge comes in two types, which we call positive and negative. Then a charge (the object) is an object that has charge (the character trait). We use the word both ways and use the context to determine which way we're using the word each time.

When we want to include the charge in an equation, we typically use q or Q for amount of charge (the character trait) that a charge (the object) has. Either q or Q is used, but if both occur in the same problem then they refer to different amounts of charge (the trait). While q is a variable that could be positive or negative, $+q$ implies an unknown amount of positive charge and $-q$ implies an unknown amount of negative charge.

The most basic charges (the objects) are the proton and the electron. (This is not entirely true but works well until the last chapter of the book.) Protons have positive charge and electrons have the same amount of negative charge. Because the charge of a proton or electron appears so often in physics, we give it a special symbol e. e is the amount of charge on a proton or electron, so the charge of an electron is $-e$.

Charge is measured in coulombs, abbreviated C. The charge of a proton e is 1.6×10^{-19} C. Inversely, a coulomb is about 6×10^{18} protons or electrons. The charge on an electron is -1.6×10^{-19} C.

Any two charges create an electric force on each other. Two charges of the same sign repel each other, while two charges of opposite sign attract each other. The magnitude of the force they create is

$$F = \frac{k|q_1||q_2|}{r^2}$$

where q_1 and q_2 are the charges (the amount, or trait) and r is the distance between them. The constant k is

$$k = \frac{1}{4\pi\epsilon_0} = 9 \times 10^9 \text{ N} \cdot \text{m}^2/\text{C}^2$$

EXAMPLE

Three charges, of $+3$ μC, -5 μC, and $+7$ μC, are arranged along the x axis at 0, 10 cm, and 25 cm, respectively. Find the total force on the $+7$ μC charge.

Tutor: Where do we always start when dealing with forces?
Student: With a free-body diagram.
Tutor: What forces are acting on the $+7\ \mu$C charge?
Student: There is nothing in contact with it, so only gravity and electric forces.
Tutor: Often when there is an electric force we ignore gravity, because the force of gravity is much smaller than the electric force.
Student: Then just the electric force.
Tutor: What is the force that the $+3\ \mu$C charge puts on the $+7\ \mu$C charge?
Student: Is the $+3\ \mu$C charge blocked by the $-5\ \mu$C charge?
Tutor: If I hold my hand under something, does the mass of my hand block the gravity force from the mass of the Earth?
Student: No.
Tutor: So one charge does not block the force from another.
Student: The $+3\ \mu$C charge creates a force on the $+7\ \mu$C charge that is

$$F = \frac{k|q_1||q_2|}{r^2} = \frac{(9 \times 10^9\ \text{N}\cdot\ \text{m}^2/\text{C}^2)(3 \times 10^{-6}\ \text{C})(7 \times 10^{-6}\ \text{C})}{(0.25\ \text{m})^2} = 3.0\ \text{N}$$

Tutor: Is the force to the left or the right?
Student: The force is positive, so it's to the right.
Tutor: We haven't chosen axes yet, so there is no positive direction. When we did forces before, we used the diagram to get the direction of the forces right, and the equations never told us the direction.
Student: They are both positive charges, with the same sign, so they repel.
Tutor: If the $+3\ \mu$C charge repels the $+7\ \mu$C charge, what is the direction of the force on the $+7\ \mu$C charge?
Student: Away from the $+3\ \mu$C charge, or to the right.
Tutor: What is the force that the $-5\ \mu$C charge puts on the $+7\ \mu$C charge?
Student: The $-5\ \mu$C charge creates a force on the $+7\ \mu$C charge that is

$$F = \frac{k|q_1||q_2|}{r^2} = \frac{(9 \times 10^9\ \text{N}\cdot\ \text{m}^2/\text{C}^2)(5 \times 10^{-6}\ \text{C})(7 \times 10^{-6}\ \text{C})}{(0.15\ \text{m})^2} = 14.0\ \text{N}$$

Tutor: What is the direction of this 14 N force?
Student: The charges have opposite signs, so they attract. The force on the $+7\ \mu$C charge is toward the $-5\ \mu$C charge, or to the left.
Tutor: Are there any other electric forces?
Student: We did every charge except the $+7\ \mu$C charge, so we have all of the electric forces, and there are no other forces.
Tutor: Now we can finish the free-body diagram.

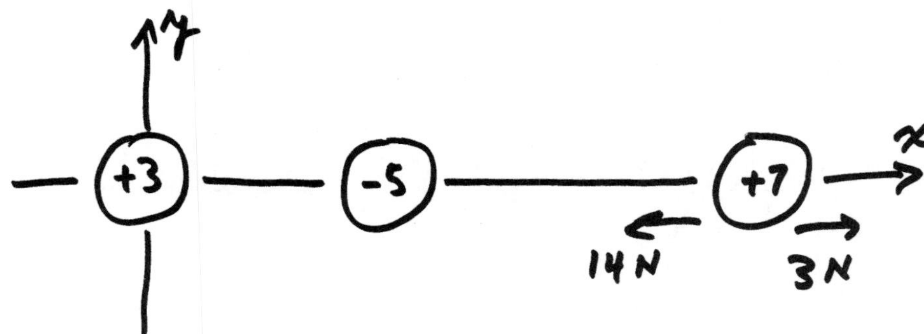

Tutor: Now you have to choose an axis.
Student: Only one axis?

Tutor: All of the forces are colinear.
Student: Since the greater force is to the left, I'll take left as the positive direction.
Tutor: What is the total force on the +7 μC charge?

$$F = (+14 \text{ N}) + (-3 \text{ N}) = +11 \text{ N}$$

Student: 11 N in the positive direction, or to the left.

EXAMPLE

Two +15 μC charges are fixed 10 cm apart on a horizontal line. A third +25 μC charge stays 12 cm above the midpoint of the first two charges. What is the mass of the third charge?

Student: How is the mass connected to the charge?
Tutor: Never mind that. Where do we start?
Student: With a free-body diagram.
Tutor: What are the forces on the +25 μC charge?
Student: There is nothing in contact with it, so gravity and electric forces.
Tutor: There's the connection. What about gravity?
Student: The gravity force is down with a magnitude of mg. We don't know the mass, so m is a variable.
Tutor: What is the electric force that the left bottom charge creates on the third charge?
Student: They are both positive, so they repel, and the force on the top charge is up and to the right:

$$r = \sqrt{(5 \text{ cm})^2 + (12 \text{ cm})^2} = 13 \text{ cm}$$

$$F = \frac{k|q_1||q_2|}{r^2} = \frac{(9 \times 10^9 \text{ N}\cdot \text{m}^2/\text{C}^2)(15 \times 10^{-6} \text{ C})(25 \times 10^{-6} \text{ C})}{(0.13 \text{ m})^2} = 200 \text{ N}$$

Tutor: What is the electric force that the right bottom charge creates on the third charge?
Student: The distance between the charges is the same as the distance we just did, and the charges are the same, so it's also 200 N. The force is up and to the left.
Tutor: Now we can finish the free-body diagram.

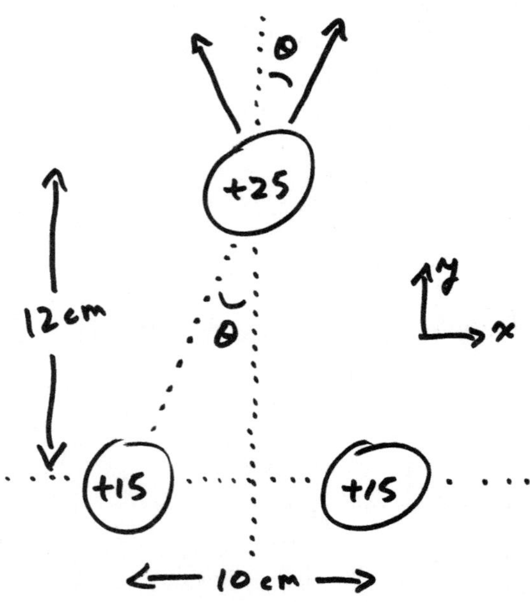

Student: Can we add the three forces, $2 \times 200\ \text{N} - mg = 0$?
Tutor: Forces are vectors. How do we add vectors?
Student: We choose axes, break the vectors into components, and add the components.
Tutor: What axes will you use?
Student: I'll choose the x axis to the right and the y axis upward.
Tutor: What is the angle between the electric force created by the bottom left charge and the y axis?
Student: It's the same as the opposite angle, so the cosine of the angle is (12/13).

$$\theta = \arccos(12/13) = 22.6°$$

Tutor: What is the x component of the electric force created by the bottom left charge?
Student: The x component is opposite to the angle, so it's sine.

$$F_{\text{L},x} = 200\ \text{N} \sin 22.6° = 77\ \text{N}$$

Tutor: What is the y component of the electric force created by the bottom left charge?
Student: The y component is adjacent to the angle, so it's cosine.

$$F_{\text{L},y} = 200\ \text{N} \cos 22.6° = 185\ \text{N}$$

Tutor: What is the x component of the electric force created by the bottom right charge?
Student: The x component is opposite to the angle, so it's sine. It goes to the left so it's negative.

$$F_{\text{R},x} = -200\ \text{N} \sin 22.6° = -77\ \text{N}$$

Tutor: What is the y component of the electric force created by the bottom right charge?
Student: The y component is adjacent to the angle, so it's cosine.

$$F_{\text{R},y} = 200\ \text{N} \cos 22.6° = 185\ \text{N}$$

Tutor: What is the x component of the gravity force?
Student: Gravity is parallel to the y axis, so there is no x component.
Tutor: Now we can write our Newton's second law equation.
Student: And the acceleration is zero.

$$\Sigma F_x = m_x$$
$$(77\ \text{N}) + (-77\ \text{N}) + (0) = 0$$

Student: What did we learn from that?
Tutor: If the x forces didn't add to zero, then the top charge would have to move.
Student: But the x forces have to add to zero because the problem is symmetric.
Tutor: If the bottom charges weren't equal, the top charge wouldn't be above the midpoint.

$$\Sigma F_y = m_y$$
$$(185\ \text{N}) + (185\ \text{N}) + (-mg) = 0$$
$$(370\ \text{N}) = mg$$
$$m = \frac{370\ \text{N}}{9.8\ \text{m/s}^2} = 38\ \text{kg}$$

EXAMPLE

Two charges, of $+2\ \mu$C and $-7\ \mu$C, are placed along the x axis at 0 and 10 cm. Where can a third charge of $+5\ \mu$C be placed so that the total electric force on it is zero?

Student: Where do we start?
Tutor: Could it be on the y axis, above the first charge?
Student: *Anywhere* on the y axis?
Tutor: If the third charge is on the y axis, what are the forces on it?
Student: The first charge creates a force up, and the second charge creates a force down and to the right.
Tutor: Could those forces add to zero?
Student: No, there would still be a component to the right.
Tutor: What if the third charge was above the midpoint between the first two?
Student: Then both of them would create forces to the right.
Tutor: If we put the third charge anywhere other than on the x axis, the first two charges will create forces that aren't colinear.
Student: So that won't cancel and won't add to zero. We have to put the third charge on the x axis.
Tutor: Could it go between them?
Student: Then the first charge creates a force to the right, and the second charge creates a force to the right, so no.
Tutor: Could it go to the right of the two charges?
Student: Then the first charge creates a force to the right, and the second charge creates a force to the left, so it could.
Tutor: If the third charge is to the right, could the electric forces have the same magnitude?
Student: The $-7\ \mu$C charge would be bigger and closer, so it would create a bigger force. The forces couldn't add to zero.
Tutor: Could it go to the left of the two charges?
Student: Then the first charge creates a force to the left, and the second charge creates a force to the right. The $-7\ \mu$C charge would be bigger, but the $+2\ \mu$C charge would be closer. How do we decide which is bigger?
Tutor: If the $+5\ \mu$C charge is really close to the $+2\ \mu$C charge, then the $+2\ \mu$C charge creates a bigger force. What if the $+5\ \mu$C charge is a long way to the left?
Student: The $+2\ \mu$C charge would still be closer.
Tutor: But the difference between $(100\text{ m})^2$ and $(100.1\text{ m})^2$ is very small.
Student: So the larger charge would create a larger force.
Tutor: Somewhere in between there must be a spot where the two charges create the same force, but in opposite directions.
Student: We've already gotten the forces in opposite directions, so that is where the magnitudes are the same.

$$\frac{\cancel{k}(2\ \mu\text{C})\cancel{(5\ \mu\text{C})}}{(r)^2} = \frac{\cancel{k}(7\ \mu\text{C})\cancel{(5\ \mu\text{C})}}{(r+0.1\text{ m})^2}$$

Student: It doesn't matter how big the third charge is, since it cancels.
Tutor: We'll see why in the next chapter.

$$\frac{(2\ \mu\text{C})}{(r)^2} = \frac{(7\ \mu\text{C})}{(r+0.1\text{ m})^2}$$

$$7(r)^2 = 2(r+0.1\text{ m})^2$$

$$\sqrt{7/2}\,(r) = (r+0.1\text{ m})$$

$$(\sqrt{7/2}-1)(r) = 0.1\text{ m}$$

$$(r) = \frac{0.1\text{ m}}{(\sqrt{7/2}-1)} = 0.115\text{ m} = 11.5\text{ cm}$$

The other important thing in this chapter is how charges behave.

Charges can move around in a conducting material, but not in an insulating material. Most metals are conductors (made of conducting material), so we make wires out of copper or aluminum. Plastic is a good insulator, so we put it on the outside of wires so that the charges don't go through us.

In dealing with charges, we often refer to "ground." Ground can supply an infinite amount of charge without becoming charged itself. If you have a charged conductor and you touch it to ground, or connect it to ground through a wire, then all of the charge can go from the conductor onto ground. Ground does not itself become charged, because ground is SO big that the charge is dispersed too small to detect. It is not true that any object connected to ground must be neutral (uncharged). If my conductor is connected to ground and a positive charge is brought close to my conductor, then negative charge will come from ground onto my conductor, attracted by the positive charge.

EXAMPLE

You have three identical conducting spheres on insulating stands, a positively charged insulating rod, and a conducting wire. You also have access to ground (the Earth). What could you do so that you can get charges of $+q$, $-\frac{1}{2}q$, and zero on the spheres? That is, you don't care how big the charges are, so long as one of the spheres has a negative charge on it, another has twice as much positive charge, and the third is neutral. Describe how you could achieve this.

Student: If I rub the rod on one of the spheres, the sphere will get some charge on it.
Tutor: Maybe some, but charge does not move easily off of the insulating rod. Is there another way to get charge onto a sphere?
Student: I could connect it to something with the wire, like ground or another sphere.
Tutor: If you connect a sphere to ground, what happens to the charge on the sphere?
Student: The sphere becomes neutral.
Tutor: Not necessarily. If the rod or another charge is nearby, opposite charges will be attracted from ground onto the sphere.
Student: So if I hold the rod nearby a sphere and connect the sphere to ground, the sphere will become positively charged.
Tutor: Yes. What happens if you connect two spheres together or touch them to each other?
Student: Charges can move from one to the other. The charges will move so that the two spheres are equal.
Tutor: The charges can move, but if the charged rod is closer to one of the spheres than the other, will the spheres be equal?
Student: No, the nearby sphere will have more negative charge than the far one.
Tutor: And if you connect a sphere to ground with nothing else nearby, it will become neutral.
Student: So I connect one sphere to ground and use the rod to pull negative charges onto it. I disconnect the sphere and move the rod away and the charges are still there. Then I connect a second sphere to ground and use the first one to pull positive charges onto it.
Tutor: That would give you two spheres with opposite charges, but it would be very difficult to get the right amount of charge on each sphere. Try a different way of getting some charge on a sphere.
Student: Okay, I connect one sphere to another and put the rod near the first sphere. I disconnect the spheres and then I have two spheres of $+q$ and $-q$.
Tutor: To make sure that the charges are equal, you need to make sure that they are both neutral beforehand. Do this by connecting each to ground with no other charges nearby. Now you need to take the $-q$ and split it in half.
Student: I'll take the negative sphere and connect it to the third sphere. But how do I know that the charge splits exactly in half?
Tutor: Using symmetry — since the spheres are identical, they have to have the same charge after the charges are done moving. That's why we're using spheres. If we used cubes, then we'd have much more difficulty getting the geometry to be symmetric.

Student: So now I have $+q$, $-\frac{1}{2}q$, and $-\frac{1}{2}q$. If I connect the third sphere to ground, with the other charges far away, then it will become neutral.

Chapter 22

Electric Fields

In mathematics, a field means that at each point in space there is a vector. Imagine walking around and measuring the normal force that the ground puts on you. On flat ground the force would be the same as your weight and upward (since there are no other forces and no acceleration). On a hillside, the normal force would be tilted, so that it was perpendicular to the hill's surface. This set of vectors would be a field. Imagine taking a single charge and moving it around among other charges, and at each point you measure the electric force on your charge. The set of vectors, one for every point in space, is a field.

Imagine now that you repeat the process, but with a greater charge, twice as big. At each point, the force that each other charge creates would be twice as big, so the total force would be twice as big. If you switched the sign of your charge, all of the forces would switch direction, so the total force would switch direction. The force on your charge is proportional to the amount of charge you use.

One way to think about electric field is that it is the force you would have on a charge if there was a charge there. The field exists whether there is a charge there or not. So the units of electric field are newtons of force for each coulomb of charge, or N/C. A 3 C charge in a 4 N/C field experiences a force of 12 N. **A point in space has an electric field, and a charge at that point experiences a force from the electric field**.

In the previous chapter we had charges creating forces on other charges. That was a simplification. Charges create electric field, and electric field creates forces on charges.

$$Q \longrightarrow E \longrightarrow F$$

Each of these steps produces a vector, either field or force. For each step, we need a way to determine the direction of the vector and an equation to determine the magnitude of the vector. Just like everything else we've done involving vectors, **the equation only tells us the magnitude of the vector**.

Imagine that a positive charge is creating the electric field. If I were to put a 1 C charge nearby, the positive charge would repel my charge, creating a force away from the positive charge. Therefore the **positive charge creates electric field away from the positive charge**. This is because the force my charge would experience if I put it nearby would be away from the positive charge. Likewise, if you imagine that a negative charge is creating the electric field, then the negative charge would attract my charge, creating a force toward the negative charge. Therefore the **negative charge creates electric field toward the negative charge**.

The electric field that a charge Q creates is

$$E = \frac{kQ}{r^2}$$

When a charge q is placed in this electric field, it experiences a force of

$$\vec{F} = q\vec{E}$$

Notice that this is a vector equation. If q is negative, then **the force on the negative charge is in the opposite direction as the electric field**. (It is possible to write the first equation as a vector equation, but not as helpful at this time.)

Why do we bother with electric fields at all? Electric fields can do things between the time they are created and the time they act on a charge. Also, there are some types of calculations that are only possible with electric fields. More on these later.

So we have **two ways to calculate electric fields**. If we have the charge(s) that create the electric field, we look at the first step ($Q \to E$, $E = kQ/r^2$). If we have the charge that experiences the electric field, we look at the second step ($E \to F$, $F = qE$). Like any equation, we can do each step backwards if necessary.

EXAMPLE

A proton and an electron are located 2 μm apart in an electric field, with the proton to the left of the electron. The total electric force on the electron is 4×10^{-17} N to the left. What are the direction and magnitude of the external electric field?

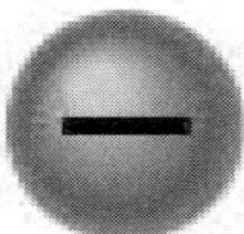

Student: The proton creates an electric field at the electron, and the electric field creates a force on the electron. We find the electric field, then find the force.
Tutor: But the force is already given.
Student: So what is there left to find?
Tutor: There is an external electric field.
Student: Meaning that it doesn't exist internally?
Tutor: Meaning that it is created by charges that we don't see — charges other than the proton. It is an externally *created* electric field, in addition to the field created by the proton.
Student: So are we doing the $Q \to E$ step or the $E \to F$ step.
Tutor: You were correct in saying that we can find the electric field that the proton creates.
Student: That would be the $Q \to E$ step, right?
Tutor: Right. That field plus the external field is the total field.
Student: How can there be more than one electric field?
Tutor: Every charge creates electric field. The proton creates electric field where the electron is, and the charges we don't see create electric field where the electron is. If we add the fields from all of the charges, we get the total field.
Student: Do we need to include the field created by the electron?
Tutor: Good question, but no, when we want the field acting on the electron we don't include the field created by the electron. Otherwise we might conclude that the charge can push itself.
Student: Okay. The proton creates an electric field

$$E_\text{p} = \frac{kQ}{r^2} = \frac{(9 \times 10^9 \text{ N} \cdot \text{ m}^2/\text{C}^2)(1.6 \times 10^{-19} \text{ C})}{(2 \times 10^{-6} \text{ m})^2} = 360 \text{ N/C}$$

Student: Is this a big field?
Tutor: No, electric fields are often 10^5 N/C or more. What is the direction of this electric field.

Student: It goes away from the proton.
Tutor: Yes, but what is the direction of this field *at the electron*?
Student: Away from the proton would be to the right.
Tutor: Correct. Now you can find the total electric field acting on the electron.
Student: How can I do that? I don't know the externally applied field.
Tutor: True, but you know the force on the electron, so you can find the field at the electron.
Student: You mean using the $E \rightarrow F$ step, only backwards.
Tutor: Yes. Or you could define a variable for the external field, write an equation for the total field, and put it into the $F = qE$ equation.
Student: Which is better?
Tutor: They're the same. It's a matter of personal preference.
Student: I'm going to find the total field. The force on the electron is

$$F = qE$$

$$(4 \times 10^{-17}\text{ N}) = (-1.6 \times 10^{-19}\text{ C})E_{\text{total}} \quad ?$$

Tutor: You need to be careful about your signs. What is your positive direction?
Student: How about to the right?
Tutor: Okay. Is E_{p} positive or negative?
Student: It was to the right, so it's positive.
Tutor: Good. Is the force on the electron positive or negative?
Student: It's to the left, so it must be negative. The charge of the electron isn't a vector, so do I include the sign?
Tutor: Yes. Because the charge q is negative, F and E_{total} will have opposite signs, indicating that they point in opposite directions.

$$(-4 \times 10^{-17}\text{ N}) = (-1.6 \times 10^{-19}\text{ C})E_{\text{total}}$$

$$E_{\text{total}} = 250\text{ N/C}$$

Student: The total electric field is positive, so it points to the right.
Tutor: Does that make sense?
Student: The electron has a negative charge, so the field and force should be in opposite directions. The force is to the left, so the field is to the right. It works.
Tutor: Good. What is the external field?
Student: The total field is the sum of the proton's field and the external field.

$$E_{\text{total}} = E_{\text{p}} + E_{\text{ext}}$$

$$(+250\text{ N/C}) = (+360\text{ N/C}) + E_{\text{ext}}$$

$$E_{\text{ext}} = -110\text{ N/C}$$

Student: The external field is to the left.

EXAMPLE

Find the electric field at the center of the square.

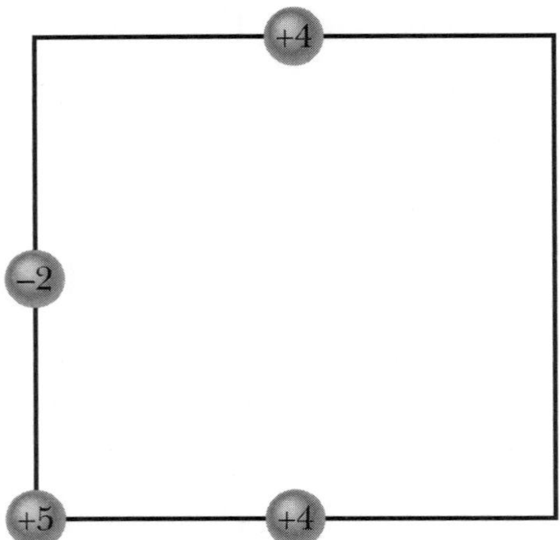

Tutor: How are you going to start?
Student: We want electric field, so we need to use either the $Q \to E$ step or the $E \to F$ step.
Tutor: True. Either we have the charges that create the electric field or we have the force created by the electric field.
Student: There is no charge at the center of the square. How can we have the force created by the electric field?
Tutor: We don't.
Student: Ah, but we have the charges that create the electric field. I find the field from each one and add them up.
Tutor: Remember to add them as vectors. Start with a diagram, except that instead of drawing forces, draw fields.
Student: Okay. The top $+4$ μC charge creates field away from it because it is a positive charge. Away from that charge is downward. The bottom $+4$ μC charge creates field away from it, or upward. The -2 μC charge creates field toward the negative charge, or to the left. And last, the $+5$ μC charge creates field away from it, or up and to the right.

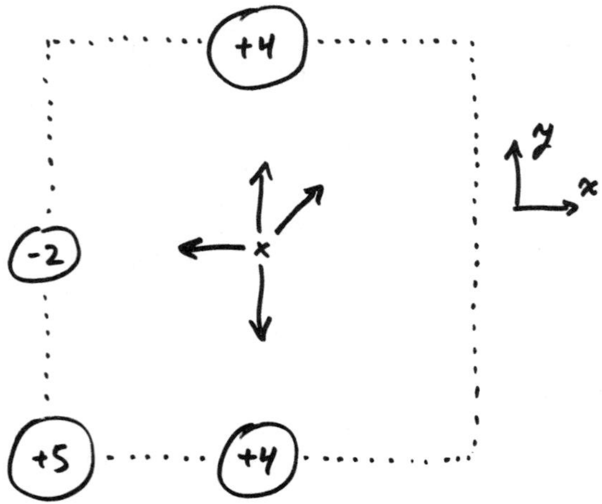

Student: Do the fields from the $+4$ μC charges cancel?
Tutor: They are in opposite directions. Do they have the same magnitude?
Student: The charges are the same, and the distances are the same, so the magnitudes should be the same. They cancel.
Tutor: They add to zero, yes. What about the other fields?

Student: I need to find the magnitudes and add them.

$$E_2 = \frac{kQ}{r^2} = \frac{(9 \times 10^9 \text{ N}\cdot \text{m}^2/\text{C}^2)(2 \times 10^{-6} \text{ C})}{(0.05 \text{ m})^2} = 7.2 \times 10^6 \text{ N/C}$$

$$E_5 = \frac{kQ}{r^2} = \frac{(9 \times 10^9 \text{ N}\cdot \text{m}^2/\text{C}^2)(5 \times 10^{-6} \text{ C})}{(0.05 \text{ m})^2} = 18 \times 10^6 \text{ N/C} \quad ?$$

Tutor: Is the +5 μC charge 5 cm from the center of the square?
Student: No, it's further. I need to use the Pythagorean theorem.
Tutor: For a 45°-45°-90° right triangle, the hypotenuse is $\sqrt{2}$ times the length of a side.
Student: Yeah, that's right. So the +5 μC charge is $\sqrt{2} \times 5$ cm from the center.

$$E_5 = \frac{kQ}{r^2} = \frac{(9 \times 10^9 \text{ N}\cdot \text{m}^2/\text{C}^2)(5 \times 10^{-6} \text{ C})}{\left(\sqrt{2}(0.05 \text{ m})\right)^2} = 9.0 \times 10^6 \text{ N/C}$$

Student: Now I decide whether to add them or subtract them.
Tutor: Are the field vectors colinear?
Student: No. Do I have to find the components like we did back in Chapter 3 or something?
Tutor: Yep. Adding vectors hasn't changed, even if the vectors are electric fields now.

$$E_x = -E_2 + E_5 \cos 45° = -(7.2 \times 10^6 \text{ N/C}) + (9.0 \times 10^6 \text{ N/C}) \cos 45° = -0.84 \times 10^6 \text{ N/C}$$

$$E_y = +E_5 \sin 45° = (9.0 \times 10^6 \text{ N/C}) \sin 45° = +6.36 \times 10^6 \text{ N/C}$$

Student: To find the magnitude of the field, I have to add the components using the Pythagorean theorem.

$$E = \sqrt{(E_x)^2 + (E_y)^2} = \sqrt{(-0.84 \times 10^6 \text{ N/C})^2 + (+6.36 \times 10^6 \text{ N/C})^2} = 6.42 \times 10^6 \text{ N/C}$$

Student: It's almost the same as the y component.
Tutor: The y component is much bigger than the x component. The angle of the vector should show that.
Student: It goes a little left and a lot up. Measured from the $-x$-axis, the angle is

$$\arctan\left(\frac{+6.36 \times 10^6 \text{ N/C}}{0.84 \times 10^6 \text{ N/C}}\right) = 82.5°$$

EXAMPLE

A charge of $Q = 12\ \mu$C is spread out uniformly over a rod of length $L = 12$ cm. The point $\mathcal{P}$ is a distance $d = 2$ cm above the midpoint of the rod. What is the electric field at point $\mathcal{P}$?

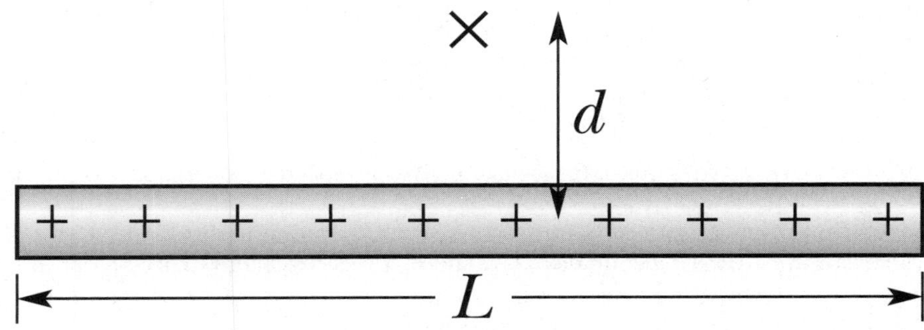

Tutor: How do we find electric fields?
Student: We use $E = kQ/r^2$, but what do I use for r?
Tutor: That is a problem, isn't it? Since not all of the charge is at the same place, you need to divide it into pieces, each so small that we can say that the whole piece is at the same place.
Student: That would have to be really tiny. There'd be hundreds of them
Tutor: Yes, maybe more. Then you find the electric field from each piece and add all the electric fields created by the pieces.
Student: But there are hundreds of pieces! Wait, this is sounding like an integral.
Tutor: Don't panic. Pick one piece, but not the first or the last.
Student: Why not the first or the last?
Tutor: We need to do all of the pieces at once, and the first and last pieces are special and aren't representative. Pick an arbitrary piece.
Student: Like, say, 1 cm to the right of the middle of the rod.
Tutor: Like, say, x to the right of the middle of the rod. x can then be anything from -6 to 6 cm.

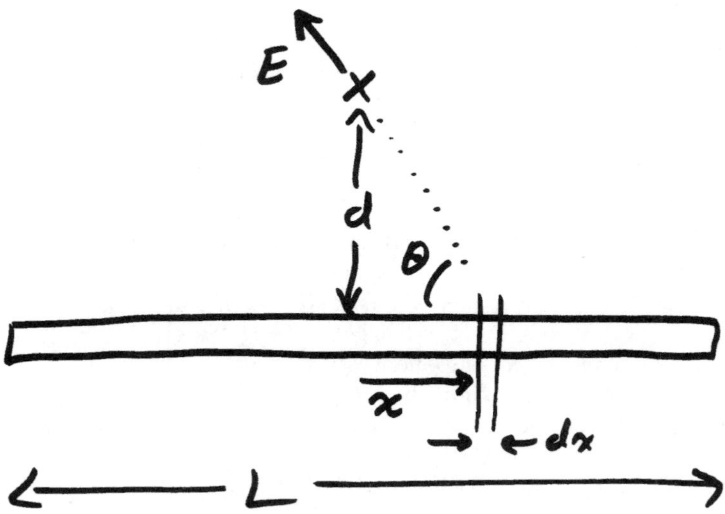

Student: So it could be any piece. Got it. Then I find the electric field from the piece using x. The magnitude of the field is

$$E = \frac{k\ dq}{(\sqrt{x^2 + d^2})^2}$$

Student: Picking x from the middle makes the denominator simpler. What's dq?
Tutor: dq is typically used to represent the charge of each piece. What is the charge dq of the piece?
Student: Well, if there are millions of pieces, then the charge of each piece is really, really small.
Tutor: Very small, but not zero. If there were exactly one million pieces, what would be the charge of each piece?
Student: The total charge is 12 μC, so a millionth of that, or 12×10^{-12} C.
Tutor: Do all of the pieces have the same charge?
Student: Uh, I think so. How can I tell?
Tutor: The charge is spread "uniformly" over the rod, so any piece of the same length has the same charge. What is the length of each piece?
Student: Is that dx?
Tutor: It is. The length of each piece is equal to the distance from one piece to the next, which is dx. You can use that to find the charge on each piece.
Student: So the number of pieces is $L/\,dx$, and the charge q of each piece is

$$dq = \frac{Q}{\text{number of pieces}} = \frac{Q}{L/\,dx} = \frac{Q\ dx}{L}$$

Tutor: Very good. Another way to get the same result is by using charge density. The charge density is the charge per length, or coulombs per meter. If we multiply the coulombs per meter (Q/L) by the meters of a piece (dx) then we get the coulombs of a piece ($Q\ dx/L$).
Student: So the field created by each piece is

$$E = \frac{k(Q\ dx/L)}{(\sqrt{x^2+d^2})^2}$$

Tutor: That is the *magnitude* of the field created by each piece. Do the electric fields created by each piece point in the same direction?
Student: No. The fields all point away from the charge but not all in the same direction.
Tutor: So we have to find the x and y components of each field.
Student: But there are millions of fields.
Tutor: Again, do it once for the piece at x and it will be the same for all of them.
Student: Okay. I need to find the angle θ, and then the x component is adjacent, so it's cos, while the y component is opposite, so it's sin.

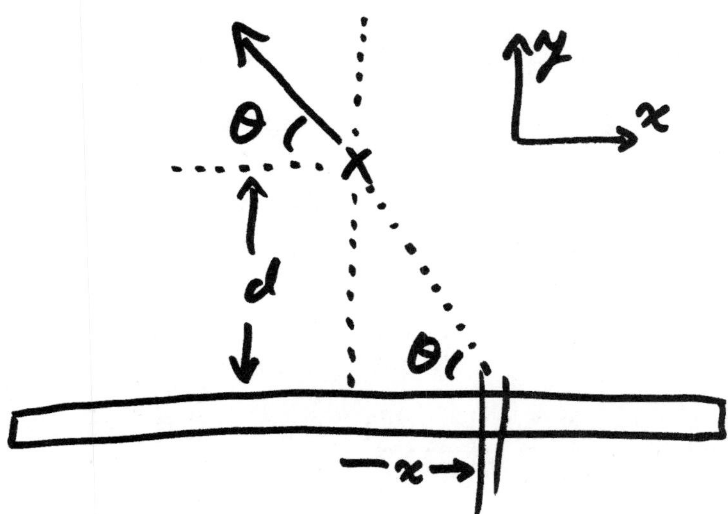

Tutor: True. But you don't really need θ, just $\cos\theta$ and $\sin\theta$. Look at the triangle you drew to find the angle. Instead of finding the angle, find the cosine and sine of the angle, using the lengths of the sides.
Student: The cosine of θ is x divided by $\sqrt{x^2+d^2}$. Is that what you mean?
Tutor: Yep. Now you can write down the x component of the electric field. Remember that if x is positive, then the x component of the electric field is negative.

$$E_x = \frac{k(Q\ dx/L)}{(\sqrt{x^2+d^2})^2} \times \frac{-x}{\sqrt{x^2+d^2}} = -\frac{kQx\ dx}{L(x^2+d^2)^{3/2}}$$

Tutor: Then the total x component is the sum of all of the individual x components. What are the limits for x?
Student: x goes from $L = -6$ cm to $+6$ cm.

$$E_x = \int_{-L/2}^{L/2} -\frac{kQx\ dx}{L(x^2+d^2)^{3/2}} = -\frac{kQ}{L}\int_{-L/2}^{L/2} \frac{x\ dx}{(x^2+d^2)^{3/2}}$$

Student: Now we have to do the integral.
Tutor: We look in the appendix of the book, or some other integral table, where we find

$$\int \frac{x\ dx}{(x^2+a^2)^{3/2}} = -\frac{1}{(x^2+a^2)^{1/2}}$$

Tutor: This is the same integral that we have, so we can use their answer.

$$E_x = -\frac{kQ}{L}\int_{-L/2}^{L/2} \frac{x\ dx}{(x^2+d^2)^{3/2}} = -\frac{kQ}{L}\left[-\frac{1}{\sqrt{x^2+d^2}}\right]_{-L/2}^{L/2}$$

$$= -\frac{kQ}{L}\left[-\frac{1}{\sqrt{(L/2)^2+d^2}} - -\frac{1}{\sqrt{(-L/2)^2+d^2}}\right] = 0$$

Student: Zero? Isn't there some easier way to do this?
Tutor: Look at the symmetry of the problem.
Student: Oh, yeah. If I flip the problem left to right, the problem is the same, so the answer has to look the same, so it has to be along the y axis.
Tutor: Good. You still need to do the y component.
Student: It should be very similar, except that it's the opposite over the hypotenuse.

$$E_y = \int_{-L/2}^{L/2} \frac{k(Q\ dx/L)}{(\sqrt{x^2+d^2})^2} \times \frac{d}{\sqrt{x^2+d^2}} = \frac{kQd}{L}\int_{-L/2}^{L/2} \frac{dx}{(x^2+d^2)^{3/2}}$$

Student: We look in the appendix of the book, where we find

$$\int \frac{dx}{(x^2+a^2)^{3/2}} = \frac{x}{a^2\ (x^2+a^2)^{1/2}}$$

$$E_y = \frac{kQd}{L}\int_{-L/2}^{L/2} \frac{dx}{(x^2+d^2)^{3/2}} = \frac{kQd}{L}\left[\frac{x}{d^2\ \sqrt{x^2+d^2}}\right]_{-L/2}^{L/2} = \frac{kQd}{Ld^2}\left[\frac{L/2}{\sqrt{(L/2)^2+d^2}} - \frac{-L/2}{\sqrt{(-L/2)^2+d^2}}\right]$$

$$= \frac{kQ}{Ld}\left[\frac{L}{\sqrt{(L/2)^2+d^2}}\right] = \frac{kQ}{d\ \sqrt{(L/2)^2+d^2}}$$

Student: Now I can put in the numbers.

$$E_y = \frac{kQ}{d\ \sqrt{(L/2)^2+d^2}} = \frac{(9\times 10^9\ \text{N}\cdot\ \text{m}^2/\text{C}^2)(12\ \mu\text{C})}{(0.02\ \text{m})\sqrt{(0.06\ \text{m})^2+(0.02\ \text{m})^2}} = 8.5\times 10^7\ \text{N/C}$$

Chapter 23

Gauss' Law

Examine the surface below. Count the number of electric field lines coming *out of* each surface. Remember to count a field line going in as a negative line going out. You should find that $N_2 = -N_1$ and $N_3 = N_4 = 0$.

This is not mere coincidence. Electric field lines start at positive charges and end at negative charges. For a field line to come out of a surface, it either has to start inside (at a positive charge) or enter somewhere else on the surface. This is Gauss' law.

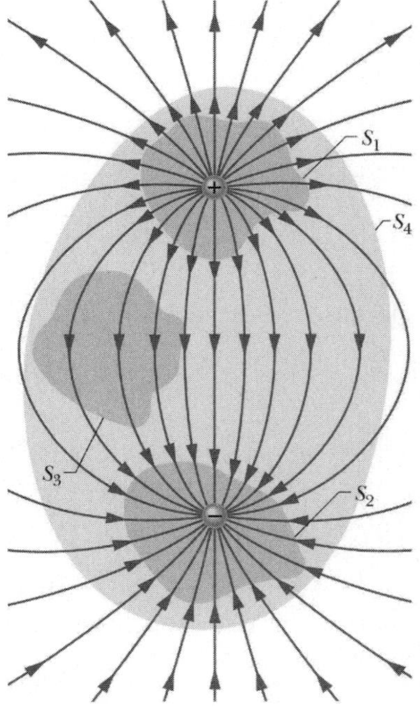

Gauss' law works for any closed surface. A closed surface is one with no exits — if you are inside the surface and you want to get out you need to pass through the surface. The surface does not have to be real — imagine or invent any surface you want and Gauss' law tells you the number of electric field lines coming out.

The mathematical expression for Gauss' law is

$$\Phi_E = \frac{q_{\text{net}}}{\epsilon_0}$$

where Φ_E is the electric field flux coming out of the area, q_{net} is the net charge inside the area, and ϵ_0 is $4\pi k = 8.85 \times 10^{-12}\ \text{C}^2/\text{N m}^2$.

What is electric field flux Φ_E? Imagine that you are out in the rain holding a bucket. Flux would be the rate at which raindrops enter the bucket. You could reduce this rate by using a smaller bucket, by going to a place with less rain, or by tilting the bucket sideways. The electric field flux Φ_E is effectively a count of the number of electric field lines passing through a surface. Mathematically,

$$\Phi_E = \int \vec{E} \cdot d\vec{A}$$

which says to divide the surface into tiny pieces, at each piece find the component of the electric field that is perpendicularly out of the surface (normal to the surface), multiply times the area of the tiny piece, and add the results for all of the tiny pieces.

Despite the scary appearance of this integral, it's not so bad. We only do the integral when it's dirt simple, when not doing this integral leads to a much nastier integral.

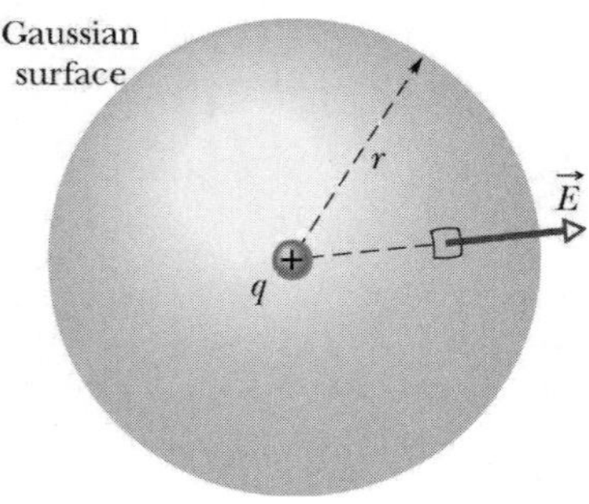

Let's see how it works. We want to find the electric field created by a single point charge $q = +2\ \mu$C. What is the electric field a distance r away? We can invent any surface we want to use with Gauss' law, so we'll use a sphere, centered on the charge q with a radius of r (r is a variable, it could be anything).

We want to do the integral to find the flux. To do that we need to know the electric field at each spot on our imaginary surface. We don't know the field — that's what we're trying to find. What we do know is that the field on one side of the imaginary sphere is the same as the field on the other side. Pick any two spots on the imaginary sphere, and the electric field is the same. It has to be because of the symmetry of the problem — it looks the same from any angle, so the electric field is the same at any angle.

At any spot on our imaginary sphere the electric field is perpendicularly out of the surface and has the same magnitude. Because the field is perpendicularly out,

$$\vec{E} \cdot d\vec{A} = E\ dA\ \cos 0° = E\ dA$$

Because the field E is the same everywhere on our imaginary "Gaussian" surface

$$\Phi_E = \int E\ dA = E \int dA$$

The integral that remains is to add up the area of each tiny piece, so it's equal to the area of the imaginary surface.

$$\Phi_E = E \int dA = EA$$

Gauss' law tells us that the flux is equal to the charge "enclosed" by the imaginary surface divided by ϵ_0.

$$EA = \Phi_E = \frac{q_{\text{net}}}{\epsilon_0}$$

Because the surface is a sphere, the area is $4\pi r^2$

$$E(4\pi r^2) = \frac{q_{\text{net}}}{\epsilon_0}$$

$$E = \frac{q_{\text{net}}}{4\pi r^2 \epsilon_0} = \frac{q}{(4\pi\epsilon_0) r^2} = \frac{k(+2\ \mu\text{C})}{r^2}$$

This is the same thing we got in the last chapter (whew).

What if we had used a negative charge $q = -2\ \mu$C? Everything would have been the same until we put the negative in and got a negative electric field. We always measure the field out from the surface, never in, so a negative field is inward or toward the negative charge.

The goal is to use Gauss' law to calculate the electric field. We can only do that if the symmetry in the problem lets us do the critical step above — the electric field, whatever it is, is the same everywhere on the imaginary surface so we can pull it out of the integral.

$$\overbrace{EA = \int \vec{E} \cdot d\vec{A}}^{\text{symmetry}} = \overbrace{\Phi_E = \frac{q_{\text{net}}}{\epsilon_0}}^{\text{Gauss' law}}$$

$$\underbrace{\int \vec{E} \cdot d\vec{A} = \Phi_E}_{\text{definition of flux}}$$

EXAMPLE

A charge of $Q = +4\ \mu$C is placed on a hollow conducting sphere of radius $R = 5$ cm. Find the electric field magnitude everywhere.

Tutor: How do we find the electric field?
Student: We divide the charge into little pieces, find the electric field from each, and add the electric fields (as vectors).
Tutor: We could do that, but the resulting integral would be very difficult. Instead, let's try using Gauss' law.
Student: So we need some kind of "Gaussian surface."
Tutor: Yes, but that's not complicated. Imagine that you want the electric field at a point that is a distance r from the middle of the sphere. Draw an imaginary sphere that is concentric with the conducting sphere and has radius r.
Student: What's r? Isn't it the same as R?
Tutor: R is the size of the sphere. r is the distance to the point where we want the electric field.
Student: But we want the electric field everywhere. Don't we have to pick a value for r?
Tutor: We could, but the beauty of leaving it as a variable is that it works for any value. Then we only have to do the problem once.
Student: So the electric field is the same everywhere?
Tutor: Not necessarily. Our result might be a function of r, so that at different values of r we get different electric fields.
Student: Okay, but why a sphere?
Tutor: Because of the symmetry. Look at the problem. You can rotate it any way you want around the center, and the problem looks the same. Therefore, the answer also has to look the same (remember doing this when we did center of mass?). We don't know what the electric field is, but everywhere on a concentric sphere the electric field is the same. If we chose some other shape, when we rotated the Gaussian surface with the sphere, the problem would look different, and we couldn't say that the electric field is the same everywhere. This is the key to using Gauss' law: choosing a surface where the electric field, whatever it is, is the same everywhere.

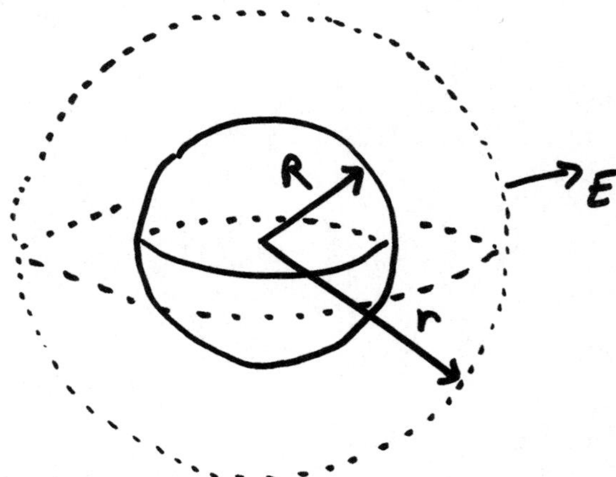

Student: Why does the field have to be the same everywhere?
Tutor: So that we can calculate the flux. We divide the area of the surface into tiny pieces, find the electric field through each, multiply by the (tiny) area, and add them. If the field is the same everywhere, then this is the same as EA.
Student: Okay, so the flux Φ is equal to EA, and we don't know E. Is A the area of the conducting sphere or the imaginary Gaussian surface?
Tutor: It's the area of the Gaussian surface $4\pi r^2$.
Student: But as r gets bigger, the area increases and the flux increases, but the number of field lines doesn't increase.
Tutor: It is good that you're thinking ahead. As r gets bigger, the area gets bigger, but the number of field lines doesn't increase. Therefore, the electric field must get smaller so that the flux stays the same.
Student: Ah, so E does depend on r.
Tutor: Correct. Once you have chosen your Gaussian surface, write down Gauss' law.

$$\frac{Q_{\rm enc}}{\epsilon_0} = \Phi_E = EA$$

Student: I thought that we had to do vector dot products and stuff like that.

Tutor: We do, but we need the field to be at least *piecewise-constant* in order to use Gauss' law. More on that in the next example. Look at the Gaussian surface. At each point on the surface, what is the direction of the electric field?
Student: Directly away from the sphere.
Tutor: Correct. Because of the symmetry, the field points directly away from the sphere. How much of the field is perpendicular to the Gaussian surface, parallel to the normal vector?
Student: All of it.
Tutor: Correct. You just did the vector dot product. All of the field is parallel to the normal, so the $\cos\theta$ is 1.
Student: And the field is the same everywhere, so the integral equals EA.
Tutor: Yes. Now, what is the area of the Gaussian surface?
Student: The area is the area of a sphere, $4\pi r^2$.

$$\frac{Q_{\text{enc}}}{\epsilon_0} = \Phi_E = E(4\pi r^2)$$

Tutor: How much charge is enclosed?
Student: All of it, $Q = +4\ \mu\text{C}$.
Tutor: If $r > R$ and the Gaussian surface encloses the sphere, then yes. What if $r < R$?
Student: Then none of the charge is enclosed.
Tutor: Correct. Let's start inside. If none if the charge is enclosed, what is the electric field?

$$\frac{\cancelto{0}{Q_{\text{enc}}}}{\epsilon_0} = \Phi_E = E(4\pi r^2)$$
$$0 = E(4\pi r^2)$$
$$E = 0$$

Student: There won't be any electric field.
Tutor: Correct. What about outside the sphere?

$$\frac{Q_{\text{enc}}}{\epsilon_0} = \Phi_E = E(4\pi r^2)$$
$$\frac{Q}{\epsilon_0} = E(4\pi r^2)$$
$$E = \frac{Q}{4\pi\epsilon_0 r^2} = \frac{kQ}{r^2}$$

Student: That's the same thing we had before!
Tutor: Before it was for a point charge. Now it is for a sphere.
Student: So any sphere has the same electric field as a point charge?
Tutor: Keep in mind that the purpose of the example is not to reach such a conclusion, but to be able to use Gauss' law.
Student: Yes, I know. Is the electric field for a sphere the same as a point charge, though?
Tutor: As long as you have spherical symmetry, and all of the charge is enclosed, then everything we did works.
Student: And by asking what the field is everywhere, it means both inside and outside the sphere?
Tutor: Yes, and because the field isn't constant, our electric field depended on r.

EXAMPLE

A long, solid, insulating cylinder of radius $R = 6$ cm has a uniform charge density of $\lambda = -3$ C/m. Find the electric field magnitude everywhere.

Student: So the symmetry, combined with the difficulty of dividing the charge into little pieces, tips me off to use Gauss' law.
Tutor: Very good. Of course, it tips you off to *try* Gauss' law. We're never entirely sure something will work until we can see the end.
Student: Okay. I want a Gaussian surface with the same symmetry as the charge, so I choose a cylinder, concentric with the cylinder of charge. My imaginary cylinder has a radius of r, which might be more of less than R.
Tutor: All very good.

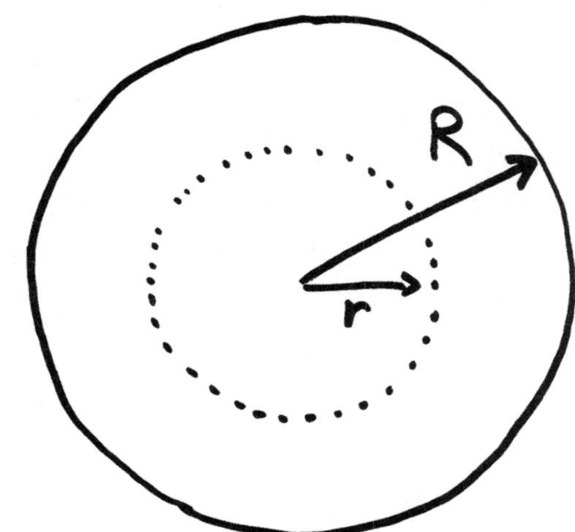

Student: Then I write down Gauss' law.

$$\frac{Q_{\text{enc}}}{\epsilon_0} = \Phi_E = EA$$

Student: The area is the area of my cylinder, which is $(2\pi rL) + 2(\pi r^2)$, including the ends, where L is the length of my imaginary Gaussian surface.
Tutor: Correct. Now it's time to think about the electric field. What is the direction of the electric field?
Student: It points outward from the cylinder of charge.
Tutor: Radially outward. What is the flux through the top and bottom of the Gaussian surface?
Student: There isn't any. The field skims the surface but doesn't go through.
Tutor: So Gauss' law becomes

$$\frac{Q_{\text{enc}}}{\epsilon_0} = \Phi_E = E_{\text{side}}(2\pi rL)(1) + E_{\text{end}}(2\pi r^2)(0)$$

Student: So the 1 and the 0 are the $\cos\theta$'s?
Tutor: Correct.
Student: Then E_{end} disappears from the equation.
Tutor: Yes, and that's good because the field wasn't constant as it moves out from the cylinder along the ends.
Student: If $r > R$, the charge enclosed is all of it, but they don't tell us the total charge.
Tutor: The longer the cylinder is, the more charge it has. The charge is -3 coulombs for each meter of length. Multiply the length enclosed by the charge per length to get the charge enclosed.

Student: So $Q_{\text{enc}} = \lambda L$?

$$\frac{\lambda L}{\epsilon_0} = \Phi_E = E(2\pi r L)$$

$$E = \frac{\lambda \cancel{L}}{2\pi r \cancel{L} \epsilon_0} = \frac{\lambda}{2\pi \epsilon_0 r} = \frac{2k\lambda}{r}$$

Tutor: Good. As we get further from the cylinder, the field gets weaker as the field lines spread out. What if $r < R$?
Student: Then the enclosed charge is zero, so the field is zero.
Tutor: Not true this time. If the cylinder were made of a conducting material, then the charges would move to the surface and that would be true. Now the charge is spread throughout the cylinder.
Student: So if $r < R$, some of the charge is enclosed?
Tutor: Correct. How much of the charge is enclosed?
Student: I don't know.
Tutor: How much of the volume is enclosed?
Student: What does that matter?
Tutor: If half of the volume is enclosed, then half of the charge is enclosed.
Student: So I find the volume enclosed, divide by the total volume to get the fraction enclosed, and multiply by the charge?
Tutor: Essentially.
Student: The volume enclosed is $\pi r^2 L$. The total volume is $\pi R^2 L$. The charge enclosed is

$$Q_{\text{enc}} = \lambda \times \frac{\pi r^2 L}{\pi R^2 L} = \lambda \times \frac{r^2}{R^2} \quad ?$$

Tutor: Almost. The charge in a volume $\pi R^2 L$ is λL.

$$Q_{\text{enc}} = \lambda L \times \frac{\pi r^2}{\pi R^2}$$

Tutor: Now back to Gauss' law.

$$\frac{1}{\epsilon_0} \times \left((\lambda L) \times \frac{r^2}{R^2} \right) = \Phi_E = E(2\pi r L)$$

$$\frac{\lambda r^2}{\epsilon_0 R^2} = E(2\pi r)$$

$$E = \frac{\lambda r}{2\pi \epsilon_0 R^2} = \frac{2k\lambda r}{R^2}$$

Student: The field increases as we move away from the cylinder.
Tutor: The field increases as we move away from the center, so long as we are still inside the cylinder. As we move away, the charge enclosed increases faster than the area, so the field increases. What is the electric field at $r = 0$?
Student: It is zero.
Tutor: Which it has to be because of symmetry.
Student: Ah, if it wasn't zero, and we rotated the cylinder around its axis, it would be different.
Tutor: Note that λ is negative, so the field is going into the Gaussian surface.
Student: Good — the field should go towards the negative charge.

EXAMPLE

A $d = 6$ cm thick plane of insulating material contains a uniform charge density $\rho = 7\ \text{C/m}^3$. Find the electric field magnitude everywhere.

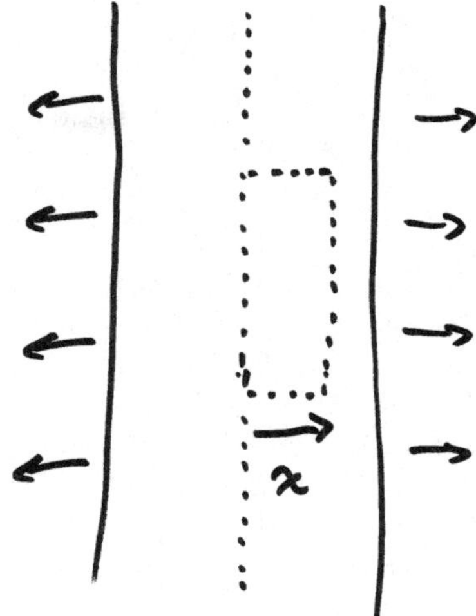

Student: I need a Gaussian surface that keeps the symmetry of the charge.
Tutor: Correct. What does the electric field look like?
Student: It points perpendicular to the plane of charge.
Tutor: Does it go in both directions?
Student: Can we use symmetry to determine that? If I flip the problem left-to-right, the problem looks the same, so it goes both ways and has to have the same field each way.
Tutor: Good. Draw a Gaussian cuboid that starts at the center and goes out a distance x.
Student: Then I write down Gauss' law.

$$\frac{Q_{\text{enc}}}{\epsilon_0} = \Phi_E = EA$$

Tutor: What is the area?
Student: I only include the area that has field coming out of it. Nothing comes out of the paper or goes into the paper, or goes up or down the paper, because it skims the surface rather than going through. The field at the center is zero, so the only side that has field is the right side. But what is the area of the right side?
Tutor: It's your Gaussian surface, so you get to choose.
Student: Okay. The area is A, whatever that is. It better cancel, right?
Tutor: Correct. The field better not depend on the size of your imaginary surface — I might imagine a different surface but that wouldn't change the field. What is the charge enclosed?
Student: It's the volume in meters^3 times the density in coulombs per meter^3.
Tutor: Good — what's the volume?
Student: The width times the height times the depth.
Tutor: The height times the depth is your area A, and the width is x.
Student: So the volume is xA, and the charge is $xA\rho$.
Tutor: As long as the right side is inside the plane ($x < d/2$).

$$\frac{xA\rho}{\epsilon_0} = \Phi_E = EA$$

Student: The area does cancel.

$$E = \frac{x\rho}{\epsilon_0}$$

Tutor: What if $x > d/2$, and the right side is outside the plane?
Student: The area is still the same.
Tutor: Yes, but not all of the Gaussian surface contains charge.

Student: Oh, so the enclosed charge stops increasing once we leave the plane.

$$\frac{(d/2)A\rho}{\epsilon_0} = \Phi_E = EA$$

$$E = \frac{\rho d}{2\epsilon_0}$$

Student: I saw the equations $E = \sigma/\epsilon_0$ and $E = \sigma/2\epsilon_0$. Is this the 2 in the equations?
Tutor: Somewhat. If the field goes in both directions, then you use the equation with the 2. If the field goes in only one direction, then you don't use the 2. This happens if you have a conducting plane with charge on the surface, since there won't be an electric field inside the conductor. The field here goes in both directions, and the 2 appears because, by measuring out from the center, we include only half the charge.
Student: So what's with the Greek letters λ and ρ for charge?
Tutor: We still use Q or q for charge, but we use Greek letters for charge density. Greek lambda λ is typically used for coulombs per meter, Greek sigma σ for coulombs per meter^2, and Greek rho ρ (pronounced "row" as in row, row, row your boat) for coulombs per meter^3. It's a tip-off or reminder of how the charge is distributed.

Chapter 24

Electric Potential

We've learned to use forces when we deal with charged objects. Can we use conservation of energy and conservation of momentum? Certainly energy and momentum are both conserved, but can we use them effectively to solve problems? Conservation of momentum doesn't help us because some of the charges are usually "glued down," and we don't know the force on them so we don't know the impulse. Conservation of energy is so effective in dealing with charges that we do it all the time, often without noticing.

How did we use conservation of energy earlier? The energy you start with plus the energy you put in (work) is the energy you end with.

$$KE_i + PE_i + W = KE_f + PE_f$$

The work done by some forces (gravity and springs) is easier to deal with by using potential energy, and then we don't include them in the work term. Work done by electric force is the other work that is easier to do with potential energy.

How did potential energy from gravity work? **Potential energy is work that we've done that we might get back out**. When we raise a rock we exert a force equal to gravity mg (a little more at the beginning to get it going) while we raise it a height h. Since we push in the same direction as the motion we do positive work. Gravity pushes in the opposite direction as the motion and does negative work as the potential energy increases. The work that we do that is stored as potential energy is mgh. If we tied a string to the rock beforehand and wrapped the other end around a generator, then as the rock rolled down the hill we could get electricity to power our stereo and make tunes, which is the purpose of physics. The work that gravity does during the process is the same no matter how we get the rock to the top.

If we moved a rock that was twice as big then the potential energy would be twice as big. The potential energy is the product of the size of the rock (m) and something that depends on the position (gh). This latter part, the part that depends on position, is the potential. The units of gh are $\mathrm{m^2/s^2}$ or J/kg — how much work it takes to move one kilogram of rock.

With electricity, the potential energy is equal to the product of the size (charge q) of the thing we move times the potential V, which depends on the position.

$$U = PE = qV$$

(U is often used for potential energy.) The potential energy in joules is equal to the charge in coulombs times the potential, or joules per coulomb. One joule per coulomb is a volt and is sometimes called voltage.

$$1\ \mathrm{V} = 1\ \frac{\mathrm{J}}{\mathrm{C}}$$

The only significant difference is that the charge can be negative, while with gravity the mass could only be positive.

Remember that potential and voltage are the same thing, but potential and potential energy are not the same. A location has potential, and a charge at the location has potential energy.

The question then is: how do we determine the potential.

When we push a charge against the electrical force we do positive work, increasing the potential energy. When the charge moves on its own and we hold it back, we do negative work (or get work out) while the potential energy decreases (some of the potential energy is turned into work). To calculate the electrical potential energy we have to find the work we do pushing against the electrical force, or the negative work done by the electrical force.

$$\Delta PE = -\vec{F} \cdot \vec{x} = -q\vec{E} \cdot d\vec{x} = -qEx\cos\theta$$

Often the electric field is not constant as we move the charge. Therefore we divide the path into tiny pieces – so tiny that we can say that the electric field is constant for the whole tiny path. Then we add the work we've done (increase in potential energy) for all of the tiny pieces of the path.

$$\Delta PE = q\int -\vec{E} \cdot d\vec{x}$$

where $\vec{E} \cdot d\vec{x}$ is the component of the electric field parallel to the tiny piece $d\vec{x}$ times the length of the tiny piece.

The potential energy is the product of two pieces, the amount of charge that we move and something that depends only on the path along which we move it. The part that depends on the path (and not the charge) is the potential.

$$\Delta V = \int -\vec{E} \cdot d\vec{x}$$

You might conclude that we're going to be doing lots of integrals, but this is not so. There are two cases for which it's reasonable to work out the integral and reuse it. The cases are point charge and constant field. Since these come up regularly, using potential often *keeps* us from needing to do an integral.

If the field is constant, then the integral becomes

$$\Delta V_{\text{constant}} = \int -\vec{E} \cdot d\vec{x} = -\vec{E} \cdot \int d\vec{x} = -\vec{E} \cdot \Delta\vec{x}$$

if the field is created by a point charge, then the potential is

$$V_{\text{point charge}} = \frac{kq}{R}$$

This assumes that the potential is equal to zero at infinity.

EXAMPLE

What is the work needed to move a proton from $\mathcal{P}$ to ∞?

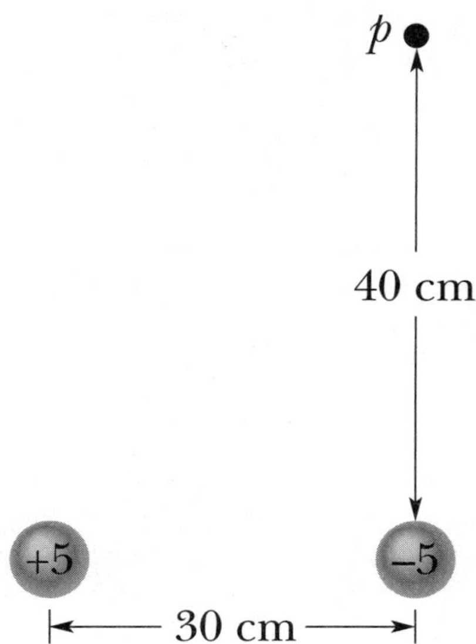

Student: The problem asks for work, so that means we'll need to use energy.
Tutor: Good. What is the equation for conservation of energy?
Student: Energy before plus energy in equals energy after.

$$KE_i + PE_i + W = KE_f + PE_f$$

Tutor: What is the kinetic energy before?
Student: The proton isn't moving, is it?
Tutor: The problem doesn't say, so that implies that it isn't.
Student: So the kinetic energy before is zero. Is the proton moving when it gets to infinity?
Tutor: If it was, then it would have a positive kinetic energy, and we'd have to do more work. We don't want to do that.
Student: So the kinetic energy afterward is zero. How can a proton move to infinity?
Tutor: It can't really, of course, because the universe is finite. It takes something like half of the work to move the proton the first 40 cm, and half of the remaining work to move it the next 80 cm, and half of the remaining to move it the next 160 ...
Student: How do you come up with that so easily?
Tutor: If we double the distance, we cut the potential energy in half.
Student: So by the time we've moved it three meters, we've already done 90% of the work needed to move it to infinity?
Tutor: Yes, and we can calculate how much work it would take to get to infinity, but we'd never *really* get it there.
Student: Okay. Both kinetic energies are zero. We need the potential energies.
Tutor: True. The potential energy due to electric forces is qV.
Student: So...

$$\cancelto{0}{KE_i} + qV_{\mathcal{P}} + W = \cancelto{0}{KE_f} + qV_\infty$$

Tutor: Very good. Now we need to calculate the potentials at $\mathcal{P}$ and ∞.
Student: So I find the potential at $\mathcal{P}$ due to each charge and add the potentials?

Tutor: Yes.
Student: The potential at $\mathcal{P}$ due to the $-5\ \mu$C charge is

$$V = \frac{kQ}{r} = \frac{k(-5\ \mu\text{C})}{(40\ \text{cm})} = -11.25 \times 10^4\ \text{N·m/C}$$

Tutor: Yes. A newton times a meter is a joule, and a joule per coulomb is a volt.
Student: Okay. The potential at $\mathcal{P}$ due to the $+5\ \mu$C charge is

$$V = \frac{kQ}{r} = \frac{k(+5\ \mu\text{C})}{\sqrt{(30\ \text{cm})^2 + (40\ \text{cm})^2}} = 9 \times 10^4\ \text{V}$$

Tutor: Correct.
Student: Now I need to find the components and add the x and y components together.
Tutor: No. Potential doesn't have direction. The values you just found are the potentials created by each charge, and you just add the potentials together.
Student: No vector stuff at all? Great.

$$V_{\mathcal{P}} = (9 \times 10^4\ \text{V}) + (-11.25 \times 10^4\ \text{V}) = -22,500\ \text{V} = -22.5\ \text{kV}$$

Tutor: That's one of the reasons to use potential. If we had been doing a vector, then you should have used the absolute value of the charge, so that you got a positive value for the magnitude. Since potential isn't a vector, you need to include the minus sign so that a negative charge creates a negative potential.

$$(1.6 \times 10^{-19}\ \text{C})\,(-22.5\ \text{kV}) + W = (1.6 \times 10^{-19}\ \text{C})V_\infty$$

Student: Now I need to find the potential at infinity.
Tutor: Yep. What is the potential if r is ∞?
Student: Uh, it's zero.
Tutor: Yes. We chose the potential to be zero at infinity. If we had picked anywhere else, then the equation for potential would have been more complicated and harder to calculate.
Student: So the potential is *always* zero at infinity?
Tutor: Whenever we use $V = kQ/r$ for point charges, the potential is zero at infinity. The other common situation is a constant field, and then we can't take the potential to be zero at infinity. We'll do one of those next.
Student: Oh good! Anyway, now I can find the work.

$$(1.6 \times 10^{-19}\ \text{C})\,(-22.5\ \text{kV}) + W = (1.6 \times 10^{-19}\ \text{C})\, \cancelto{0}{V_\infty}$$
$$W = -(1.6 \times 10^{-19}\ \text{C})\,(-22.5\ \text{kV}) = 3.6 \times 10^{-15}\ \text{J}$$

EXAMPLE

What is the work needed to move an electron from $\mathcal{C}$ to $\mathcal{D}$?

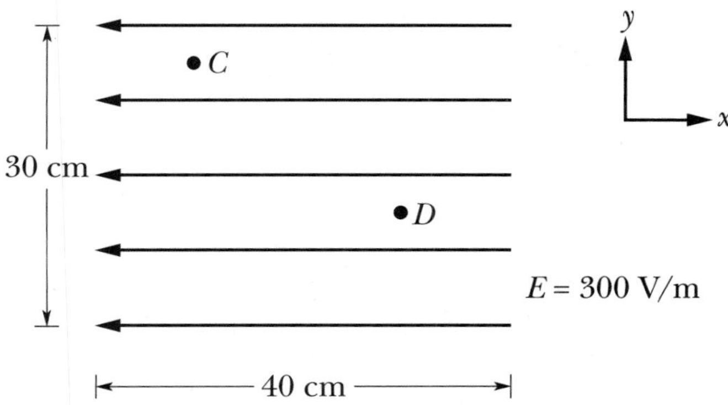

Student: It asks for work again, so we'll be using conservation of energy again.

$$KE_i + PE_i + W = KE_f + PE_f$$

Student: Again, the particle isn't moving before or after, and again, the potential energy is qV.

$$\cancelto{0}{KE_i} + qV_{\mathcal{C}} + W = \cancelto{0}{KE_f} + qV_{\mathcal{D}}$$

Tutor: You're doing well. What is the potential at $\mathcal{C}$?
Student: Well, that is a problem. Where are the charges that create the electric field?
Tutor: We can't see them.
Student: So I have to start at infinity, where the potential is zero, and integrate the electric field as I come in to $\mathcal{C}$?
Tutor: That would work, in theory. The problem is that the field is a constant, and integrating a constant over an infinite distance gives an infinite value. We didn't have this problem with a point charge, because the field got smaller as we went further away, and the integral converged.
Student: Ah, big math term. So if I can't see the charges that create the field, and I can't integral the electric field from infinity, how can I find the potential at $\mathcal{C}$?
Tutor: It's kind of a trick question. You don't need the potential at infinity, you need only the difference in potentials between $\mathcal{C}$ and $\mathcal{D}$.

$$W = qV_{\mathcal{D}} - qV_{\mathcal{C}} = q\left(V_{\mathcal{D}} - V_{\mathcal{C}}\right)$$

Student: So I have to integrate only from $\mathcal{C}$ to $\mathcal{D}$.
Tutor: Correct. In fact, it is important that you integrate from $\mathcal{C}$ to $\mathcal{D}$ and not from $\mathcal{D}$ to $\mathcal{C}$.
Student: How can you tell?
Tutor: Because a change is always the final minus the initial, so the final is $\mathcal{D}$ and the initial is $\mathcal{C}$. Fortunately, the electric field is constant, and the integral is easy.
Student: Oh, right. The potential difference is

$$\Delta V = E \times d = (300 \text{ V/m})(0.50 \text{ m}) = 150 \text{ V} \quad ?$$

Student: I thought that the units for electric field were newtons per coulomb.
Tutor: They are. Volts per meter are equal to newtons per coulombs. Also, remember that only the part of the displacement that is parallel to the electric field matters.
Student: Oh, so

$$\Delta V = E \times d = (300 \text{ V/m})(0.40 \text{ m}) = 120 \text{ V} \quad ?$$

Tutor: Now you need to make sure that you have the sign right.
Student: How do I do that?
Tutor: One way is to follow the math. As you go from $\mathcal{C}$ to $\mathcal{D}$, the displacement is to the right. The electric field is to the left. The displacement in the direction of the electric field is −40 cm, so

$$\Delta V = -\vec{E} \cdot \Delta \vec{x} = -(300 \text{ V/m})(-0.40 \text{ m}) = 120 \text{ V}$$

Tutor: Another way is to look at the electric field. Electric field always goes from higher potential to lower potential.
Student: So $\mathcal{D}$ must be at a higher potential than $\mathcal{C}$.
Tutor: Yes, and therefore $V_{\mathcal{D}} - V_{\mathcal{C}}$ is...
Student: ...positive.
Tutor: Is the potential at $\mathcal{D}$ positive or negative?
Student: It's greater than the potential at $\mathcal{C}$, so it must be positive.
Tutor: But −4 is greater than −7, and −4 isn't positive.
Student: So $V_{\mathcal{D}}$ is negative?
Tutor: We don't know, and we don't care. We do know that $V_{\mathcal{D}} > V_{\mathcal{C}}$, and that's what's important.
Student: Okay. I can finish the problem now.

$$W = q\,(V_{\mathcal{D}} - V_{\mathcal{C}}) = (-1.6 \times 10^{-19} \text{ C})(120 \text{ V}) = -1.92 \times 10^{-17} \text{ J}$$

Tutor: Does it make sense that the work you need to do is negative?
Student: The electric force will push the electron to the right, so it will go there on its own. I won't need to do any work.
Tutor: Negative work means that we can get work out of the process as the electron moves.

EXAMPLE

A charge of $Q = +4\ \mu\text{C}$ is spread uniformly over a solid sphere of radius $R = 5$ cm. Find the potential at the center of the sphere, taking the potential to be zero at infinity.

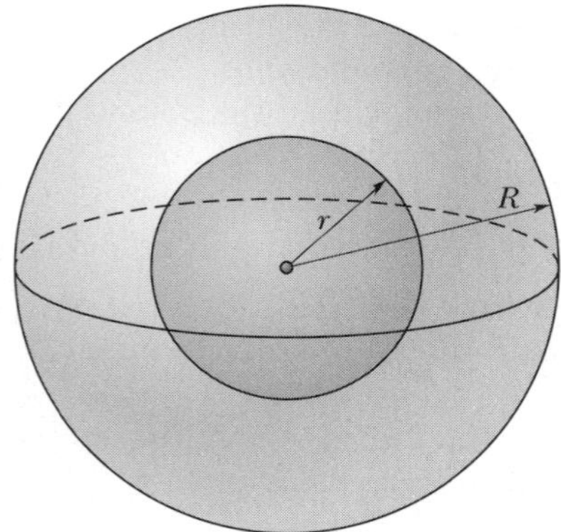

Student: Nothing about energy, just find the potential.
Tutor: Yep. We have three ways to do that. If the field is constant, then the potential difference is the displacement times the parallel component of the field.
Student: The field isn't a constant. We found the field in the last chapter.
Tutor: Not quite. We found the field due to a hollow conducting sphere. Because the solid sphere is also spherically symmetric, the field outside the sphere is the same as for the hollow sphere. Our second way to find a potential is that if the field is created by point charges, then we can find the potential from each point

charge and add the potentials.
Student: The charge is spread out over the sphere, so we would have an infinite number of point charges.
Tutor: True. The third way is to start at a point where we know the potential, and integrate the electric field to the point we want.
Student: The only place where we know the potential is at infinity.
Tutor: So we start there and integrate in.

$$\Delta V = \int -\vec{E} \cdot d\vec{x}$$

$$V_R = \cancelto{0}{V_\infty} + \int_\infty^R -\vec{E} \cdot d\vec{x}$$

Tutor: I have trouble with minus signs when integrating from a higher value to a lower one, so I'm going to reverse the limits and add a minus sign.

$$V_R = -\int_R^\infty -\vec{E} \cdot d\vec{x}$$

Student: The field outside the sphere is kQ/r^2.
Tutor: As we move from R to ∞, we are using x to represent our current position.
Student: So the field outside the sphere is kQ/x^2?
Tutor: Correct. What is the direction of the electric field?
Student: The field points away from the sphere.
Tutor: What is the direction of $d\vec{x}$?
Student: What is $d\vec{x}$?
Tutor: $d\vec{x}$ is the infinitesimally small length of the steps that we take as we move out to ∞.
Student: If it's infinitesimally small, then it doesn't have direction.
Tutor: It may be really small, but it isn't zero. When we add up all of the steps, we have to get the displacement from R to ∞.
Student: So the direction of each step, small though it is, is outward from the sphere.
Tutor: Good. How much of the electric field is in the direction of the step?
Student: They are in the same direction.
Tutor: Yes. How much of the electric field is in the direction of the step?
Student: All of it.
Tutor: Yes, so

$$\vec{E} \cdot d\vec{x} = E \ dx$$

Student: That's how we deal with the vector stuff.
Tutor: Yes. Now the potential V_R at R is

$$V_R = \int_R^\infty E\,dx = \int_R^\infty \frac{kQ}{x^2}\,dx = kQ\left[-\frac{1}{x}\right]_R^\infty = kQ\left[-\frac{1}{\infty} - -\frac{1}{R}\right] = \frac{kQ}{R}$$

Student: That's the same as the potential for a point charge!
Tutor: The field is the same as for a point charge, so the integral of the field is also the same.
Student: Now we do the same for the field inside the sphere.
Tutor: Yes. What is the electric field inside the sphere?
Student: We need to use Gauss' law, picking an imaginary sphere to keep the symmetry of the charge. Only part of the charge is enclosed, so we need to find Q_{enc} like we did in the cylinder example from the last chapter.

$$\frac{Q_{\text{enc}}}{\epsilon_0} = \Phi_E = EA$$

$$\frac{1}{\epsilon_0}\left(\frac{\text{volume enclosed}}{\text{volume of sphere}} \times (\text{charge of sphere})\right) = E(4\pi r^2)$$

$$\frac{1}{\epsilon_0}\left(\frac{\frac{4}{3}\pi r^3}{\frac{4}{3}\pi R^3} \times Q\right) = E(4\pi r^2)$$

$$\frac{Qr^3}{\epsilon_0 R^3} = E(4\pi r^2)$$

$$E = \frac{Qr^3}{4\pi\epsilon_0 R^3 r^2} = \frac{Qr}{4\pi\epsilon_0 R^3} = \frac{kQr}{R^3}$$

Tutor: Wonderful. Now we can do the potential integral.

$$V_0 = V_R + \int_R^0 -E\ dx$$

Tutor: Again, I prefer to reverse the limits, because otherwise I'm likely to make a sign error.

$$V_0 = V_R + -\int_0^R -E\ dx = \frac{kQ}{R} + \int_0^R \frac{kQr}{R^3}\ dx \quad ?$$

Student: What's wrong with that?
Tutor: The field inside the sphere is a function of how far we are from the center of the sphere. In using Gauss' law to find the field, you used r for that distance, but in the potential integral we're using x.
Student: So the field inside the sphere is $\frac{kQx}{R^3}$ when we do the integral.
Tutor: Yes, as x goes from 0 to R.

$$V_0 = \frac{kQ}{R} + \int_0^R \frac{kQx}{R^3}\ dx = \frac{kQ}{R} + \frac{kQ}{R^3}\int_0^R x\ dx$$

$$V_0 = \frac{kQ}{R} + \frac{kQ}{R^3}\left[x^2\right]_0^R = \frac{kQ}{R} + \frac{kQ}{R^3}\left[R^2 - 0^2\right] = \frac{kQ}{R} + \frac{kQ}{R}$$

$$V_R = \frac{2kQ}{R} = \frac{2(9\times 10^9\ \text{N}\cdot\ \text{m}^2/\text{C}^2)(+4\ \mu\text{C})}{(0.05\ \text{m})} = 1.44\times 10^6\ \text{V} = 1.44\ \text{MV}$$

EXAMPLE

A charge Q is spread uniformly over a length L. What is the potential at $\mathcal{P}$, a distance d away from the center of the line of charge?

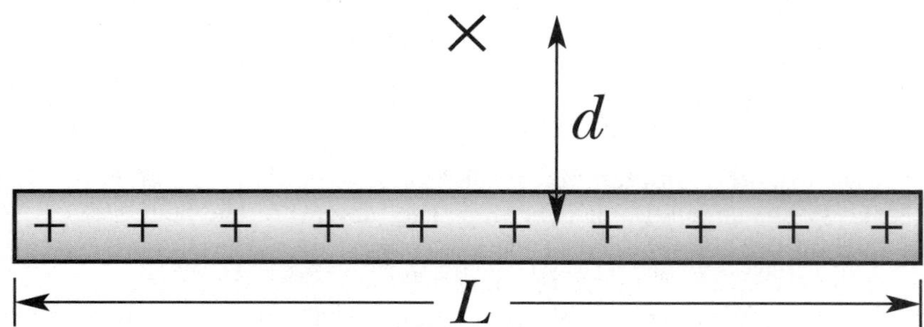

Student: Let me see if I have this right. We have three ways to find the potential. If the field is constant, then we use $\Delta V_{\text{constant}} = -\vec{E}\cdot\Delta\vec{x}$. If the field is from point charges, we use $V_{\text{point charge}} = \frac{kq}{R}$. If we can't do either of those, then we integrate the electric field $\Delta V = \int -\vec{E}\cdot d\vec{x}$ from someplace where we do know the potential.
Tutor: Very good. Which way are you going to try here?

Student: The field isn't a constant, so we can't do that. The charge isn't a point charge, but we could divide it into point charges and add the potentials using an integral. Or we could integrate the field. We have to do an integral either way. Which way is easier?
Tutor: We can do it both ways for practice. It's probably easier to integrate the charge.
Student: We divide the rod into tiny bits, each dx long. As before, the charge of each piece is

$$dQ = \frac{\text{(length of piece)}}{\text{(length of rod)}} \times \text{(charge of rod)} = \frac{dx}{L}Q$$

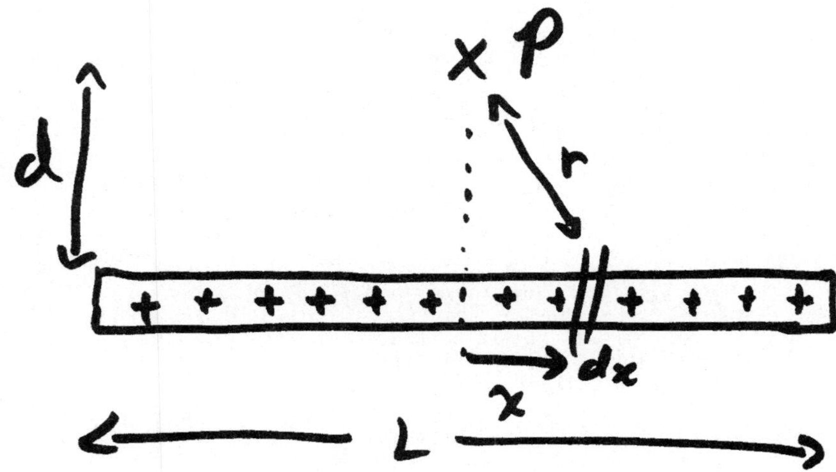

Tutor: Good. What is the potential created by each tiny piece?
Student: k times the charge divided by the distance.

$$V_{\mathcal{P}} = \int \frac{k\,(Q\,dx/L)}{\sqrt{d^2 + x^2}}$$

Tutor: What are the limits for the integral? What are the minimum and maximum values for x?
Student: x goes from $-L/2$ to $+L/2$.

$$V_{\mathcal{P}} = \int_{-\frac{L}{2}}^{+\frac{L}{2}} \frac{k\,(Q\,dx/L)}{\sqrt{d^2 + x^2}}$$

Student: Can I do the integral now?
Tutor: Sure, everything there is a constant or x, the variable of integration.

$$V_{\mathcal{P}} = \int_{-L/2}^{+L/2} \frac{k\,(Q\,dx/L)}{\sqrt{d^2 + x^2}}$$

$$V_{\mathcal{P}} = \frac{kQ}{L} \int_{-L/2}^{+L/2} \frac{dx}{\sqrt{d^2 + x^2}}$$

$$V_{\mathcal{P}} = \frac{kQ}{L} \left[\ln \left(x + \sqrt{d^2 + x^2} \right) \right]_{-L/2}^{+L/2}$$

$$V_{\mathcal{P}} = \frac{kQ}{L} \left[\ln \left((L/2) + \sqrt{d^2 + (L/2)^2} \right) - \ln \left((-L/2) + \sqrt{d^2 + (-L/2)^2} \right) \right]$$

Tutor: We can do algebra to simplify this, but the physics of the problem is done.

$$V_{\mathcal{P}} = \frac{kQ}{L} \left[\ln \left(\frac{(L/2) + \sqrt{d^2 + (L/2)^2}}{(-L/2) + \sqrt{d^2 + (L/2)^2}} \right) \right]$$

$$V_{\mathcal{P}} = \frac{kQ}{L} \ln \left(\frac{(L/2) + \sqrt{d^2 + (L/2)^2}}{(-L/2) + \sqrt{d^2 + (L/2)^2}} \cdot \frac{(L/2) + \sqrt{d^2 + (L/2)^2}}{(L/2) + \sqrt{d^2 + (L/2)^2}} \right)$$

$$V_{\mathcal{P}} = \frac{kQ}{L} \ln \left(\frac{\left((L/2) + \sqrt{d^2 + (L/2)^2}\right)^2}{-(L/2)^2 + (\sqrt{d^2 + (L/2)^2})^2} \right)$$

$$V_{\mathcal{P}} = \frac{kQ}{L} \ln \left(\frac{\left((L/2) + \sqrt{d^2 + (L/2)^2}\right)^2}{d^2} \right)$$

$$V_{\mathcal{P}} = \frac{2kQ}{L} \ln \left(\frac{(L/2) + \sqrt{d^2 + (L/2)^2}}{d} \right)$$

Student: Now you want to integrate the field for practice?
Tutor: Sure. To get the potential at $\mathcal{P}$, start at ∞ and go to $\mathcal{P}$, integrating the electric field.

$$V_{\mathcal{P}} = V_\infty + \int_\infty^{\mathcal{P}} -\vec{E} \cdot d\vec{x}$$

Student: But then you want to integrate the other way.
Tutor: It's easier to get the sign right that way.

$$\cancelto{0}{V_\infty} = V_{\mathcal{P}} + \int_{\mathcal{P}}^{\infty} -\vec{E} \cdot d\vec{x}$$

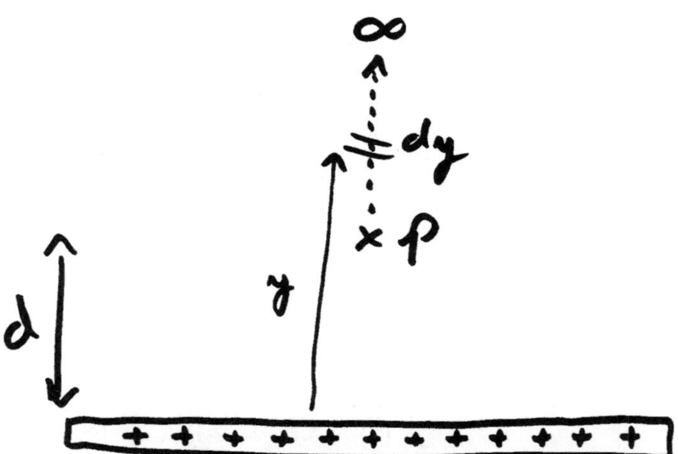

Student: The electric field is kQ/x^2, since x is our variable of integration.
Tutor: Just because x is the variable you're integrating over doesn't mean that the field depends on x.
Student: True. But the field does depend on where we are. As x moves from $\mathcal{P}$ to ∞, we need the field at a distance x.
Tutor: Good. But the field is not created by a point charge.
Student: Arrg! We have to integrate over the charge to find the field?
Tutor: Yes and no. Yes, we have to find the field, and that means doing an integral. No, we don't have to now, because we did that in an earlier example.

$$E_y = \frac{kQ}{d\sqrt{d^2 + (L/2)^2}}$$

Student: There's no r.
Tutor: No. d was the distance from the rod to the point where we found the field.
Student: So we replace d with x.

$$E_y(x) = \frac{kQ}{x\sqrt{x^2 + (L/2)^2}}$$

Tutor: Yes, then we let x go from d to ∞.

$$0 = V_{\mathcal{P}} + \int_d^\infty - \left(\frac{kQ}{x\sqrt{x^2 + (L/2)^2}} \right) dx$$

$$0 = V_{\mathcal{P}} + \int_d^\infty - \frac{kQ}{x\sqrt{x^2 + (L/2)^2}} \, dx$$

$$V_{\mathcal{P}} = kQ \int_d^\infty \frac{dx}{x\sqrt{x^2 + (L/2)^2}}$$

$$V_{\mathcal{P}} = kQ \left[-\frac{1}{(L/2)} \ln \left(\frac{(L/2) + \sqrt{x^2 + (L/2)^2}}{x} \right) \right]_d^\infty$$

$$V_{\mathcal{P}} = \frac{2kQ}{L} \left[-\ln \left(\frac{(L/2) + \sqrt{\infty^2 + (L/2)^2}}{\infty} \right) + \ln \left(\frac{(L/2) + \sqrt{d^2 + (L/2)^2}}{d} \right) \right]$$

$$V_{\mathcal{P}} = \frac{2kQ}{L} \left[-\ln \left(\frac{(L/2) + \infty}{\infty} \right) + \ln \left(\frac{(L/2) + \sqrt{d^2 + (L/2)^2}}{d} \right) \right]$$

$$V_{\mathcal{P}} = \frac{2kQ}{L} \ln \left(\frac{(L/2) + \sqrt{d^2 + (L/2)^2}}{d} \right)$$

Student: So we get the same answers either way.

Chapter 25

Capacitance

Imagine taking two large flat pieces of metal, such as cookie sheets, and placing them parallel to each other but not touching. Then connect the two ends of a battery to the two cookie sheets. Positive charge flows out of the positive end of the battery to the first cookie sheet (really it's negative charges flowing out of the negative end of the battery onto the other cookie sheet, but the effect is the same). When the positive charges get to the first conductor (cookie sheet), they can go no further so they collect there. Other positive charges from the second conductor take the place of the previous charges, going into the negative terminal of the battery and leaving behind a net negative charge. After a while (which might be only microseconds) this process slows down and stops.

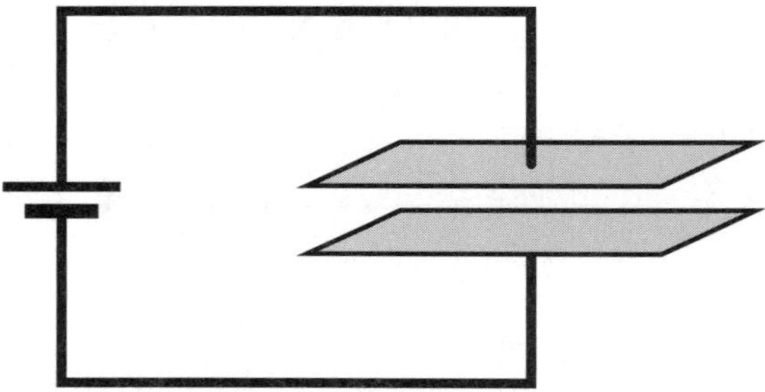

What if we were now to disconnect the battery from the cookie sheets? Could the charges on the conductors foresee what we're about to do and rush back into the battery? No, the charges are trapped on the conductors. We could now connect the conductors to the two ends of a light bulb, and the excess positive charge would go through the bulb, creating a momentary current, and meet up with the excess negative charge on the other conductor. We have a device for storing small amounts of charge. The amount of charge Q that can be stored on a capacitor with capacitance C is

$$Q = CV$$

Equal amounts of positive and negative charge are stored on the two plates of the capacitor, and Q is the absolute value of one of these two charges. The units of capacitance are farads (F).

It takes work to store charge on a capacitor. Since this energy can all be extracted later, we can express it as a potential energy

$$U_{\text{CAP}} = \frac{1}{2}QV$$

This is similar to the equation we had for the potential energy of a charge ($U = qV$) except for the one-half. One way to understand this difference is to imagine stacking books. If you start with a bunch of (equally sized) books on the floor, it doesn't take any work to put the first book on the floor. The second book takes a little work, since it must be moved up by the thickness of the first book. The third book takes even more work, and so on, until the last book takes the most work. We could add the work needed for each book, or multiply the number of books by the *average* work for each book. The average is the work for the middle book, which is lifted to half the height of the pile. Likewise it takes no work to put the first charge on the capacitor because with no charge it has no voltage. The second charge takes some work, the third more, and the last charge takes the most work. The total work is the charge times the average voltage, which is half of the final voltage. Use $U = qV$ for individual charges, but use $U = \frac{1}{2}QV$ when each successive charge takes more work than the next.

We have four variables (C, V, Q, and U) and two equations. **If we know any two of the four variables, we can find the other two using our two equations**.

EXAMPLE

What size capacitor would you need to store 2 mJ when connected to a 12 V battery?

Student: If we know any two of the four, we can solve for the others, but I need an equation with U, V, and C.
Tutor: It's true that if you know two of the four, you can find the others. It's also true that we don't have an equation with those three. What can you find?
Student: I can solve for the charge Q.

$$U_{\text{CAP}} = \frac{1}{2}QV$$
$$(0.002\text{ J}) = \frac{1}{2}Q(12\text{ V})$$
$$Q = 333\ \mu\text{C}$$

Student: That seems like a pretty small charge, but then 2 mJ seems like a pretty small amount of energy.
Tutor: Most capacitors aren't very big. They store only a little charge.
Student: Now that I have the charge, I guess I can find the capacitance after all.

$$Q = CV$$
$$333\ \mu\text{C} = C(12\text{ V})$$
$$C = 28\ \mu\text{F}$$

Tutor: One farad is a big capacitance, so many capacitors are measured in microfarads or even picofarads.
Student: So why don't we come up with another equation?
Tutor: We could ...

$$Q = \frac{2U}{V} \quad \longrightarrow \quad \frac{2U}{V} = CV \quad \longrightarrow \quad U = \frac{1}{2}CV^2$$

Tutor: We could come up with an equation for every problem, but then there would be an infinite number of equations. It's better to remember fewer equations and to combine them when necessary.
Student: An infinite number of equations *would* be hard to remember.

The next difficulty comes when we have two or more capacitors at the same time. If two capacitors are connected in series or in parallel, then we have ways of dealing with them. **It is not true that any two capacitors must be either in series or parallel**, but if they are then we can do something with them.

When two devices are connected so that all of the charge that leaves one must go through the other, we say they are in series. Likewise when two devices are connected so that they must have the same voltage, we say they are in parallel.

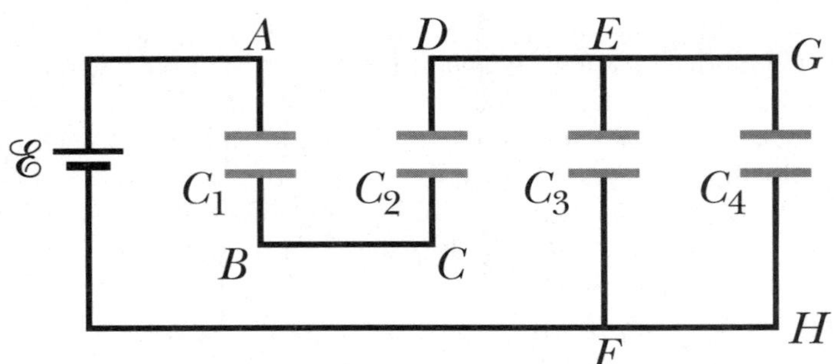

How do we determine whether two capacitors are in series? Consider C_1 and C_2 above. Charge passes from A through C_1 to B. From there the charges can only go to C and through C_2 to D. C_1 and C_2 are in series because the charges that pass through C_1 must then go through C_2, and the charges that pass through the two must be equal. Charges that pass through C_2 to point D can go either through C_3 to F or through C_4 to H. The charges through C_2 and C_3 do not have to be equal, so C_2 and C_3 are not in series.

How do we determine whether two capacitors are in parallel? Consider C_3 and C_4 above. Take a finger from each hand, and place them at each end of C_3. The voltage difference between your fingers is equal to the voltage difference between the ends of C_3, or the voltage across C_3. Now try moving your fingers, but only along wires and not across any circuit element, such as a capacitor or battery. If by doing this you can get your fingers to lie at the two ends of C_4, then C_3 and C_4 have the same voltage across them and they are in parallel (they are, and you should be able to do this). Are C_3 and C_2 in parallel? We can move a finger from E to D but we can't get from F to C. We could only do this by jumping from one wire to another nearby wire (not allowed), or by going through the battery and C_1 (also not allowed). Therefore, C_3 and C_2 are not in parallel. Even though both are drawn vertically, so that they are geometrically parallel, they are not electrically parallel. Any two capacitors do not have to be either in series or in parallel.

If two capacitors are in series, then they can be treated as if they were a single capacitor. The capacitance of this replacement or "equivalent" capacitor is

$$\frac{1}{C_S} = \frac{1}{C_1} + \frac{1}{C_2}$$

If two capacitors are in parallel, then they can be treated as if they were a single capacitor. The capacitance of this replacement or "equivalent" capacitor is

$$C_P = C_1 + C_2$$

EXAMPLE

Find the voltage across each capacitor.

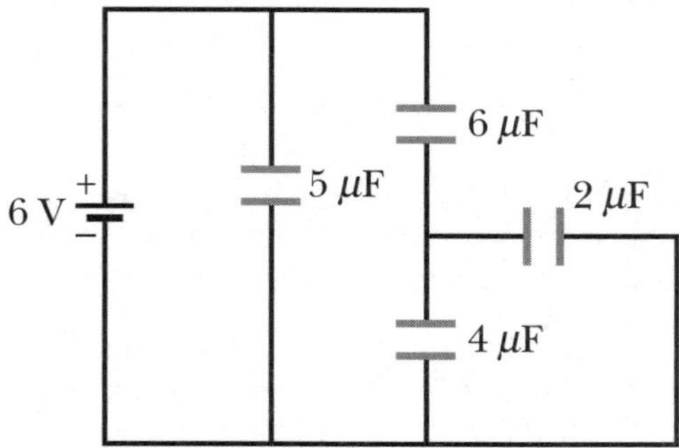

Student: The voltage on the 5 μF capacitor is 6 volts.
Tutor: Rather than "voltage on," use "voltage across." The voltage across is the difference in the voltages, or potentials, at the two ends.
Student: Okay. The voltage across the 5 μF capacitor is 6 volts.
Tutor: How do you know?
Student: Because it's a 6 volt battery.
Tutor: That's not enough. Is the 5 μF capacitor in parallel with the battery?
Student: Why does it matter?
Tutor: Two things in parallel have the same voltage across them.
Student: So if the 5 μF capacitor is in parallel with the battery, then it has the same voltage across it as the battery does?
Tutor: Yes. How do you check?
Student: I do the "finger test." I put one finger on each side of the 5 μF capacitor and try to move them to the ends of the battery. I slide the top one over, and I slide the bottom one over, and I'm there. The 5 μF capacitor is in parallel with the battery.
Tutor: How do you know that the voltage across the battery is 6 volts?
Student: Uh, because it's a 6 volt battery?
Tutor: Yes. A battery creates a potential difference, or voltage difference. A 6 volt battery creates a 6 volt potential difference.
Student: So the voltage across the 5 μF capacitor is the same as the voltage across the battery, and that's 6 volts.
Tutor: Good.
Student: The voltage across the 6 μF capacitor is 6 volts.
Tutor: How do you know?
Student: Because it's a 6 volt battery.
Tutor: That's not enough. Is the 6 μF capacitor in parallel with the battery?
Student: Here we go again.
Tutor: It gets faster with experience.
Student: If the 6 μF capacitor is in parallel with the battery, then it has the same potential across it as the battery does. I put one finger on each side of the 6 μF capacitor and try to move them to the two sides of the battery. The top one slides over to the top of the battery. The bottom one has to jump over the 2 μF capacitor or the 4 μF capacitor...
Tutor: ...and it can't do that.
Student: So the 6 μF capacitor is not in parallel with the battery.
Tutor: Correct.
Student: So the voltage across the 6 μF capacitor isn't 6 volts.
Tutor: It isn't necessarily 6 volts. It might be 6 volts, but we don't know that yet (and it won't be).
Student: So what can I do? Neither the 2 μF capacitor nor the 4 μF capacitor is in parallel with the battery either.
Tutor: If you can find two capacitors in series or parallel, then you can replace them with a single capacitor.
Student: So I can replace the 6 μF capacitor and the 4 μF capacitor with a 10 μF capacitor, and the 8 μF capacitor is in parallel with the battery.
Tutor: Slow down. You just did three steps, all of which we need to check. Are the 6 μF capacitor and the 4 μF capacitor in series or parallel?
Student: They are in series. Charge goes from the 6 to the 4.
Tutor: Let's check. Does *all* of the charge that comes out of the bottom of the 6 μF capacitor have to go into the 4 μF capacitor, or could some of it go somewhere else?
Student: Some of the charge could go into the 2 μF capacitor.
Tutor: To be in series, *all* of the charge from one must go into the other.
Student: So the 6 μF capacitor and the 4 μF capacitor *aren't* in series.
Tutor: Correct.
Student: And the 6 μF capacitor isn't in series with the 2 μF capacitor either.
Tutor: Also correct.
Student: Nor is the 4 μF capacitor in series with the 2 μF capacitor.

Tutor: Correct.
Student: Are some of them in parallel?
Tutor: Check.
Student: Okay. I'll check to see if the 4 is in parallel with the 6. I put one finger on each side of the 4 μF capacitor, and try to move them to the two sides of the 6 μF capacitor. I can move one from the top of the 4 to the bottom of the 6, but I can't move the other from the bottom of the 4 to the top of the 6.
Tutor: But you can move them to the two sides of the 2 μF capacitor.
Student: Oh. How can the 4 μF capacitor and the 2 μF capacitor be in parallel when one is vertical and the other is horizontal?
Tutor: There's a difference between electrically parallel and graphically parallel. Often two capacitors that are graphically parallel are also electrically parallel, but not always, so you need to check for electrically parallel. The 6 μF capacitor and the 5 μF capacitor are graphically parallel, but not electrically parallel.
Student: So I can replace the 4 μF capacitor and the 2 μF capacitor with a single capacitor?
Tutor: Yes. How do we add capacitors in parallel?
Student: We add the capacitances.
Tutor: Yes. Earlier you tried to add capacitors that you thought were in series.
Student: Oops.

$$C_{\mathrm{P}} = C_1 + C_2$$

$$C_{4\&2} = C_4 + C_2 = (4\ \mu\mathrm{F}) + (2\ \mu\mathrm{F}) = 6\ \mu\mathrm{F}$$

Tutor: Good.
Student: Is the replacement capacitor in series with the 6 μF capacitor?
Tutor: Check.
Student: The charge that comes out of the bottom of the 6 μF capacitor can go into the top of the 4 μF capacitor, or into the left side of the 2 μF capacitor, but nowhere else.
Tutor: So the 6 μF capacitor is in series with the replacement capacitor, or with the 4-2 pair.
Student: So I can replace them with a single capacitor.

$$\frac{1}{C_{\mathrm{S}}} = \frac{1}{C_1} + \frac{1}{C_2}$$

$$\frac{1}{C_{6\&4\&2}} = \frac{1}{C_6} + \frac{1}{C_{4\&2}} = \frac{1}{(6\ \mu\mathrm{F})} + \frac{1}{(6\ \mu\mathrm{F})}$$

Tutor: Be careful that you don't do something silly, like $1/6 + 1/6 = 1/12$.
Student: I had almost forgotten how much I enjoyed fractions.
Tutor: They aren't that hard, but do be careful.

$$\frac{1}{C_{6\&4\&2}} = \frac{1}{(6\ \mu\mathrm{F})} + \frac{1}{(6\ \mu\mathrm{F})} = \frac{2}{(6\ \mu\mathrm{F})} = \frac{1}{(3\ \mu\mathrm{F})}$$

Tutor: The replacement capacitor is not $1/3$ μF. One over the replacement capacitor is $1/3$ μF.
Student: So the replacement capacitor is 3 μF.
Tutor: Yes. Sometimes we call it the equivalent capacitance of the trio of capacitors.
Student: Now the trio of capacitors is in parallel with the 5 μF capacitor.
Tutor: True, but not necessary.
Student: I don't need to combine the 5 μF capacitor with the other trio?
Tutor: No. The trio is in parallel with the battery.
Student: It is. I can move one finger from the top of the 6 to the top of the battery, and the other from the bottom of the trio to the bottom of the battery. Isn't the 5 μF capacitor in the way?
Tutor: It may be between the battery and the trio, but it doesn't affect whether or not the trio is in parallel with the battery. What does it mean that the trio is in parallel with the battery?
Student: They have the same voltage, so the voltage on the trio is 6 volts, so the voltage on each of the three capacitors is 6 volts.
Tutor: Right and wrong. The voltage *across* the trio is 6 volts, but that doesn't mean that the voltage on each capacitor in the trio is 6 volts.

Student: No?
Tutor: One volt is one joule per coulomb, so each coulomb of charge leaves the battery with 6 joules of energy. First the charges go through the 6 μF capacitor, then through the 4-2 pair (which is also 6 μF). How much energy does the coulomb of charge "spend" to get through each capacitor?
Student: Does it spend 3 joules and 3 joules?
Tutor: Yes.
Student: But then the three capacitors in the trio each have 3 volts across them. That's 9 volts!
Tutor: Each coulomb of charge goes through first the 6 μF capacitor and then *either* the 4 μF capacitor or the 2 μF capacitor. So for each charge it's 3 volts plus 3 volts, which is 6 volts.
Student: How can we be sure that it's 3 and 3?
Tutor: Because the two capacitors (6 and 4-2 pair) have the same capacitance, so they have to have the same voltage across them.
Student: What if the 6 and the 4-2 pair didn't have the same capacitance?
Tutor: Then they wouldn't have the same voltage. Let's do that in the next example.

EXAMPLE

Find the voltage across each capacitor.

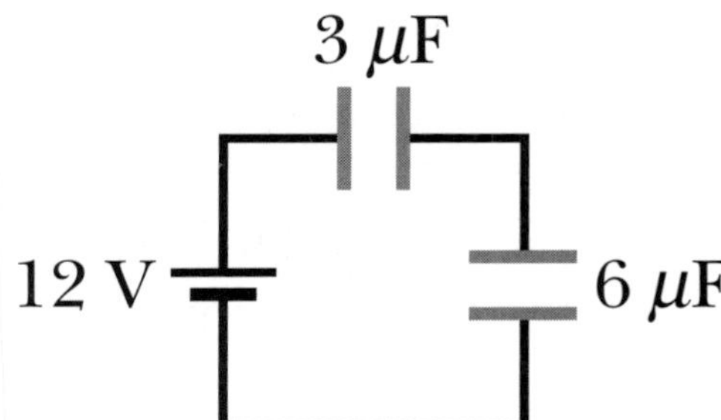

Student: Neither capacitor is in parallel with the battery, so the voltage across the capacitors isn't 12 volts.
Tutor: Good.
Student: But if the capacitors are in series or parallel, then I can replace them with a single capacitor.
Tutor: Yes.
Student: I think that they're in series, so I'll check that first. When charge comes out of the 3 μF capacitor, it has nowhere to go other than into the 6 μF capacitor, so they are in series.
Tutor: Good. How do we combine capacitors in series?
Student: By adding their reciprocals.

$$\frac{1}{C_S} = \frac{1}{C_1} + \frac{1}{C_2}$$

$$\frac{1}{C_{\text{pair}}} = \frac{1}{C_3} + \frac{1}{C_6} = \frac{1}{(3\ \mu\text{F})} + \frac{1}{(6\ \mu\text{F})} = \frac{3}{(6\ \mu\text{F})}$$

$$C_{\text{pair}} = 2\ \mu\text{F}$$

Tutor: What is the voltage across the 2 μF pair?
Student: The pair is in parallel with the battery, so it has the same voltage as the battery, so 12 volts.
Tutor: What is the charge on the equivalent capacitor?
Student: We're trying to find voltage. Why are you asking about the charge?

Tutor: Because that's how we get to voltage. Remember that $Q = CV$, so if we know the charge we can find the voltage.
Student: The charge on the equivalent capacitor is

$$Q_{\text{pair}} = C_{\text{pair}} V_{\text{pair}} = (2\ \mu\text{F})(12\ \text{V}) = 24\ \mu\text{C}$$

Tutor: How much of that charge is on each capacitor?
Student: Half on each?
Tutor: Think about what happens to the charge. 24 μC of charge comes out of the battery and goes into the first capacitor. There's a gap, so the charge can't go any further, and it stops. 24 μC of charge takes it's place, coming out of the bottom of the first capacitor and going into the top of the second capacitor. There it stops, and 24 μC of charge from the bottom of the second capacitor goes into the bottom of the battery.
Student: So there's 24 μC of charge on *each* capacitor?!
Tutor: Yes. When capacitors are in series, they each have all of the charge.
Student: In the last example, when the 4-2 pair was in series with the 6, the 4-2 pair would have the same charge as the 6?
Tutor: Yes. The 4 and the 2 would not each have the same charge as the 6, but their combined charge would be the same as the 6.
Student: So capacitors in parallel share charge, but in series they each get the whole charge.
Tutor: Yes.
Student: Now that we know the charges on the capacitors, we can find the voltages.

$$Q = CV \quad \rightarrow \quad V = \frac{Q}{C}$$

$$V_3 = \frac{Q_3}{C_3} = \frac{(24\ \mu\text{C})}{(3\ \mu\text{F})} = 8\ \text{V}$$

$$V_6 = \frac{Q_6}{C_6} = \frac{(24\ \mu\text{C})}{(6\ \mu\text{F})} = 4\ \text{V}$$

Student: Look! They add up to 12 volts.
Tutor: They need to, because each coulomb of charge uses up 12 joules as it goes through the two capacitors. The rules for adding capacitors in series and parallel were invented so that the voltages would add up to 12 volts.
Student: But you said a moment ago that the charges can't go through a capacitor.
Tutor: True. Positive charges stop on one plate, and different positive charges leave the other plate, leaving behind an excess negative charge. But because all charges dress the same, we can't tell them apart, so we sometimes say that the charges go through the capacitor, even though we know that they really don't.

Chapter 26

Current and Resistance

So far, we have not had moving charges or electric fields in conducting materials. These two are connected. **An electric field in conducting material causes current — moving charges.** So far we have simply waited until the charges quit moving and the electric field disappeared. Not anymore.

An electric field in a conducting material causes a current density $\vec{J}$.

$$\vec{J} = \frac{\vec{i}}{A} = (ne)\vec{v}_d = \sigma\vec{E} = \frac{\vec{E}}{\rho}$$

The current density J is the current i divided by the cross-sectional area A. Most of the time we care more about the current than the current density, but an electric field causes current density. The current appears because the valence electrons drift in the conducting material, and the current density is equal to the density n of charges times the charge e of each times the average drift velocity v_d. σ is the conductivity of the material, and ρ is the resistivity of the material ($\sigma = 1/\rho$).

Of more practical importance is Ohm's law.

$$V = iR$$

The difference in voltage (or potential) between the two sides of a resistor is equal to the current through the resistor times the resistance of the resistor. Current is measured in amperes (1 A = 1 C/s). Resistance is measured in ohms. To avoid confusing an uppercase "O" with a zero, the symbol for ohms is the Greek letter uppercase omega (1 Ω = 1 V/A).

Consider also the energy used in the resistor. Each charge uses an energy that depends on the change in the potential — if the voltage difference between the two sides of the resistor is 6 volts, or 6 joules per coulomb, then each coulomb of charge uses 6 joules of energy getting through the resistor. The more current, the faster coulombs of charge go through, the quicker energy is turned into heat in the resistor. So the power, or rate of energy used in the resistor, is

$$P = iV$$

The units of power used earlier work here — volts times amps equals watts.

$$(1\ \text{V})(1\ \text{A}) = (1\ \text{J}/\not{\text{C}})(1\ \not{\text{C}}/\text{s}) = 1\ \text{J/s} = 1\ \text{W}$$

We have four variables (R, V, i, and P) and two equations. If we know any two of the four variables, we can find the other two using our two equations.

To find the resistance of a wire, we use

$$R = \frac{\rho L}{A}$$

where ρ and A are the resistivity of the material and the cross-sectional area above, and L is the length. It is not necessary for the wire to be round or cylindrical, just so long as it has the same cross section over its whole length.

EXAMPLE

A light bulb uses 100 watts of power when attached to 120 volts. What is the resistance of the light bulb?

$$V = IR \text{ and } P = IV$$

$$R = \frac{V}{I} = \frac{V}{P/V} = \frac{V^2}{P}$$

$$R = \frac{(120 \text{ V})^2}{(100 \text{ W})} = 144\ \Omega$$

Student: If I know two of the four, I can solve for the other two. I know $P = 100$ W and $V = 120$ V. It seems strange to say that $V = 120V$.
Tutor: Be careful. Remember that variables are in italics and units in regular type. Your first statement said that the potential difference (or voltage) was equal to 120 volts, which is correct. Your second statement said that the potential difference was equal to 120 times the potential difference, which is not correct.
Student: Yes, I'm going to have to be careful. I do have two of them, so I should be able to find the resistance. I can use Ohm's law $V = iR$, but I don't know the current.
Tutor: True. You'll need to find the current in order to use Ohm's law.
Student: I could use $P = iV$, but I can't use it to solve for the resistance. I could use it to find the current, then put that current back into Ohm's law to find the resistance.
Tutor: Very good.

$$P = iV$$

$$100 \text{ W} = i(120 \text{ V})$$

$$i = \frac{100 \text{ W}}{120 \text{ V}} = 0.83 \text{ A}$$

$$V = iR$$

$$(120 \text{ V}) = (0.83 \text{ A})R$$

$$R = \frac{120 \text{ V}}{0.83 \text{ A}} = 144\ \Omega$$

Student: It seems like we could invent a new formula for this.
Tutor: We could.

$$P = iV = \left(\frac{V}{R}\right) V = \frac{V^2}{R}$$

Tutor: We could invent a formula for *every* situation, but after a while the number of formulas would become too large. Better to be able to use a few formulas many ways.

EXAMPLE

A power plant delivers 100 MW of electrical power to the big city at 50 kV. The power travels through wires with a total resistance of 2 Ω. How much power is lost in the wires?

Student: I have the power and the voltage and the resistance, so with three of the four this should be easy. In fact, I have the power and I want the power, so what's the problem?
Tutor: The problem is that there are two resistors. One is the resistance of the wires to the big city, and the second is the resistance of the big city.

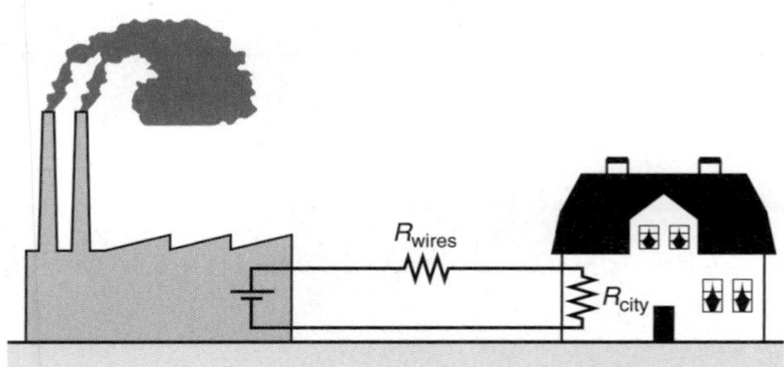

Tutor: What do you know about the resistance of the wires?
Student: The resistance is 2 Ω. I want the power. The current isn't mentioned. Is the voltage 50 kV?
Tutor: 50 kV is the voltage across the resistance of the big city.
Student: So I don't know the voltage across the wires. I know only one of the four, and I can't do the problem.
Tutor: It is true that you know only one of the four, but don't give up. What do you know about the resistance of the big city?
Student: The power is 100 MW and the voltage is 50 kV. I don't know the current or the resistance.
Tutor: But you could find them. You know two of the four for the big city, so you could find the other two if you wanted.

	Wire	City
V		50 kV
i		
R	2 Ω	
P	?	100 MW

Student: Yes, I suppose I could.
Tutor: What happens to the charges after they go through the wire?
Student: They go through the big city.
Tutor: Is the number of charges passing through the wire equal to the number of charges passing through the big city?
Student: You mean, are the two resistors in series?
Tutor: We'll handle series and parallel resistors in the next chapter, but yes, that's what I mean.
Student: They are in series. So the currents are the same.
Tutor: So if you found the current through the big city, then that would be the current through the wires. You would know two of the four and could solve for the power in the wires.
Student: I could do that.

$$P_{\text{city}} = i_{\text{city}} V_{\text{city}}$$

$$100 \text{ MW} = i_{\text{city}} \text{ (50 kV)}$$

$$i_{\text{city}} = \frac{100 \text{ MW}}{50 \text{ kV}} = \frac{100 \times 10^6 \text{ W}}{50 \times 10^3 \text{ V}} = 2000 \text{ A}$$

$$P_{\text{wire}} = i_{\text{wire}} V_{\text{wire}} = i_{\text{wire}} (i_{\text{wire}} R_{\text{wire}}) = i_{\text{wire}}^2 R_{\text{wire}} = (2000 \text{ A})^2 (2\ \Omega) = 8 \times 10^6 \text{ W} = 8 \text{ MW}$$

Tutor: About 8 % of the power is lost in the wires.
Student: So my electric bill is 8% higher because of energy lost in the wires?
Tutor: To avoid this, the electric company delivers electricity at much higher voltages, like 350 kV. At seven times the voltage, the current is one-seventh as much, and the power in the wires is 1/49 as much or 98% less — only 0.2% is lost.
Student: That's better.

EXAMPLE

If the length of a wire is increased by a factor of 7, by what factor must we change the diameter of the wire so that the resistance remains unchanged?

Student: No numbers. This is one of those "scaling" problems.
Tutor: It is. How do you deal with scaling problems?
Student: I work out an equation for each case, before and after, and divide or substitute equations.
Tutor: Very good. What is the appropriate equation?
Student: Isn't it just ...

$$R = \frac{\rho L}{A}$$

Tutor: Almost. You want to get the diameter involved.
Student: The area is

$$A = \pi r^2 = \pi \left(\frac{d}{2}\right)^2$$

$$R = \frac{4\rho L}{\pi d^2}$$

Student: So I'll use R_1 for the original resistance, and R_2 for the resistance afterward, and likewise everything else.

$$R_1 = \frac{4\rho_1 L_1}{\pi d_1^2}$$

$$R_2 = \frac{4\rho_2 L_2}{\pi d_2^2}$$

Tutor: Excellent. What do you know about the two resistances?
Student: They are the same. We change the diameter so that the resistances are the same.

$$\frac{4\rho_1 L_1}{\pi d_1^2} = R_1 = R_2 = \frac{4\rho_2 L_2}{\pi d_2^2}$$

Tutor: The 4 and the π cancel. What do you know about the two resistivities?
Student: If the wires are made of the same material, then the resistivities are the same.

$$\frac{\cancel{\rho_1} L_1}{d_1^2} = \frac{\cancel{\rho_2} L_2}{d_2^2}$$

Tutor: What is the length of the wire afterward (L_2)?
Student: The problem doesn't say.
Tutor: It is true that we don't have a value, but we do know something about it.
Student: We know that the length afterward is seven times the length beforehand.

$$L_2 = 7L_1$$

$$\frac{L_1}{d_1^2} = \frac{(7L_1)}{d_2^2}$$

Tutor: So the length cancels, with the factor of seven left behind.

$$\frac{1}{d_1^2} = \frac{(7)}{d_2^2}$$

Student: I still can't solve and get a value for d_2.
Tutor: Solve for d_2 anyway.

$$d_2 = \sqrt{7d_1^2} = \sqrt{7}\ d_1$$

Tutor: The question was: by what factor do we need to change the diameter.
Student: And we need to multiply it by $\sqrt{7}$.
Tutor: And that's true regardless of the specific values.

Chapter 27

Circuits

What happens if you have two (or more) resistors in the same circuit? We did this with capacitors, and now we'll do it with resistors.

Like before, resistors can be in series, or in parallel, or neither. When two devices are connected so that they must have the same current, we say they are in series. Two resistors in series act the same as a single resistor that has resistance

$$R_S = R_1 + R_2$$

Likewise, when two devices are connected so that they must have the same voltage, we say they are in parallel. Two resistors in parallel act the same as a single resistor that has resistance

$$\frac{1}{R_P} = \frac{1}{R_1} + \frac{1}{R_2}$$

Note that these rules are the reverse of the rules for capacitors.

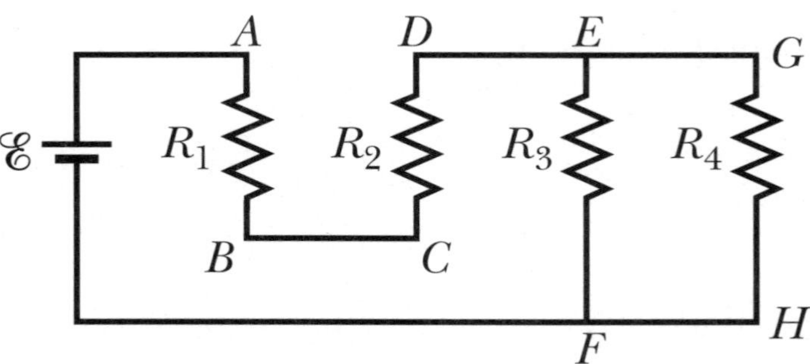

How do we determine whether two resistors are in series? Consider R_1 and R_2 above. Current passes from A through R_1 to B. From there the charges can only go to C and through R_2 to D. R_1 and R_2 are in series because the charges that pass through R_1 must then go through R_2, and the currents through the two must be equal. Charges that pass through R_2 to point D can go either through R_3 to F or through R_4 to H. The currents through R_2 and R_3 do not have to be equal, so R_2 and R_3 are not in series.

How do we determine whether two resistors (or capacitors) are in parallel? Consider R_3 and R_4 above. Place a finger at each end of R_3. The voltage difference between your fingers is equal to the voltage difference between the ends of R_3, or the voltage across R_3. Now try moving your fingers, but only along wires and not across any circuit element, such as a resistor, capacitor, or battery. If by doing this you can get your fingers to lie at the two ends of R_4, then R_3 and R_4 have the same voltage across them and they are in parallel (they are, and you should be able to do this). Are R_3 and R_2 in parallel? We can move a finger from E to D but we can't get from F to C. We could only do this by jumping from one wire to another nearby wire (not allowed), or by going through the battery and R_1 (also not allowed). Therefore, R_3 and R_2 are not in parallel. Even though both are drawn vertically, so that they are geometrically parallel, they are not electrically parallel.

We have found that R_2 and R_3 are neither in series nor in parallel. **Two resistors do not have to be either in series or in parallel.**

When doing more complicated circuits, we often do the series and parallel thing many times. Ohm's law ($V = IR$) can be applied to any resistor, but it can also be applied to a pair in series, or to any group of resistors. Likewise $Q = CV$ can be applied to any capacitor, but it can also be applied to a pair in series, or to any group of capacitors. It's important to keep track of what value applies to what circuit element or group of elements.

EXAMPLE

Find the voltage across each resistor.

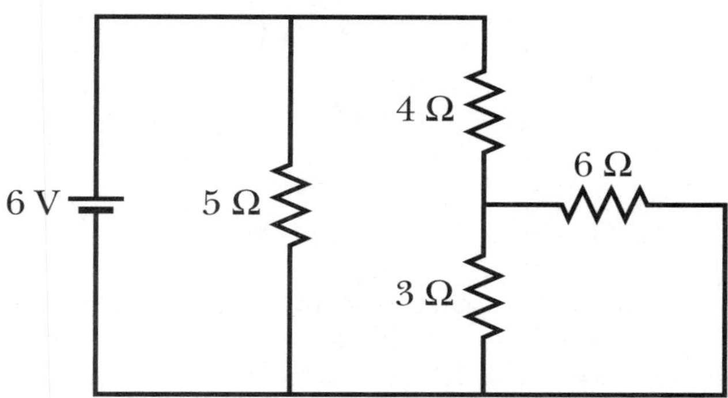

Student: This looks a lot like the capacitor circuit example from a couple of chapter ago.
Tutor: It is a lot like that example, and the techniques involved are very similar. Let's see how much you remember.
Student: The voltage *across* the 5 Ω resistor is 6 volts.
Tutor: How do you know?
Student: Because it's in parallel with the 6 volt battery. Two things in parallel have the same voltage across them.
Tutor: Yes. How do you check?
Student: I do the "finger test." I put one finger on each side of the 5 Ω resistor and try to move them to the ends of the battery. I slide the top one over, and I slide the bottom one over, and I'm there. The 5 Ω resistor is in parallel with the battery.
Tutor: How do you know that the voltage across the battery is 6 volts?
Student: Because a battery creates a potential difference, or voltage difference. A 6 volt battery creates a 6 volt potential difference.
Tutor: All good.
Student: The voltage across the 4 Ω resistor is 6 volts.
Tutor: How do you know?
Student: Because it's a 6 volt battery.
Tutor: That's not enough. Is the 4 Ω resistor in parallel with the battery?
Student: Oops. If the 4 Ω resistor is in parallel with the battery, then it has the same potential across it as the battery does. I put one finger on each side of the 4 Ω resistor and try to move them to the two sides of the battery. The top one slides over to the top of the battery. The bottom one has to jump over the 3 Ω

resistor or the 6 Ω resistor. It can't do that. So the 4 Ωresistor is not in parallel with the battery.
Tutor: Correct. So the voltage across the 4 Ω resistor isn't 6 volts.
Tutor: If I can find two resistors in series or parallel, then I can replace them with a single resistor.
Tutor: Can you find two resistors in either series or parallel? Check to see if the 4 Ω resistor and the 3 Ω resistor are in series.
Student: Does *all* of the current that comes out of the bottom of the 4 Ω resistor have to go into the 3 Ω resistor, or could some of it go somewhere else? Some of the current could go into the 6 Ω resistor.
Tutor: To be in series, *all* of the charge or current from one must go into the other.
Student: So the 4 Ω resistor and the 3 Ω resistor *aren't* in series.
Tutor: Correct.
Student: I believe that the 3 Ω resistor is in parallel with the 6 Ω resistor. I put one finger on each side of the 3 Ω resistor, and try to move them to the two sides of the 6 Ω resistor. I can move one from the top of the 3 to the top of the 6, and I can move the other from the bottom of the 3 to the bottom of the 6. They are in parallel electrically, even if they are not graphically parallel.
Tutor: Doing well. How do we add resistors in parallel?
Student: We add the reciprocals of the resistances.

$$\frac{1}{R_{\text{P}}} = \frac{1}{R_1} + \frac{1}{R_2}$$
$$\frac{1}{R_{3\&6}} = \frac{1}{R_3} + \frac{1}{R_6} = \frac{1}{3\ \Omega} + \frac{1}{6\ \Omega} = \frac{3}{6\ \Omega} = \frac{1}{2\ \Omega}$$

Student: The replacement resistor is not 1/2 Ω, but the reciprocal, 2 Ω.
Tutor: Yes. Sometimes we call it the equivalent resistance of the pair of resistors.

$$R_{3\&6} = 2\ \Omega$$

Student: The replacement resistor is in series with the 4 Ω resistor.
Tutor: Check to be sure.
Student: The current that comes out of the bottom of the 4 Ω resistor can go into the top of the 3 Ω resistor, or into the left side of the 6 Ω resistor, but nowhere else.
Tutor: So the 4 Ω resistor is in series with the replacement resistor, or with the 3-6 pair.
Student: Then I can replace them with a single resistor.

$$R_{\text{S}} = R_1 + R_2$$
$$R_{\text{trio}} = R_4 + R_{3\&6} = (4\ \Omega) + (2\ \Omega) = 6\ \Omega$$

Student: The trio of resistors has an equivalent resistance of 6 Ω. This trio of resistors is in parallel with the 5 Ω resistor.
Tutor: True, but not necessary. The trio is in parallel with the battery, so we can determine the voltage on the trio.
Student: It is. I can move one finger from the top of the 4 to the top of the battery, and the other from the bottom of the trio to the bottom of the battery. Isn't the 5 Ω resistor in the way?
Tutor: It may be between the battery and the trio, but it doesn't affect whether or not the trio is in parallel with the battery. What does it mean that the trio is in parallel with the battery?
Student: They have the same voltage, so the voltage across the trio is 6 volts. That doesn't mean that the voltage on each of the three resistors is 6 volts.
Tutor: Excellent. One volt is one joule per coulomb, so each coulomb of charge leaves the battery with 6 joules of energy. First the charges go through the 4 Ω resistor, then through the 3-6 pair (which is 2 Ω). It takes more joules for charges to get through the 4 Ω resistor, so the voltage (joules per coulomb) on the 4 Ω resistor is higher.
Student: So how do I find the voltage across each of the resistors in the trio?
Tutor: You need to find the current through the trio.
Student: Like finding the charge in the trio a couple chapters ago.

$$V_{\text{trio}} = i_{\text{trio}} R_{\text{trio}}$$

$$6 \text{ V} = i_{\text{trio}}(6\ \Omega)$$

$$i_{\text{trio}} = 1 \text{ A}$$

Student: One amp of current goes through the trio.
Tutor: Yes. This is true regardless of the 5 Ω resistor, because the voltage across the trio is not affected by the presence of the 5 Ω resistor.
Student: Is there a current through the 5 Ω resistor?
Tutor: Oh yes. There is a current of 1.2 A, so that the total current from the battery is 2.2 A.
Student: What do I do, now that I know the current through the trio?
Tutor: What is the current through the 4 Ω resistor?
Student: All of the current that goes through the trio goes through the 4 Ω resistor first, and then through the 3-6 pair. The current through the 4 Ω resistor is 1 A, and the current through the pair is 1 A.
Tutor: Now you can find the voltage across the 4 Ω resistor.
Student: Using Ohm's law.

$$V_4 = i_4 R_4$$

$$V_4 = (1 \text{ A})(4\ \Omega) = 4 \text{ V}$$

Tutor: What is the current through the 3-6 pair?
Student: All of the current that goes through the trio goes through the 4 Ω resistor first, and then through the 3-6 pair. The current through the pair is 1 A.
Tutor: Correct. Now you can use Ohm's law to find the voltage across the pair.

$$V_{\text{pair}} = i_{\text{pair}} R_{\text{pair}}$$

$$V_{\text{pair}} = (1 \text{ A})(2\ \Omega) = 2 \text{ V}$$

Student: The voltage across the pair is 2 volts.
Tutor: Good. What is the voltage across the 3 Ω resistor?
Student: Doesn't the 3 Ω resistor get all of the 2 volts?
Tutor: It does. The voltage difference across the pair is the same as the voltage across the 3 Ω resistor and the same as the voltage across the 6 Ω resistor.
Student: So $V_3 = V_6 = V_{\text{pair}} = 2$ V. And it's not a problem that $4 + 2 + 2 \neq 6$.
Tutor: No, because none of the charges go through all three resistors of the trio. They go through the 4 and the 3, or through the 4 and the 6, and either way the sum of the voltages is 6 volts, or 6 joules for each coulomb of charge.
Student: Isn't there some shortcut?
Tutor: In general, no. If the numbers are nice, then sometimes. Look at the 4 Ω resistor and the 2 Ω pair. The are in series, so they have the same current. The 4 Ω resistor has twice the resistance as the 3-6 pair, so it must have twice the voltage. Also, the voltage must add up to 6 volts. Can you think of two numbers, one twice as big as the other, so that the two numbers add up to 6?
Student: Four and two.
Tutor: Yes. When the numbers are nice, sometimes you can do that. Usually the numbers aren't nice, and often they aren't even integers.

In more complicated circuits, equivalent resistance is not enough. We need more powerful tools.

Kirchhoff has two rules for dealing with circuits:

- The node rule: The currents into any point or element in the circuit is the same as the current out of that point or element.
- The loop rule: The sum of the voltage changes along any closed path in the circuit is zero.

Kirchhoff's node rule says that charge cannot collect anywhere. This is the same idea we have used to say that two devices in series had to have the same current.

Kirchhoff's loop rule says that the energy that each charge has (12 joules for each coulomb if it comes from a 12 volt battery) must be exactly used up by the charge as it goes from the positive to the negative terminal of the battery. This is why in a complex circuit the sum of the resistor voltages is not equal to the battery voltage. Instead the sum of the voltages along the path that any charge takes adds up to be equal to the battery voltage.

In practice, we follow complete loops adding up the voltage changes. A complete loop is one where we start and stop at the same place. When we get back to the starting place, the sum of the voltage changes must be zero. (This is essentially conservation of energy — if you took a mountain hike, adding all of the changes in elevation, when you returned to your starting spot the sum of the changes would be zero.)

In the process, we usually come across resistors. The voltage across the resistors depends on the current, and we don't know the current. We do what we always do in physics when we want to put a value into an equation and we don't have a value — invent a variable. This variable or symbol represents the current through the resistor, and the voltage across the resistor is IR. Because the current could go in either direction, we choose a direction to be positive current through the resistor. It is not important that our choice be correct — a negative value indicates that the current is opposite to our choice. It is important that we use our choice consistently.

EXAMPLE

What is the current through the 5 Ω resistor?

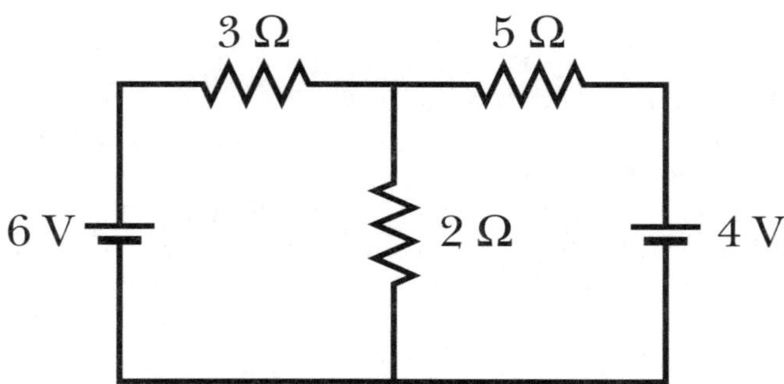

Tutor: Are any of the resistors in series or parallel?
Student: Is the 3 Ω resistor in series with the 2 Ω resistor? I'll check. Current that goes through the 3 Ω resistor left to right could go down through the 2 Ω resistor, but it could also go right through the 5 Ω resistor. The 3 Ω resistor is not in series with either the 2 Ω resistor or the 5 Ω resistor. I don't see anything in series or parallel.
Tutor: The 3 Ω resistor is in series with the 6 V battery, but we don't have a rule for replacing a battery and a resistor in series.
Student: So we need to use the advanced tools.
Tutor: Pick a point, and go around a loop adding the voltage changes.
Student: Okay, I'll start at the bottom left corner. First I go through the 6 volt battery, and the change is 6 volts.
Tutor: Is the change positive or negative?
Student: How do I tell?
Tutor: Which end of the battery is at a higher potential?
Student: The end with the longer line on the symbol, which is the top end in this circuit.
Tutor: So as you move up over the battery, do you go to a higher or a lower potential?

Student: I go to a higher potential.
Tutor: Yes, you go from a lower to a higher potential. Is the change in potential positive or negative?
Student: The change is to increase the potential, so it is positive.
Tutor: Good.
Student: Now I go through the 3 Ω resistor. The voltage across the 3 Ω resistor is equal to the resistance times the current. I don't know the current through the resistor, so I need to invent a variable. I choose i to be the current.
Tutor: Careful. We're going to have more than one current that we don't know, so we'll need more than one current variable. How about using i_3?
Student: The three must be for the 3 Ω resistor. So the voltage across the 3 Ω resistor is $(3\ \Omega)(i_3)$.
Tutor: Yes. Is the change positive or negative?
Student: I'm not sure. How do I tell?
Tutor: Current goes from higher potential to lower potential. If you go with the current, you're going to lower potential, and if you go against the current, you're going to higher potential.
Student: Which way is the current going?
Tutor: You get to decide.
Student: Wait a minute. The current is certainly going one way or the other. How can I be the one to decide?
Tutor: True, the current is going one way or the other, determined by the circuit configuration. You get to choose which way to call positive. When a car was moving on the highway, you got to decide whether the car was moving in the positive or negative direction. Your choice of axis did not affect the motion of the car, but only the sign you used to represent the motion of the car. Here you are choosing an axis, and your choice does not change the actual current but rather the sign you use to represent the current. Pick either direction, it doesn't matter as long as you use it consistently.
Student: Okay. The current is going left to right. What if I'm wrong?
Tutor: Then the value of i_3 will be negative. Is the voltage change positive or negative?
Student: The current is going left to right, so the higher potential is at the left and the lower potential is at the right. But if I choose the current direction wrong then those will be reversed.
Tutor: And then the negative value of i_3 will make everything work out right. Draw an arrow and the symbol on the circuit so that you'll remember what you chose.
Student: Okay, I draw on the circuit, and all I have to do is use the direction I chose consistently. Okay. I'm moving across the resistor from left to right, from higher potential to lower potential, so the change in potential is negative.
Tutor: Good. Now do the next circuit element.
Student: I go down across the 2 Ω resistor. The voltage across the 2 Ω resistor is equal to the resistance times the current.
Tutor: Is the current through the 2 Ω resistor equal to the current through the 3 Ω resistor?
Student: If it is, then they are in series. We already determined that they aren't in series, so no, they are not equal.
Tutor: So you need a different variable name for the current through the 2 Ω resistor.
Student: I'll use i_2. I need to pick a direction, so I'll choose downward as positive i_2. The higher potential is at the top and the lower potential is at the bottom. I'm going downward, toward lower potential, with the current, so the change in voltage is negative.
Tutor: Good. Draw an arrow and the symbol on the circuit. Now you're back where you started.
Student: So I add all of the voltage changes and they have to add up to zero.

$$+6\ \text{V} - (3\ \Omega)i_3 - (2\ \Omega)i_2 = 0$$

Tutor: Good. Can you solve this equation to find the current through the 5 Ω resistor?
Student: That current isn't even in this equation. We have two variables and one equation, so we can't solve it for anything.
Tutor: So you need to do another loop.
Student: I'll do the right-hand loop, starting at the bottom middle. First I go through the 2 Ω resistor. I need to choose a variable and direction for the current through the 2 Ω resistor.
Tutor: You already chose a variable and direction.

Student: I don't get to choose the direction again?
Tutor: No, you have to use the direction you chose earlier.
Student: So that's what you mean about using the direction consistently. I'm going upward, against the current and toward higher potential, so the change is positive $(2\ \Omega)(i_2)$.
Tutor: Good.
Student: Now I go across the 5 Ω resistor. I don't have a variable for that yet, so I choose i_5 and left to right is the positive direction. I draw an arrow and the symbol on the circuit. I'm going with the current, toward lower potential, so the change in voltage is negative.

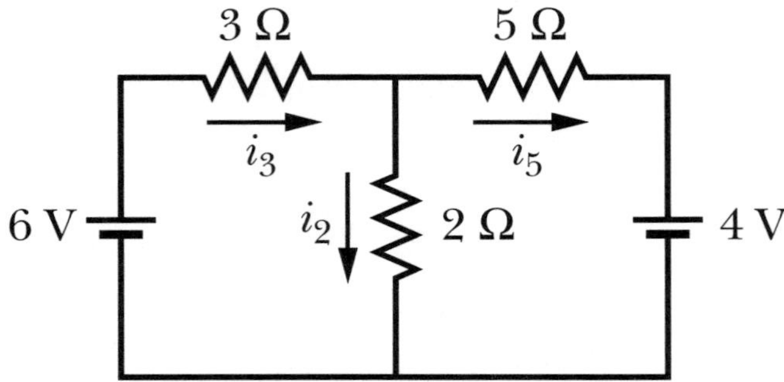

Tutor: You may be getting the hang of this.
Student: Now I go across the battery. Do I need a variable for the current through the battery?
Tutor: No. The voltage difference across the battery is the same, regardless of the current through the battery.
Student: Okay. The voltage difference on the battery is 4 volts. I'm going toward lower potential, so the change is negative. Now I'm back where I started, so the sum of the voltage changes is zero.

$$+(2\ \Omega)i_2 - (5\ \Omega)i_5 - 4\ \text{V} = 0$$

Tutor: Good. Can you solve this equation to find the voltage across the 5 Ω resistor?
Student: No, because there are two unknowns in the equation.
Tutor: Can you solve the two equations together?
Student: Then I have two equations but three different unknowns, so again I can't solve yet. I need to do another loop. Maybe around the outside.
Tutor: Tempting though that might be, it won't help. Going around the outside is the same as doing both of the loops you've already done — 6, 3, 2, back through the 2, 5, 4. The resulting equation would be the same as if we added the two equations you already have.
Student: But I'd have three equations and three unknowns. I could solve them.
Tutor: You wouldn't have three *independent* equations. When you solved them, you would cancel stuff until you got to something like 6 = 6. That wouldn't help.
Student: So I need another equation, but I can't do another loop.
Tutor: Correct. You need to do a "node rule" equation.
Student: How does that work?
Tutor: Look at the spot midway between the 3 Ω resistor and the 5 Ω resistor, where the connection to the 2 Ω resistor comes in. Charge can't build up there, so the current going into that spot has to equal the current going out of that spot.
Student: So i_3 in has to equal $i_2 + i_5$ out?
Tutor: Correct. That is your third equation.

$$i_3 = i_2 + i_5$$

Student: Now I have three equations and three unknowns. I can solve for i_5.
Tutor: Is i_5 what you're trying to find?

Student: No, I need the voltage across the 5 Ω resistor. But if I know i_5, then I can find the voltage.

$$\begin{cases} +6\text{ V} - (3\ \Omega)i_3 - (2\ \Omega)i_2 = 0 \\ +(2\ \Omega)i_2 - (5\ \Omega)i_5 - 4\text{ V} = 0 \\ i_3 = i_2 + i_5 \end{cases}$$

$$+6\text{ V} - (3\ \Omega)(i_2 + i_5) - (2\ \Omega)i_2 = 0$$

$$+6\text{ V} - (3\ \Omega)i_5 - (5\ \Omega)i_2 = 0$$

$$+(5\ \Omega)i_2 = +6\text{ V} - (3\ \Omega)i_5$$

$$+(2\ \Omega)\left(\frac{+6\text{ V} - (3\ \Omega)i_5}{+(5\ \Omega)}\right) - (5\ \Omega)i_5 - 4\text{ V} = 0$$

$$+2.4\text{ V} - (1.2\ \Omega)i_5 - (5\ \Omega)i_5 - 4\text{ V} = 0$$

$$-(6.2\ \Omega)i_5 = 1.6\text{ V}$$

$$i_5 = -0.26\text{ A}$$

Student: It came out negative. That means that the current through the 5 Ω resistor is really right to left.
Tutor: Very good. What is the voltage across the 5 Ω resistor?
Student: I use Ohm's law and the current to find the voltage.

$$V_5 = i_5R_5 = (-0.26\text{ A})(5\ \Omega) = -1.3\text{ V}$$

Tutor: Remember that the negative indicates direction.
Student: So I'd say that the voltage is really 1.3 volts, right?
Tutor: Right. Which side of the 5 Ω resistor is at the higher voltage?
Student: The current really goes from right to left, so the right side is at the higher potential.
Tutor: Correct.
Student: Since the current ended up being negative, shouldn't I reverse the direction on my drawing?
Tutor: No. Then you'd get confused over the direction of the current. It's perfectly okay now, with positive i_5 being left to right and i_5 having a negative value.

EXAMPLE

Find the current through each resistor.

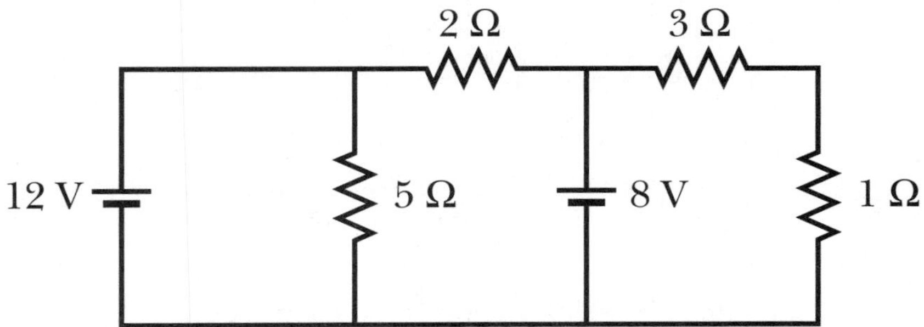

Student: I need "loop rule" equations around each of the loops, and then "node rule" equations for each of the nodes.
Tutor: Slow down. There is a tendency to use the most recently learned technique for everything. It might

not be necessary.
Student: But I can't combine all of the resistors using series and parallel. The 5 Ω resistor isn't in series or parallel with any of the other resistors.
Tutor: But it is in parallel with the 12 volt battery.
Student: It is. I can put one finger on each end of the 5 Ω resistor and move one from the top of the resistor to the top of the battery, and the other from the bottom of the resistor to the bottom of the battery.
Tutor: What does that mean?
Student: Because they are in parallel, they have to have the same voltage. The voltage across the 5 Ω resistor is 12 volts. The current through the 5 Ω resistor is

$$i_5 = \frac{V_5}{R_5} = \frac{12\ \text{V}}{5\ \Omega} = 2.4\ \text{A}$$

Tutor: One down. What about the 2 Ω resistor?
Student: It's not in parallel with either the 12 volt battery or the 8 volt battery.
Tutor: Look at the two batteries. The batteries and the 2 Ω resistor form a loop.
Student: So does the 2 Ω resistor, the 5 Ω resistor, and the 8 volt battery. Why are you so excited about your loop rather than mine?
Tutor: How many variables are in your loop?
Student: One for each resistor, so two.
Tutor: How many variables are in my loop?
Student: One for the 2 Ω resistor. I see. We can solve the equation from your loop but not mine.

$$\begin{aligned} +12\ \text{V} - (2\ \Omega)i_2 - 8\ \text{V} &= 0 \\ 4\ \text{V} &= (2\ \Omega)i_2 \\ i_2 &= 2\ \text{A} \end{aligned}$$

Tutor: Two down, two to go. What can you tell me about the 3 Ω resistor and the 1 Ω resistor?
Student: They are in series. All of the current that goes through the 3 Ω resistor has to go through the 1 Ω resistor — it has nowhere else to go. I can replace them with a single resistor of 4 Ω. That equivalent resistor is in parallel with the 8 volt battery.

$$i_3 = i_1 = i_{\text{pair}} = \frac{V_{\text{pair}}}{R_{\text{pair}}} = \frac{8\ \text{V}}{4\ \Omega} = 2\ \text{A}$$

Student: So if we haven't learned anything new, why did we do this example?
Tutor: To make the point that you don't have to use your "best" tool for every job.
Student: So look for series and parallel pieces and simple loops before resorting to lots of loop equations.
Tutor: Exactly.

Chapter 28

Magnetic Fields

Charges create electric field, and electric field creates forces on charges. Moving charges create magnetic field, and magnetic field creates forces on moving charges.

When we did electric field, for each step we needed a way to find the direction and a formula to find the magnitude. For magnetic field, for each step we need a way to find the direction and a formula to find the magnitude. This chapter is about finding the magnetic force created by the magnetic field. The next chapter is about finding the magnetic field created by the moving charges.

The equation for magnetic force is

$$\vec{F_B} = q\vec{v} \times \vec{b}$$

Because it is a vector equation, this equation contains both the magnitude and direction information. We don't typically use the vector equation. Instead, the magnitude of the magnetic force is

$$F_B = |q|vB\sin\phi$$

where ϕ is the angle between the velocity of the moving charge and the magnetic field. When lots of charges are moving in a current, the magnetic force a the wire is

$$F_B = iLB\sin\phi$$

where L is the length of the wire.

Then we need to find the direction of the magnetic force. **The magnetic force on a moving charge is always perpendicular to the magnetic field and is always perpendicular to the velocity of the charged particle**. Imagine a proton moving out of the page in a magnetic field that points left to right. Is there any direction that is perpendicular to both out of the page and right at the same time? Only up the page and down the page. Note that any magnetic force problem must be three-dimensional.

To determine whether the magnetic force is into the page or out of the page we use the right-hand rule. Hold your hand so that the fingers point out of the page (as in the figure) — keep your fingers pointing out of the page and rotate your hand so that the palm points right — curl your fingers from out to right — where does your thumb point? Up the page.

If q is negative, then that reverses the direction of the magnetic force. Also, if v and B are parallel, then the angle between them is zero and there is no magnetic force.

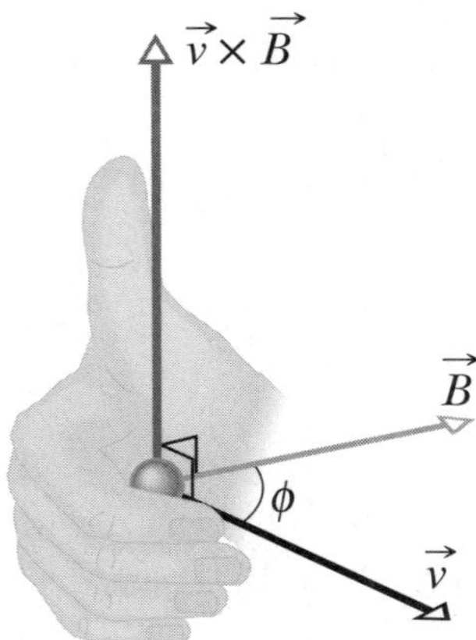

Drawing a vector up the page or right is easy, but drawing one into or out of the page is less so. To describe a vector out of the page we use the dot of the arrow coming at us ($\odot$ symbol), like in a 3-D movie. To describe a vector into the page we use the tailfeathers of the arrow going away from us ($\otimes$ or $\times$ symbol).

Magnetic field is measured in tesla:

$$1 \text{ T} = 1 \text{ N / A m} = 1 \text{ N s/ C m}$$

A tesla is a large magnetic field — a strong handheld magnet typically produces a magnetic field strength of 0.01 T to 0.1 T.

EXAMPLE

A +10 μC charge moves from left to right at a speed of 5×10^6 m/s. There is a 0.04 T magnetic field toward one o'clock, if the page were a clock. Find the magnetic force on the charge.

Tutor: How are you going to do this?
Student: We have a formula to tell us the magnetic force.

$$F_B = |q|vB \sin\phi$$

Student: I know the charge, the speed, and the magnetic field.

$$F_B = (10\ \mu\text{C})(5 \times 10^6 \text{ m/s})(0.04 \text{ T}) \sin\phi$$

Tutor: What is the angle ϕ?
Student: It's the angle between the velocity and the magnetic field. Here it's less than 90°.
Tutor: If the magnetic field pointed to three o'clock, it would be parallel to the velocity. What would the angle be if the magnetic field pointed toward twelve o'clock?
Student: Then it would be 90°. Oh, so the angle is two-thirds of the way from 0° to 90°, or 60°.

$$F_B = (10\ \mu\text{C})(5 \times 10^6 \text{ m/s})(0.04 \text{ T}) \sin 60° = 17.3 \text{ N}$$

Tutor: What is the direction of the magnetic force?
Student: I have to find the direction too?
Tutor: Force is a vector and has direction.
Student: I need to use the right-hand rule. How do I do that?
Tutor: Begin by pointing your fingers in the direction of the velocity.
Student: So I point my fingers to the right.
Tutor: Now, while keeping your fingers pointed in that direction, rotate your hand so that the palm points in the direction of the magnetic field.

Student: If I rotate my hand, the fingers move. They no longer point to the right.
Tutor: Draw an imaginary line to the right. This is the axis that you rotate your hand around, until the palm points up.
Student: But the magnetic field isn't up, it's up and to the right.
Tutor: The rightward component of the magnetic field doesn't create any force, so we care only about the upward component.
Student: Is that what the $\sin\phi$ does?
Tutor: Yes, $B\sin\phi$ is the component of the magnetic field that is perpendicular to the velocity of the charge.
Student: Now my palm is up and my fingers point to the right. What now?
Tutor: Curl your fingers toward the magnetic field.
Student: Just how far should I curl my fingers?
Tutor: Basically 90°. What direction does your thumb point?
Student: If points toward me.
Tutor: What direction does your thumb point, relative to the page?
Student: It points out of the page.
Tutor: That's the direction of the magnetic force.
Student: And if the charge was a negative charge, then the force would be the other way, into the page?
Tutor: Correct. Now let's check. The direction of the magnetic force has to be perpendicular to both the velocity of the charge and the magnetic field. Is it?
Student: Yes, but the magnetic field and the velocity are not perpendicular.

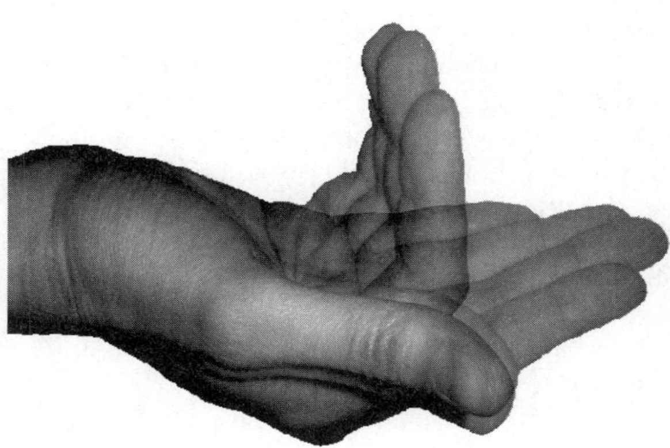

Tutor: Those don't have to be. We can create both of those, so they can be in any direction. But nature decides what direction the magnetic force is, and nature makes the force perpendicular to the velocity and the field.
Student: What if I use my left hand?
Tutor: Then you'll get the opposite answer. Also, you have to do the velocity first and then the field, or you'll get the opposite answer.
Student: But if I do both ...
Tutor: There are many ways to determine the correct direction of the magnetic force. It is important that you have a systematic approach, so that you can always get the right answer. This is the way we teach you to do it.

EXAMPLE

The Earth's magnetic field is about 50 μT northward. Find the velocity that an electron would have to have so that the magnetic force balances gravity.

Student: So this time I already know the direction of the magnetic force. The force is upward.
Tutor: Yes, but what is the direction of the velocity?
Student: Oh. So I have to decide whether it's into the page or whatever?
Tutor: There is no "page" here. When we do 3-D problems with the page as a reference, our directions are up, down, left, right, into, and out of. When we use the compass as our reference, our directions are north, south east, west, up, and down. Up compared to north is not the same as up the page.
Student: So the velocity is either north, south, east, west, up, or down.
Tutor: It could be a combination, say southeast. Could the velocity be up?
Student: How can I decide?
Tutor: If the velocity is up, could the magnetic force be up?

Student: No, the force is perpendicular to the velocity. The force is up, so the velocity can't be up. Or down.
Tutor: Could the velocity be southward?
Student: Then the field and the velocity would be parallel. The magnetic force would be zero.
Tutor: When two vectors point in opposite directions, we sometimes call them antiparallel.
Student: Okay, but the velocity has to be east or west.
Tutor: It could be southwest. The west component would cause a magnetic force that balances the weight, and the south component wouldn't create any force, so the weight would still be balanced.
Student: So there are two answers?
Tutor: There are many answers, but the east-west component is the same for all of them. Let's keep it simple and assume that the velocity is either east or west.
Student: Which one is it?
Tutor: Guess, then see if you're right.
Student: Okay. I'll guess eastward. I point my fingers to the east, with my palm to the north. Then I curl my fingers toward the magnetic field, or northward. My thumb points up, so the force is up.
Tutor: The force on a positive charge would be up. What is the charge on an electron?
Student: Oh, it's negative, so the force on an electron would be down. I need an upward magnetic force, so the velocity is the other way, or westward.
Tutor: Now you can find the speed.
Student: Yep. I look up the mass of the electron in the inside back cover of the book.

$$F_B = |q|vB\sin\phi$$

$$mg = |-e|vB\sin 90^\circ$$

Student: I could use some other angle, though. Then I'd get a different velocity.
Tutor: True, but the westward component of the velocity would be the same.
Student: I'll stick with directly westward.

$$v = \frac{mg}{eB} = \frac{(9.1\times10^{-31}\text{ kg})(9.8\text{ m/s}^2)}{(1.6\times10^{-19}\text{ C})(50\times10^{-6}\text{ T})} = 1.1\times10^{-6}\text{ m/s}$$

Student: That's a really small speed.
Tutor: Yes. Magnetic forces are much stronger than gravitational forces. We often ignore gravity if there's a magnetic field around.
Student: But we couldn't ignore it here.
Tutor: No, because we were comparing to gravity.
Student: So if moving charges create magnetic fields that create forces on moving charges, how do permanent magnets pick up steel?
Tutor: The magnet is made of atoms, and every atom has little electrons moving inside of it. Steel is also made of atoms with moving electrons. That comes up in a later chapter.

EXAMPLE

What magnetic field strength is necessary so that an electron moving with a speed of 5×10^5 m/s travels in a circle of radius 1 m?

Student: Why would the electron go in a circle?
Tutor: The magnetic force is perpendicular to the velocity, so it causes the electron to change direction. Then the force is perpendicular to the new velocity, and so on. When something was going in a circle, what was the direction of the acceleration?
Student: In toward the middle of the circle.
Tutor: And therefore perpendicular to the velocity. A charged particle in a magnetic field goes in a circle.

Student: So now we need a new formula.
Tutor: Not necessarily. We can use two that we already have. What is the magnitude of the magnetic force?
Student: The magnetic force is $F_B = qvB\sin\phi$.
Tutor: Let's take ϕ to be 90°, with the magnetic field perpendicular to the velocity. What is the force on something going in a circle?
Student: Looking way back, it's $F = mv^2/r$. Are these equal?
Tutor: The force needed to get the particle to go in a circle is provided by the magnetic force, so yes.

$$qvB = \frac{mv^2}{r}$$

$$qB = \frac{mv}{r}$$

Student: I need to find the magnetic field strength B.

$$B = \frac{mv}{rq} = \frac{(9.1\times10^{-31}\text{ kg})(5\times10^{5}\text{ m/s})}{(1\text{ m})(1.6\times10^{-19}\text{ C})} = 2.8\times10^{-6}\text{ T} = 2.8\ \mu\text{T}$$

Tutor: Does it matter what direction the magnetic field is?
Student: The direction of the field will determine which way the charge goes around the circle, whether it's clockwise or counterclockwise, but it won't affect the size of the circle.

Consider the loop of current shown, in a magnetic field that points to the right. There is no force on the top or bottom, because these currents are parallel to the magnetic field. The force on the left side is into the page, and the force on the right side is out of the page. With the same length and current, and in the same magnetic field, these add to a net force of zero. But they do create a torque on the loop, rotating it clockwise as seen from the top of the page.

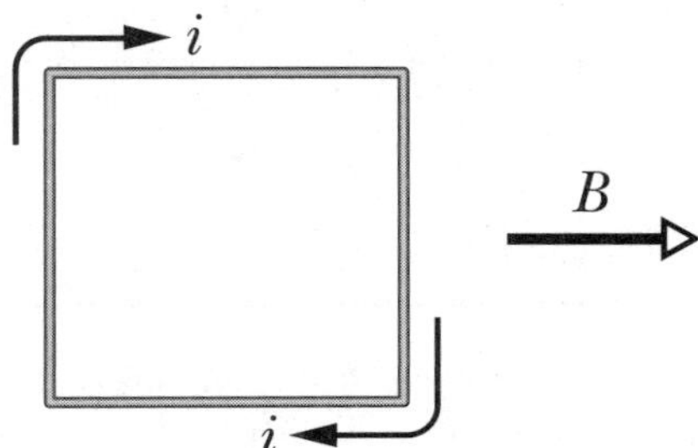

Magnetic fields do not always create torque on current loops. What if the magnetic field had been out of the page? The magnetic force on each side would be inward, toward the center, and there would be no torque.

To find the torque, start by finding the magnetic moment. Anything that creates a magnetic field has a magnetic moment. Magnetic moments are vectors. The magnitude of the magnetic moment of a loop of current is

$$|\vec{\mu}| = NiA$$

where a loop of N turns of current i has an area A. To find the direction of the magnetic moment vector, curl the fingers of your right hand around the loop in the direction of the current. Then your right thumb points in the direction of the magnetic moment.

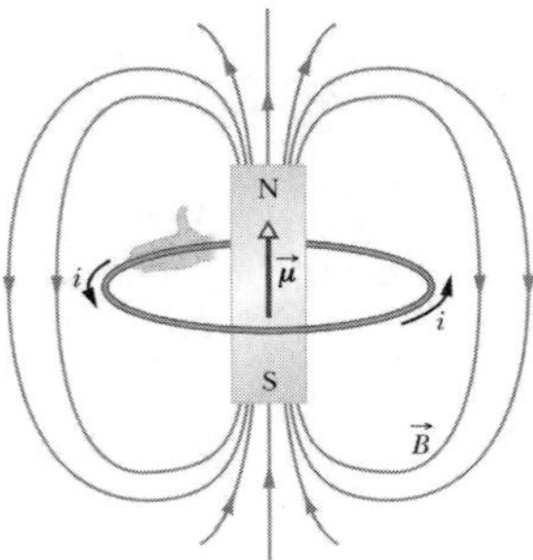

The torque on a magnetic moment in a magnetic field is

$$\vec{\tau} = \vec{\mu} \times \vec{B}$$

The magnitude of a "vector cross product" is equal to the magnitude of one vector, times the component of the second that is perpendicular to the first, found with the sine of the angle between them.

$$\tau = \mu B \sin\theta$$

The direction of the torque is such that the magnetic moment rotates to align parallel to the external field.

This is how a compass works. The needle is a small magnet, free to rotate. The torque from the Earth's magnetic field causes the compass needle to rotate until it lines up with the Earth's field. The magnetic field from the compass comes out of the north pole, and aligns with the magnetic field from the Earth, which comes out of the Antarctic and toward the Arctic. Thus the geographic north pole of the Earth (where Santa's workshop is) is the magnetic south pole of the Earth.

Magnetic moments also create magnetic fields. Along the axis of the current loop, or magnetic moment, the field is

$$\vec{B} = \frac{\mu_0}{2\pi} \frac{\vec{\mu}}{z^3}$$

where z is the distance from the magnetic moment. This equation does not apply close to the current loop, but only at large distances. As always, large means "compared to other distances in the problem," or the radius of the current loop.

EXAMPLE

A single rectangular loop of wire 14 cm × 20 cm carries a current of 18 A in the direction shown. The loop has a mass of 47 grams and is hinged at the upper-left (darker) longer side so that it can rotate up and down. A magnetic field of 0.24 T points vertically up. At what angle ϕ will the loop reach equilibrium? That is, at what angle ϕ will the magnetic and gravitational torques cancel?

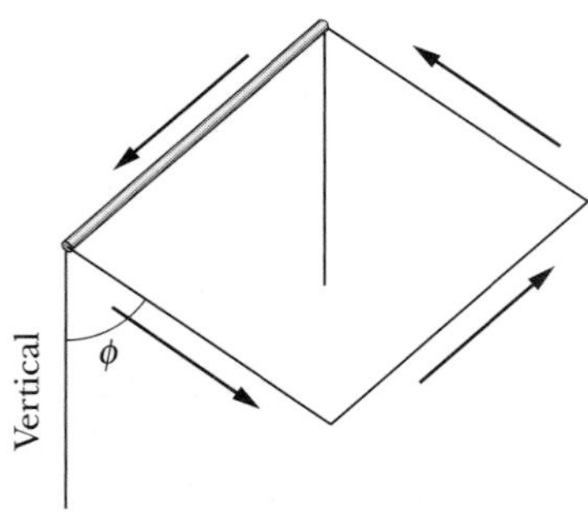

Student: It doesn't seem like torque is the issue here. Should we be more interested in force?
Tutor: The end where the loop is attached can provide all the vertical force we need, but that upward force is not through the center of mass of the loop, so it causes the loop to turn.
Student: If there's no current then the loop falls, doesn't it?
Tutor: That's one way to look at it, but if it's stationary then we can look at it from a torque standpoint. What is the direction of the magnetic moment of the loop?
Student: I point my right-hand fingers around the loop in the direction of the current, and my thumb points up and to the right. 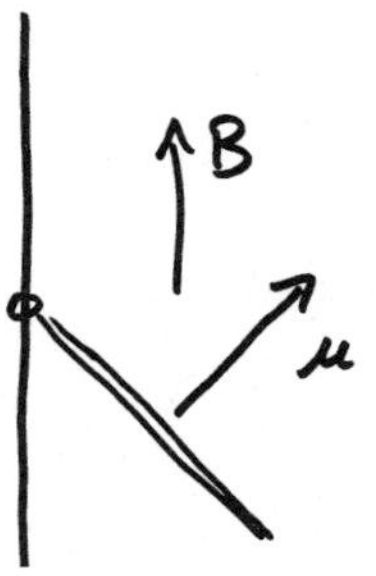
Tutor: Correct. The torque from the magnetic field causes the loop to try to align with the magnetic field.
Student: And that turns it counterclockwise, in the opposite direction as the torque from gravity.
Tutor: Good.
Student: So we just find the two torques and set them equal?
Tutor: Yep.
Student: The magnetic moment is

$$\mu = NiA = (1)(18\text{ A})\,[(0.14\text{ m})(0.20\text{ m})] = 0.50\text{ Am}^2$$

Student: The torque is

$$\tau = \mu B \sin\theta = (0.50\text{ Am}^2)(0.24\text{ T})\sin\phi \quad ?$$

Tutor: In the equation for torque, θ is the angle between the magnetic moment and the magnetic field. When they are aligned, the angle is zero and, because $\sin 0° = 0$, there is no torque. When the magnetic moment is vertically up and aligned with the field, the angle ϕ in the diagram is 90°, so the equation would give maximum torque.
Student: The angle ϕ isn't the same angle as θ?
Tutor: No.
Student: That's nasty. The angle $\theta = 90° - \phi$.
Tutor: Yes, and $\sin(90° - \alpha) = \cos\alpha$.
Student: Is that always true?
Tutor: It's one of the trig identities. In a right triangle, α is one angle and $90° - \alpha$ is the other. The side that is adjacent to α is opposite to $90° - \alpha$.
Student: Okay, the torque from the magnetic field is $(0.121\text{ Am}^2\text{T})(\cos\phi)$.
Tutor: $F = iLB$, so a newton is an amp meter tesla.
Student: Then the torque is $(0.121\text{ N·m})(\cos\phi)$, and we have the units we expect. Now we find the torque from gravity. The moment arm is 14 cm $\times \sin\phi$, so

$$(0.121\text{ Nm})(\cos\phi) = (0.047\text{ kg})(9.8\text{ m/s}^2)(0.14\text{ m})(\times\sin\phi)$$

Tutor: Remember that sine over cosine is tangent.
Student: Yes, so

$$\tan\phi = \frac{\sin\phi}{\cos\phi} = \frac{(0.121\text{ Nm})}{(0.047\text{ kg})(9.8\text{ m/s}^2)(0.14\text{ m})} = 1.88$$

$$\phi = \arctan 1.88 = 62°$$

Chapter 29

Magnetic Fields Due to Currents

Moving charges, or currents, create magnetic fields. A single moving charge creates only a tiny magnetic field, so we'll only worry about the fields created by currents. We need a way to determine the direction and magnitude of the magnetic field created by a current. We do this with the Biot-Savart law (they were French, so the t's are silent).

$$d\vec{B} = \frac{\mu_0}{4\pi}\frac{i\,d\vec{s}\times\vec{r}}{r^3}$$

where $\mu_0 = 4\pi \times 10^{-7}$ T m/A.

Let's try to explain this. Divide the current up into tiny pieces of length $d\vec{s}$. The vector r is from the tiny piece of current to the spot where we're finding the magnetic field. The cross product $d\vec{s}\times\vec{r}$ is found just like we did in the last section, using the right-hand rule. Multiply by the other stuff and you get the tiny amount of magnetic field created by that tiny piece of current. Add up all the tiny magnetic fields (and there's a lot of such tiny pieces) and you get the magnetic field.

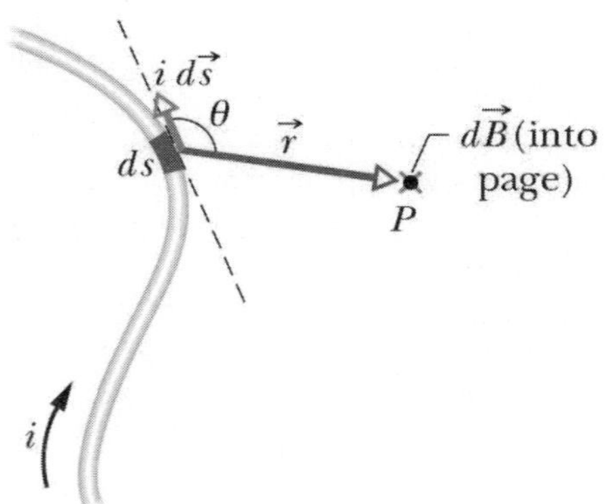

Finding electric fields wasn't this complicated, at least at the beginning. The difference is that each piece of the current is usually a different distance from the spot of interest, and so we have to use calculus to do this.

Doing this calculation is a long (or painful, if you prefer) process, so we take a few common situations to work it out and remember them. The magnetic field created by a long straight wire is

$$B_{\text{wire}} = \frac{\mu_0 i}{2\pi R}$$

where R is the distance from the wire to the point. How long is long? The length of the wire must be much greater than the distance to the wire. If the distance to the wire is 10 cm, then the wire would need to be a meter or so long. The shorter the wire is, the greater the error in pretending that it is infinitely long.

How do we find the direction of the magnetic field? Magnetic field lines never end, but always circle back onto themselves. They circle around a long straight wire. If the wire is coming at you, out of the page, then the magnetic field lines go around the wire counterclockwise. How do we know whether they go clockwise or counterclockwise? We use another right-hand rule. Take your thumb, like you're trying to "hitch" a ride, and point the thumb in the direction of the current. The fingers go around the wire in the same direction as the magnetic field.

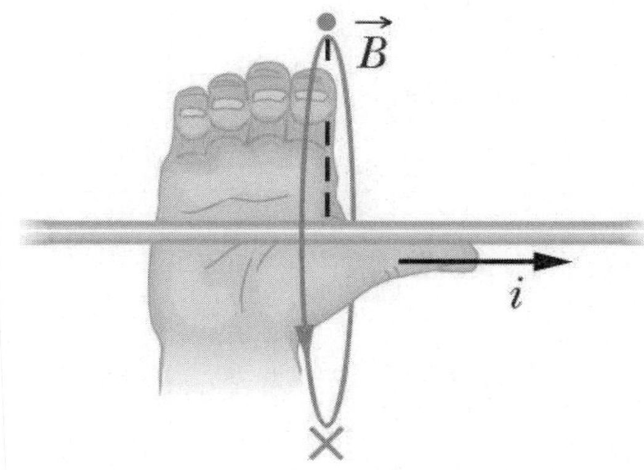

Another common situation is a circular loop of wire. The magnetic field at the center of the loop is

$$B_{\text{loop}} = \frac{\mu_0 i}{2R}$$

To get the direction of the magnetic field, pick a spot on the wire loop and use the right-hand rule: thumb in the direction of the current and curl your fingers. Alternatively, curl your fingers with the current (counterclockwise) and your thumb will point in the direction of the magnetic field in the center of the loop (out of the page). Either method works, as long as you don't use your left hand (easy to do if your pencil is in your right hand).

EXAMPLE

A rectangular loop of current 6 A is beneath, and in the same plane as, a long straight wire with a current of 8 A. Find the magnetic force that the long straight wire exerts on the rectangular loop of current.

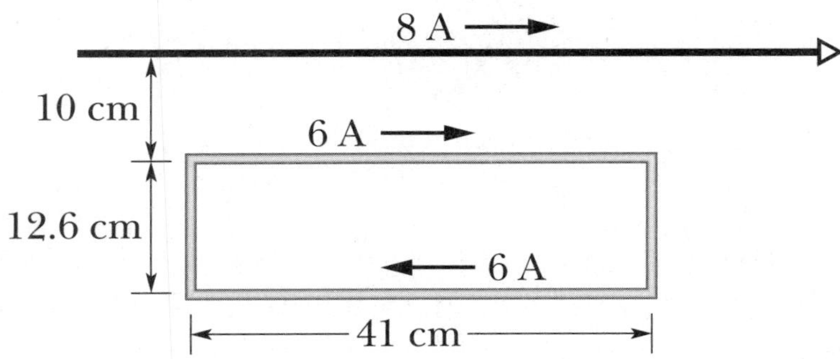

Student: Whoa! This looks complicated.
Tutor: It's not. What is the magnetic field that the long straight wire creates?
Student: The magnetic field from a wire is $B_{\text{wire}} = \mu_0 i/2\pi R$.
Tutor: Okay. What's i?

Student: That's the current in the long straight wire, 8 amperes.
Tutor: Good. What's R?
Student: That's a good question. Some of the rectangular loop is 10 cm away and some is 26 cm away. If I pick a different point on the long wire, then the distances are even greater. What do I use?
Tutor: We always use the closest point, or shortest distance to the long straight wire, so you do use 10 cm and 26 cm. You need to use both. To find the force on the top of the rectangle, you need to find the magnetic field at the top of the rectangle.
Student: So

$$B_{\text{top}} = \frac{\mu_0 i}{2\pi R} = \frac{\mu_0(8\text{ A})}{2\pi(0.10\text{ m})}$$

$$B_{\text{bottom}} = \frac{\mu_0 i}{2\pi R} = \frac{\mu_0(8\text{ A})}{2\pi(0.126\text{ m})}$$

Tutor: Yes. What is the force on the top side of the rectangle?
Student: The force is

$$F_{\text{top}} = (6\text{ A})(0.41\text{ m})B_{\text{top}}\sin\phi$$

Tutor: Good so far. What is ϕ?
Student: It's the angle between the current and the magnetic field.
Tutor: So we need to know the direction of the magnetic field.
Student: Right-hand rule: I point my thumb to the right, and my fingers go down. The magnetic field is down.
Tutor: I don't think you're doing it right. The magnetic field goes around the 8 A current, and down doesn't go around it. Take your hand the way you had it, and bend your fingers at the knuckles.
Student: My fingers go into the page.
Tutor: That's the direction of the magnetic field.
Student: So my thumb goes with the current, and my hand goes toward the point where I want the field, and then when I bend my fingers at the knuckles, the fingertips point in the direction of the field.

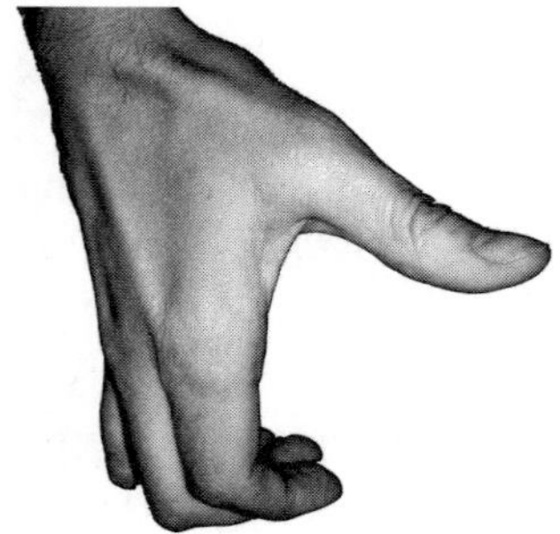

Tutor: Yes.
Student: Got it. The field is into the page, and the 6 A current is to the right, so the angle ϕ is 90°.
Tutor: When you go to find the direction of the force, remember to do the current first, then the field.
Student: Oh, the first right-hand rule. Fingers to the right with the current, bend them into the page, and the thumb points up. The force on the top is up.
Tutor: What is the force on the bottom?
Student: The magnetic field at the bottom is also into the page, but weaker. To find the magnetic force, I point my fingers to the left, bend them into the page, and the thumb points down. The forces are opposite, so I need to subtract.
Tutor: What about the sides?
Student: For the left side, the fingers point up the page, and bend into the page, and the thumb points

left. But the field changes along the length of the side. How do I find the magnitude of the force?
Tutor: We'll have to do an integral. What about the right side?
Student: The force on the right side is to the right.
Tutor: How do the magnitudes of the forces on the sides compare?
Student: Well, the fields are the same for the left and right, though they change along the length of the sides. The currents are the same and the lengths are the same, so the forces are the same. They cancel.
Tutor: To find those individual forces we would need to do an integral, but ...
Student: They'll add to zero, so we don't need to.
Tutor: Now you can finish adding the forces.
Student: Right. Taking upward as positive.

$$F = F_{\text{top}} - F_{\text{bottom}} = (6\text{ A})(0.41\text{ m})B_{\text{top}}\sin 90^\circ - (6\text{ A})(0.41\text{ m})B_{\text{bottom}}\sin 90^\circ$$

$$F = (6\text{ A})(0.41\text{ m})\left(B_{\text{top}} - B_{\text{bottom}}\right)$$

$$F = (6\text{ A})(0.41\text{ m})\left(\frac{\mu_0(8\text{ A})}{2\pi(0.10\text{ m})} - \frac{\mu_0(8\text{ A})}{2\pi(0.26\text{ m})}\right)$$

$$F = (4\pi\times10^{-7}\text{ T m/A})(6\text{ A})(8\text{ A})(0.41\text{ m})\frac{1}{2\pi}\left(\frac{1}{(0.10\text{ m})} - \frac{1}{(0.26\text{ m})}\right)$$

$$F = 2.4\times10^{-5}\text{ N}$$

Student: That seems like a small force.
Tutor: Magnetic forces are generally weaker than electric forces ...
Student: ... but stronger than gravity ...
Tutor: ... and we have two magnetic forces that partially cancel, so it's a small force.

EXAMPLE

The long wire was a single turn of radius 12 cm. What must the current be so that the magnetic field at the center of the loop is 100 μT into the page?

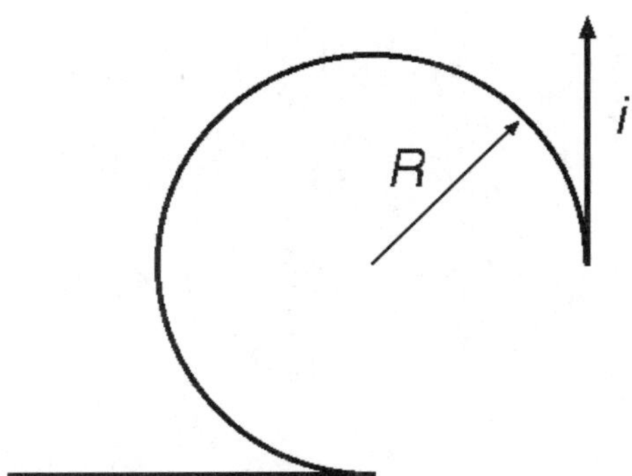

Student: I need to find the magnetic field from the long straight wire and from the loop and add them.
Tutor: Correct.
Student: But the wire is split in two.
Tutor: Is each piece exactly one-half of a long straight wire?
Student: What do you mean "exactly" half?
Tutor: If I take a long straight wire and cut it into two equal pieces, the dividing point is the closest point on the wire. If I split it somewhere else, the two pieces wouldn't be symmetric.

Student: I see, and symmetric is good?
Tutor: If the two pieces are symmetric, then they each create the same magnetic field — half of a long straight wire.
Student: Okay. Each of the two half-long wires ends at the closest point along the line, so they are each half of a long straight wire. I can use $\frac{1}{2}\frac{\mu_0 i}{2\pi R}$ to find the field from each of them.
Tutor: Now you have to deal with the loop.
Student: I know that the magnetic field from a full loop is $B_{\text{loop}} = \frac{\mu_0 i}{2R}$, but I don't have a full loop.
Tutor: Each piece of the loop contributes the same magnetic field. If you have half a loop, you have half the field.
Student: And I have three-quarters of a loop, so I have three-quarters of the field. Now I add the two fields.

$$B = \frac{1}{2}\frac{\mu_0 i}{2\pi R} + \frac{3}{4}\frac{\mu_0 i}{2R} + \frac{1}{2}\frac{\mu_0 i}{2\pi R} \quad ?$$

Tutor: You have the right idea, but you're skipping a step. Magnetic field is a vector ...
Student: So I need to find the direction of each piece and add components.
Tutor: Something like that. What is the direction of the magnetic field created by the long straight wire?
Student: What direction is the current going? The problem doesn't say.
Tutor: So pick a direction as positive, and if you end up with a negative current then the current is really the other way.
Student: Like doing circuits.
Tutor: Or anything in physics where you have a direction. You're simply choosing an axis for the current.
Student: I choose positive current to be in from the left, clockwise around the loop, and up out of the page. To find the field from the rightward current, I point my thumb to the right with my hand toward the top of the page. When I bend my fingers at the knuckles, they point out of the page, so the field is out of the page, and is negative.
Tutor: Good. You've picked the field you want to be positive field, and this is the other way.
Student: To do the upward-going current, I point my thumb up, hand to the left, and the fingers bend out of the page. How do I do the loop?
Tutor: Pick a spot on the loop, any spot, and find the direction from there. All pieces of the loop create field in the same direction.
Student: I pick the left edge of the loop. I point my thumb up the page and hold my hand to the right, because I want the field to the right of the piece of wire. My fingers bend into the page, so the field that the loop creates is positive.
Tutor: Yes, for the whole loop.

$$B = -\frac{1}{2}\frac{\mu_0 i}{2\pi R} + \frac{3}{4}\frac{\mu_0 i}{2R} - \frac{1}{2}\frac{\mu_0 i}{2\pi R}$$

$$B = \frac{3}{4}\frac{\mu_0 i}{2R} - \frac{\mu_0 i}{2\pi R} = \frac{\mu_0 i}{2R}\left(\frac{3}{4} - \frac{1}{\pi}\right)$$

$$100 \times 10^{-6}\ \text{T} = \frac{(4\pi \times 10^{-7}\ \text{T m/A})i}{2(0.12\ \text{m})}\left(\frac{3}{4} - \frac{1}{\pi}\right)$$

$$i = 44\ \text{A}$$

Student: That seems like a lot of current.
Tutor: It's more than you'd have in a household circuit — they typically top out at 20 amps.

When we did electric fields, we had Gauss' law. The equivalent with magnetic fields is Ampere's law.

Just as it is possible to use field lines to envision an electric field, it is possible to use field lines to envision a magnetic field. Magnetic field lines never end, but go in loops. The figure shows the magnetic field lines from a current that goes into the page. The field lines get further apart as we move away from the current because the field gets weaker.

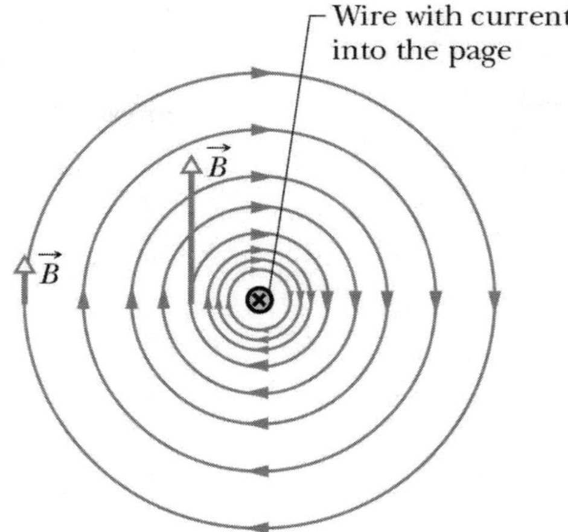

When we calculated the electric field flux, we got a total that was related to the charge inside the surface. This is because electric field lines start and end at electric charges. Magnetic field lines don't start and end, so the magnetic field flux through any closed surface is always zero. This is good to know, but it doesn't help us find the magnetic field strength.

Instead of using a closed surface, Ampere's law uses a closed loop.

$$\oint \vec{B} \cdot d\vec{s} = \mu_0 i_{\text{enc}}$$

Pick any closed loop — that is, start somewhere and go anywhere so long as you end where you started (using $\oint$ instead of $\int$ is a way to say use a closed path). As you travel along this route, at each point find the magnetic field. Take the component of the magnetic field $\vec{B}$ that is parallel to your next step $d\vec{s}$ and multiply them. When you add up the product for all of the steps, Ampere's law says that you get μ_0 times the current enclosed by the loop. This is not the current going along the path, but passing through the surface of which the path is the edge.

As with Gauss' law, we only do the integral when it's dirt simple. Let's see how this works. We want to find the magnetic field created by a long straight wire with current i. We draw an imaginary "Amperian loop" around the wire at a distance r.

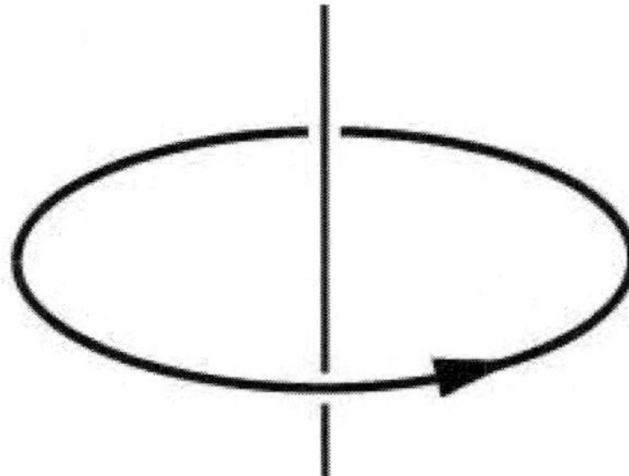

The problem has symmetry — if I rotate the problem along the axis of the wire it looks the same — so the magnetic field is the same everywhere on the imaginary path (if the wire was bent rather than straight then this would not be true). Also, from the right-hand rule this magnetic field will be parallel to the path.

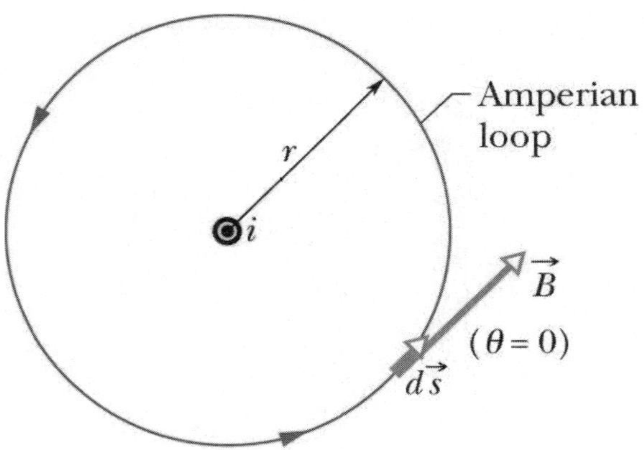

$$\oint \vec{B} \cdot d\vec{s} = \oint B\ ds = B \oint\ ds$$

The last integral $\oint\ ds$ is the sum of the lengths of each step for all steps. Because we've already dealt with the vector part (the magnetic field is parallel to the path) it is the length of the steps and not the displacement of the steps. This is the length of the path, so $\oint\ ds = L$.

$$\oint \vec{B} \cdot d\vec{s} = \oint B\ ds = B \oint\ ds = BL = B(2\pi r)$$

Ampere's law tells us that this integral is equal to $\mu_0 i_{\text{enc}}$, so

$$B(2\pi r) = \mu_0 I_{\text{enc}} = \mu_0 i$$

$$B = \frac{\mu_0 i}{2\pi r}$$

Like Gauss' law, Ampere's law is always true, but we can profitably use it only when there is symmetry in the problem. There are fewer geometries where Ampere's law is useful than there are for Gauss' law.

$$\overbrace{BL = \oint}^{\text{symmetry}}\!\!\!\!\!\!\!\!\underbrace{\oint\ \vec{B}\ \cdot\ d\vec{s} = \mu_0 i_{\text{enc}}}_{\text{Ampere's law}}$$

EXAMPLE

An extruded rectangle is curved into a toroid, so that the inner radius is 40 cm, the outer radius is 50 cm, and the height is 15 cm. A wire is wrapped evenly 140 times around the toroid and carries a current of 7 A. Find the magnetic field inside the toroid.

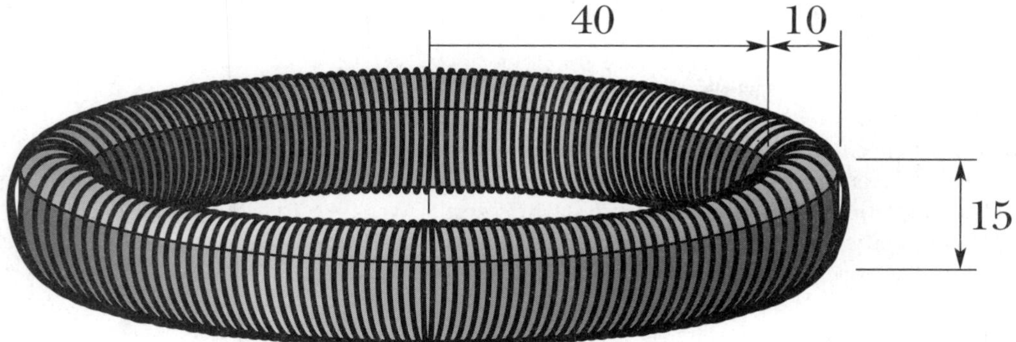

Tutor: How can you do this?
Student: We can find the magnetic field from each of the 140 loops and add them.
Tutor: We don't know how to find the field from a loop at a spot off of the axis. Even if we did, we'd have 140 fields to add, as vectors.
Student: Ouch. Presumably we can use the symmetry and Ampere's law.
Tutor: Correct. Draw an Amperian loop around the toroid.
Student: Like this?

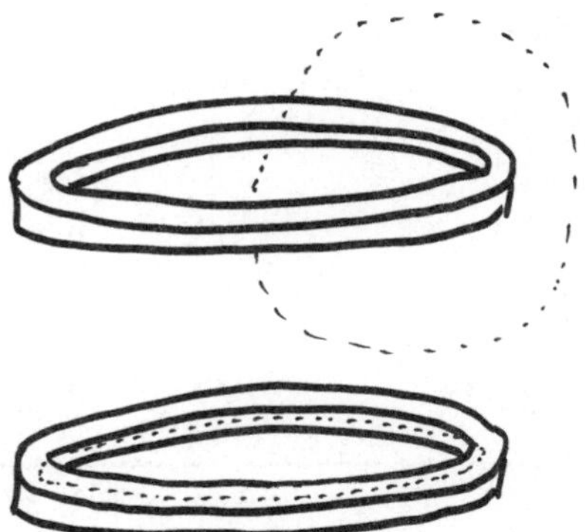

Tutor: Uh, no. We need to find a path where the magnetic field is the same everywhere on the path.
Student: But we don't know the magnetic field.
Tutor: True. We use symmetry to say that the magnetic field, whatever it is, is the same everywhere.
Student: That sounds familiar. So if I draw my path through the middle of the inside of the toroid, the field is the same everywhere?
Tutor: Use symmetry to explain how.
Student: If I rotate the whole toroid, it looks the same, so the answer has to look the same, so the field *inside* the toroid is constant.
Tutor: Careful. "Constant" means that it doesn't change with time. "Uniform" means that it's the same everywhere. The magnetic field inside the toroid isn't necessarily uniform — it could be different along the inside and outside edges. But at the same radius from the center, the magnetic field has to be the same all around the toroid.
Student: So I can use Ampere's law.

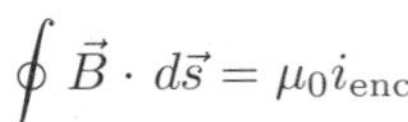

$$\oint \vec{B} \cdot d\vec{s} = \mu_0 i_{\text{enc}}$$

Tutor: What is the direction of the magnetic field inside the toroid?
Student: It goes around the inside.
Tutor: How much of the magnetic field is parallel to your Amperian path?
Student: All of it. So $\vec{B} \cdot d\vec{s} = B\,ds$.
Tutor: Very good.
Student: And B is constant, oops, the same everywhere along the path, so it comes out of the integral.

$$\oint \vec{B} \cdot d\vec{s} = \oint B\,ds = B \oint ds = B(\text{length of path}) = B2\pi r$$

Student: Is r the inside or outside radius of the toroid?
Tutor: Look at how you used it, to find the length of your path.
Student: So r is the radius of my path.
Tutor: Yes, you get to choose r.
Student: Can r be less than 40 cm or more than 50 cm?
Tutor: Sure, because the symmetry still holds.

Student: Okay. Ampere's law has given me

$$B2\pi r = \mu_0 i_{\text{enc}}$$

Student: What is the enclosed current?
Tutor: If you pick r to be less than 40 cm, then there is no enclosed current.
Student: That means that the magnetic field is zero.
Tutor: The magnetic field parallel to your path is zero. We have to use symmetry to argue that the magnetic field perpendicular to your path is also zero. What if 40 cm $< r <$ 50 cm?
Student: Then the wire goes through the loop, and the enclosed current is 7 A.
Tutor: The wire goes through the loop 140 times.
Student: Do I count every time that the wire goes through the loop?
Tutor: Yes.
Student: So the enclosed current is 140×7 A.
Tutor: Yes.
Student: And if r is greater than 50 cm, then the enclosed current is 280×7 A.
Tutor: On the outside of the toroid, the current is going the other way, so they count as negative currents.
Student: So it would be -140×7 A.
Tutor: The inside currents are still enclosed.
Student: So the inside and outside currents cancel?!
Tutor: Yes.
Student: And the field is zero if $r >$ 50 cm.
Tutor: Yes.
Student: So the only place where there is a field is inside the toroid. And that field is

$$B2\pi r = \mu_0 \left(140 \times 7 \text{ A}\right)$$
$$B = \frac{\mu_0 (140)(7 \text{ A})}{2\pi r}$$

Student: So the field gets weaker as I go further out, even inside the toroid.
Tutor: Yes, because your path becomes longer but the enclosed current stays the same.

Chapter 30

Induction and Inductance

If you connect a coil of wires to an ammeter (a current-measuring device) and place a magnet nearby, you will find that no matter how strong the magnet there is no current in the coil. However, as you move the magnet into position near the magnet there will be a current in the coil, and as you move it away from the coil there will be a current in the coil. What's happening?

Imagine a loop of wire on the edge of a magnetic field. If the loop moves down, the positive charges along the bottom of the loop experience a force to the right. They push a current counterclockwise around the loop.

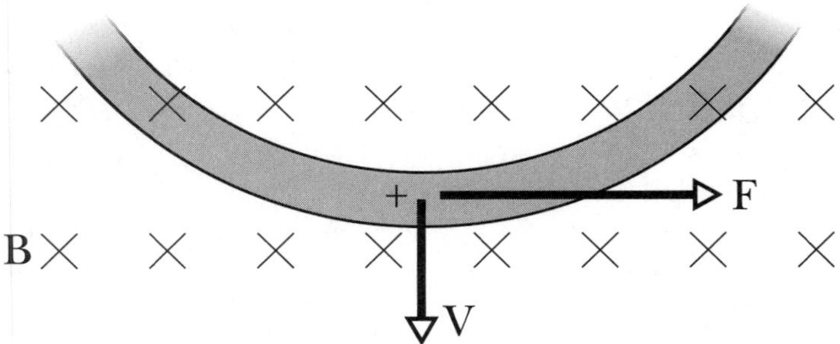

If the loop is entirely in the magnetic field (rather than partly in the magnetic field) then the positive charges at the top of the loop also feel a force to the right, and the result is no net current. This is similar to the magnet and coil, where the magnetic field itself doesn't cause a current but a movement in the magnetic field does.

The voltage induced in a loop by a magnetic field is

$$\mathcal{E} = -\frac{d\Phi_B}{dt} = -\frac{\Delta\Phi_B}{\Delta t}$$

where Φ_B is the magnetic field flux passing through the loop. This voltage causes a current if the loop is closed (complete).

The use of the word voltage here is somewhat misleading. We could say that one point in the loop had a higher voltage than the next, which had a higher voltage than the next, and so on until we got back to the original spot. This clearly can't be true, and may bring to mind Escher's waterfall. We typically use EMF $\mathcal{E}$ (electromotive force) for an electric field like this, created by a changing situation, and reserve the word voltage for static EMFs. I'll use the word voltage, but my doing so will make some physicists cringe.

What is flux? (If you remember this from Gauss' law then this is a repeat.) Imagine that you are out in the rain holding a bucket. Flux would be the rate at which raindrops enter the bucket. You could reduce this rate by using a smaller bucket, by going to a place with less rain, or by tilting the bucket sideways. The magnetic field flux is effectively a count of the number of magnetic field lines passing through a surface. Mathematically,

$$\Phi_B = \int \vec{B} \cdot d\vec{A}$$

which says to divide the surface into tiny pieces, at each piece find the component of the magnetic field that is perpendicular to the surface (normal to the surface), multiply times the area of the tiny piece, and add the results for all of the tiny pieces.

Despite the scary appearance of this integral, it's not so bad. Typically, the magnetic field is the same everywhere on the surface (or we'll pretend it is), so

$$\underbrace{\Phi_B = \int \overbrace{\vec{B} \cdot d\vec{A}}^{\text{constant field}}}_{\text{definition of flux}} = BA$$

The derivative (d) or delta (Δ) indicates that it is not magnetic field that creates an induced voltage or current. **An induced current is caused by a *change* in the magnetic flux**.

The minus sign in the equation is a reminder. The direction of the induced current is opposite to the change in the magnetic flux. How does this work? Consider our loop on the edge of the magnetic field. Initially the flux through the loop is zero since there is no magnetic field where the loop is. As we move the loop down, into the magnetic field, the magnetic flux into the page increases. The loop "wants" to keep the same flux it had (zero, in this case), so it creates magnetic field out of the page. That is, there will be an induced current in the direction that causes a magnetic field to pass through the loop out of the page.

In which direction would this current be? There are two ways to figure it out using the right-hand rules from the magnetic field section. One way is to point your thumb in the direction that you want the magnetic field through the loop to go (out of the page), then the fingers curl in the direction of the current (counterclockwise). The other is to point the fingers in the direction that you want the magnetic field through the loop to go. Then the thumb goes around the fingers in the direction of the current (counterclockwise).

EXAMPLE

Find the direction of the induced current as the coil is moved in each of the directions shown.

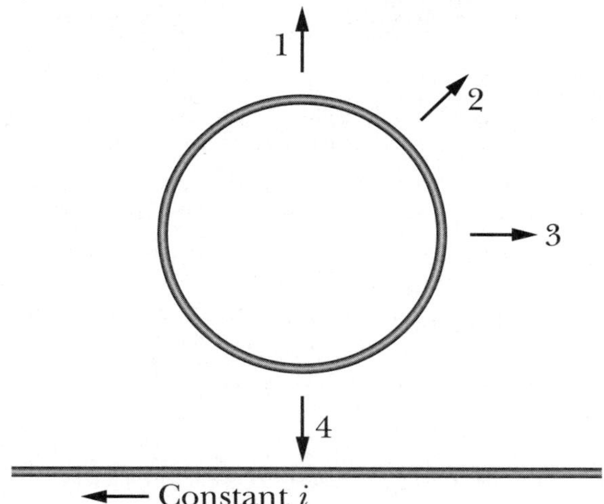

Tutor: What is the direction of the magnetic field that the long straight wire creates where the coil is?
Student: Second right-hand rule. I point my thumb to the left, put my hand up toward the coil, and bend my fingers into the page. The magnetic field goes into the page.
Tutor: So the magnetic flux is into the page.
Student: It's as easy as that?
Tutor: Field into page, flux into page. As the coil moves away from the wire, what happens to the magnetic field?
Student: The field gets weaker.
Tutor: By that you mean, the coil moves to a place where the field is weaker, so that the field through the coil is weaker.
Student: I'm sure that's what I meant.
Tutor: What happens to the magnetic flux through the coil?
Student: It gets smaller.
Tutor: It will be helpful here if you specify a direction. "Less magnetic flux into the page."
Student: Okay. The effect of moving the coil is to create less magnetic flux into the page. Happy?
Tutor: You will be. The coil wants to keep the same flux, so the loop tries to create magnetic field in which direction?
Student: Out of the page, opposite to the magnetic flux.
Tutor: It is the *change* in magnetic flux that matters. By moving the coil, we create less magnetic flux into the page.
Student: You said that.
Tutor: So the coil tries to replace the flux that was there and isn't anymore.
Student: So the coil creates magnetic flux into the page?
Tutor: Yes, to replace the flux that isn't there anymore.
Student: Even though the flux is into the page.
Tutor: Yes, because it's the change in flux that matters. We create less flux into the page, so the coil creates flux into the page, and the total flux stays the same.
Student: Okay. For the second direction, the field also gets weaker, so there's less flux into the page, and the coil creates field into the page, replacing the lost flux.
Tutor: Yes. The field that the coil creates is called the induced field. If the coil wants to create an induced field into the page, in what direction must the current in the coil go?
Student: That's the second right-hand rule again, right?

Tutor: Yes. Bend your fingers like you usually do, then point the bent fingers in the direction of the field.
Student: Into the page.
Tutor: Now keep them there and move your hand in a circle, following your thumb. Which way does your hand go?
Student: It goes clockwise.
Tutor: So a clockwise current in the coil creates a magnetic field into the page. At least in the middle of the coil.
Student: And outside the coil?
Tutor: The magnetic field circles around, since magnetic field lines never end. The field comes out of the page. But we care only about the field in the middle of the coil.
Student: Okay. In the third direction, the field doesn't change, so the flux doesn't change.
Tutor: If there is no change in the flux, what is the induced current?
Student: There isn't one.
Tutor: Correct. No matter how big the flux is, if it doesn't change then there is no induced current.
Student: In the fourth direction, the coil moves to a stronger magnetic field, so the change is "more flux into the page."
Tutor: Excellent.
Student: The coil creates magnetic field out of the page to opposite the *increase* in the magnetic flux.
Tutor: Very good.
Student: Fingers out, and the thumb follows the hand counterclockwise.

EXAMPLE

A circular coil consists of an unknown number of turns of wire, each 50 cm in diameter, in a 0.4 T magnetic field. The loop needs to rotate at 60 Hz and generate an average voltage of 120 V. How many turns must there be in the coil?

Student: If the coil rotates completely around, then the flux is back to what it started at, and there is no induced voltage.
Tutor: As the coil rotates one-fourth of the way around, the flux drops to zero.
Student: Because the magnetic field lines don't go through the coil?
Tutor: Correct. So let's look at one-fourth of a rotation.
Student: Okay. The original flux is BA, and the final flux is zero. The change happens in one-fourth of a period.

$$\mathcal{E} = N\frac{\Delta\Phi_B}{\Delta t} = N\frac{0 - BA}{\frac{1}{4}T}$$

Tutor: We need to include the factor N, since each turn induces a voltage and the voltages add. What's the period?
Student: The frequency is 60 per second. The period is one over 60.

$$\mathcal{E} = \frac{N(0.4\text{ T})A}{\frac{1}{4}(\frac{1}{60}\text{ s})}$$

Student: Is it my imagination or are we being sloppy with minus signs here?
Tutor: We are. We're not interested in the direction of the induced current, only the magnitude of the induced voltage.
Student: Okay. The radius is 25 cm, so the area is

$$A = \pi R^2 = \pi(0.25\text{ m})^2 = 0.196\text{ m}^2$$

$$120\text{ V} = \frac{N(0.4\text{ T})(0.196\text{ m}^2)}{\frac{1}{4}(\frac{1}{60}\text{ s})}$$

$$N = \frac{(120\text{ V})(\frac{1}{4})(\frac{1}{60}\text{ s})}{(0.4\text{ T})(0.196\text{ m}^2)} = 6.4$$

Tutor: What are the units for N?
Student: It shouldn't have any, should it?
Tutor: No. A tesla meter squared is flux, divided by a second is a volt.
Student: So the units all cancel. Shouldn't the number of turns be an integer?
Tutor: It should. We would have to change one of the other parameters so that N came out to be an integer.

EXAMPLE

A small circular loop (8 cm diameter) of 100 turns sits in the plane of and inside a large loop (28 cm diameter) of 100 turns. The current in the outer loop is 5 A and is increasing at a rate of 180 A/s. What is the voltage induced in the inner loop?

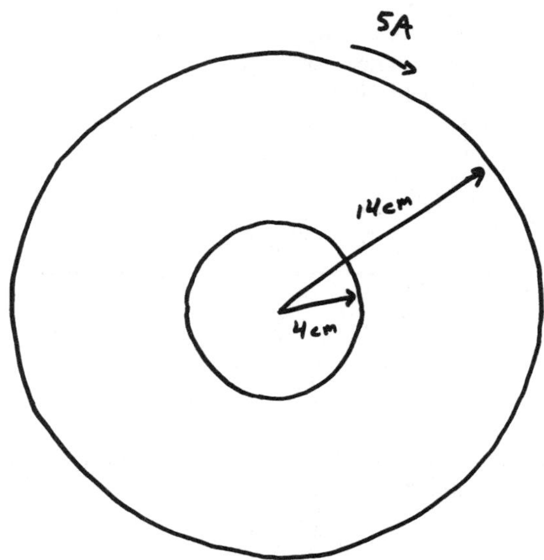

Student: The induced voltage is

$$\mathcal{E} = \frac{d\Phi_B}{dt}$$

Tutor: Good. What's the magnetic flux Φ_B?
Student: It's the magnetic field times the area.
Tutor: What is the magnetic field?
Student: We can use the magnetic field from a loop.

$$B_{\text{loop}} = \mu_0 i/2R \quad ?$$

Tutor: That's the magnetic field from each turn, and there are 100 of them.
Student: So I need a factor of 100. Is R the big loop or the small loop?
Tutor: It's the radius of the loop that is creating the magnetic field.
Student: So it's the big loop.
Tutor: Now what's the magnetic flux through the small loop?
Student: It's the magnetic field times the area πr^2.
Tutor: Which r?

Student: Uh, it's the area of the inner loop, so r is the radius of the inner loop.
Tutor: Good.

$$\mathcal{E} = \frac{d\Phi_B}{dt} = \frac{dBA}{dt} = \left(\frac{dB}{dt}\right) A$$

Student: We can pull the area outside of the derivative?
Tutor: The area isn't changing, so it is constant. $\frac{dB}{dt}$ is how fast the magnetic field is changing.
Student: But we don't know that. We know what the field is.
Tutor: So put the equation for the field into the equation for the induced voltage.

$$\mathcal{E} = \left(\frac{d}{dt}B\right) A = \left(\frac{d}{dt}\frac{N\mu_0 i}{2R}\right)(\pi r^2)$$

Tutor: N, μ_0, and R are constants, so you can pull them out of the derivative too.

$$\mathcal{E} = \left(\frac{N\mu_0}{2R}\right)\left(\frac{d}{dt}i\right)(\pi r^2)$$

Student: But I can't pull i out.
Tutor: The current i in the outer loop is changing. The change in current causes a change in magnetic field, a change in magnetic flux, and an induced voltage.
Student: So it's the *change* in the current that causes an induced voltage in the inner loop.
Tutor: Yes, and the size of the change in current that determines the magnitude of the induced voltage.

$$\mathcal{E} = \left(\frac{(100)(4\pi \times 10^{-7}\text{ T m/A})}{2(0.14\text{ m})}\right)(180\text{ A/s})\left(\pi(0.04\text{ m})^2\right) = 4.1 \times 10^{-3}\text{ V} = 4.1\text{ mV}$$

When the flux changes because of motion, then the induced voltage is

$$\mathcal{E} = BLv$$

where L is the length at the edge. The longer the length at the edge, and the faster the motion, the quicker the flux changes, and the greater the induced voltage.

EXAMPLE

Plot the induced voltage as a function of time for the single rectangular loop. Use clockwise induced current as corresponding to positive induced voltage.

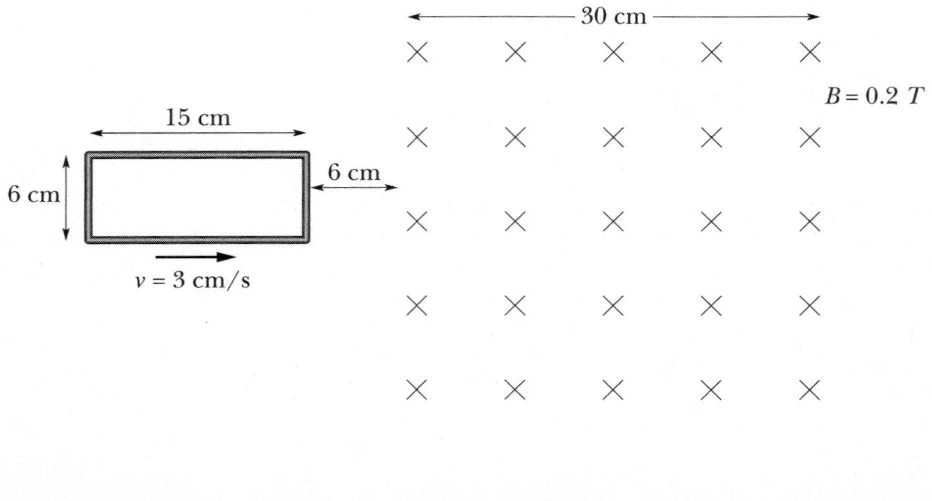

Tutor: What's the induced voltage when time begins?
Student: There's no flux, so there's no induced voltage.
Tutor: It is possible to have an induced voltage without a magnetic flux. If the flux is zero, but changing, then there will be an induced voltage.
Student: Okay. At $t = 0$ there is no flux and the flux isn't changing either. It doesn't start to change until $t = 2$ s.
Tutor: Good. How long does it take for the loop to enter the magnetic field?
Student: It has to travel 15 cm to fully enter the magnetic field, at 3 cm/s, so it takes 5 seconds. It enters from $t = 2$ s until $t = 7$ s and exits from $t = 12$ s until $t = 17$ s.
Tutor: Good. As it enters the magnetic field, what is the induced voltage?
Student: The further into the field it goes, the greater the flux, and the greater the voltage.
Tutor: More flux does not correspond to more voltage. What matters is how fast the flux is changing. Is the flux increasing more rapidly at $t = 6$ s than at $t = 3$ s?
Student: No, the flux will increase at a constant rate until the loop is inside the field.
Tutor: What's the slope of a constant rate, of a straight line?
Student: It would be constant. The induced voltage is a constant as it enters the magnetic field.
Tutor: Yes. What is that constant?
Student: Do I use $\mathcal{E} = BLv$?
Tutor: Yes.
Student: And which side is L?
Tutor: L is the width where the flux is changing. If I waited one second, the area of the loop with magnetic field in it would increase by how much?
Student: By 3 cm/s $\times$ 6 cm.
Tutor: The 3 cm/s is in v, and the 6 cm is in L.
Student: So the induced voltage is

$$\mathcal{E} = BLv = (0.2 \text{ T})(0.06 \text{ m})(0.03 \text{ m/s}) = 3.6 \times 10^{-4} \text{ V} = 0.36 \text{ mV}$$

Tutor: Yes, for the entire time that the loop is entering the magnetic field.
Student: How would you get an induced voltage that increased as it entered the field?
Tutor: I'd use a triangular loop.

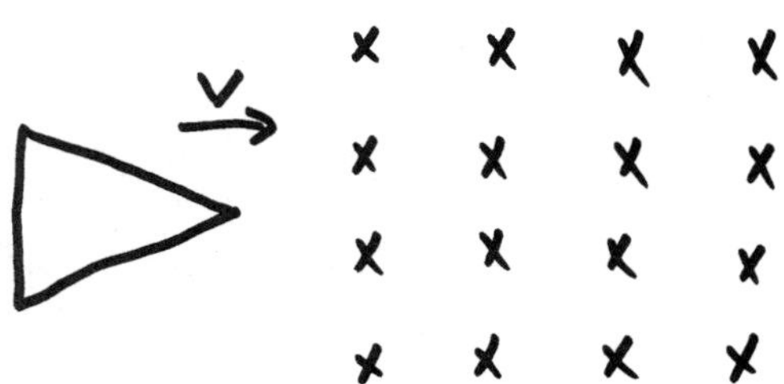

Student: Oh, so that L increased as it got further into the field.
Tutor: Yes. Is the 0.36 mV positive or negative?
Student: I need to find the direction of the induced current. When moving into the field, the change is more flux into the page. The loop tries to create field out of the page, against the change in flux, so the induced current is counterclockwise, or negative.
Tutor: So the voltage is -0.36 mV. What is the induced voltage between $t = 7$ s and $t = 12$ s, as the loop travels through the magnetic field?
Student: The flux is constant. Since it isn't changing, there is no induced voltage.
Tutor: Very good. What is the induced voltage as it exits the magnetic field?
Student: It will be just the opposite as when it enters.
Tutor: The magnitude will be the same, but check the direction.
Student: Okay. As it exits, the result is less flux into the page, so the loop creates flux into the page to

replace the flux that's disappearing, and to create flux into the page the loop needs a clockwise current. The voltage is +0.36 mV as it exits the field.

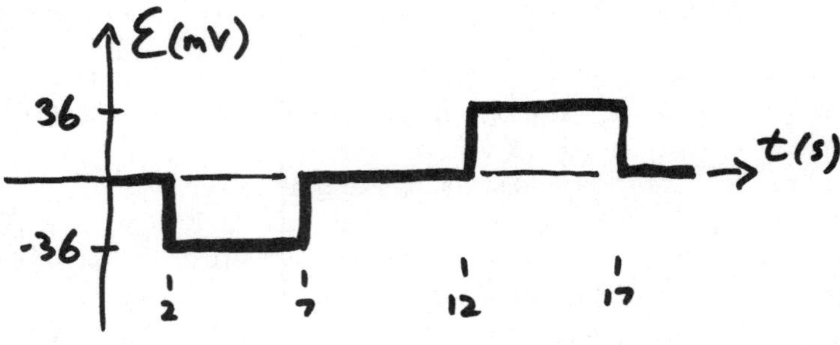

Chapter 31

Electromagnetic Oscillations and Alternating Current

A coil in a circuit is called an inductor. The voltage across an inductor is

$$V_L = L\frac{di}{dt}$$

(Some physics sources have a minus sign, left over from induced voltages, but it is incorrect to have one for an inductor.) The inductance L is measured in henries. As the current changes, the magnetic field changes, and that induces a voltage.

Consider a circuit with an inductor. A steady-state situation is one in which none of the currents or voltages are changing. If the circuit is ever to reach a steady state, then the voltage across the inductor must be zero, because a nonzero voltage indicates that the current is changing.

What if an inductor and a capacitor are placed in the same circuit? If the current is steady, then that current through the capacitor causes the charge on the capacitor to change. As the charge changes, so does the voltage, and so does the voltage on the inductor. The only possible steady state is zero — no current, no voltage, and no charge. Instead, the current through the inductor-capacitor circuit oscillates with a frequency

$$\omega = \frac{1}{\sqrt{LC}}$$

When we studied oscillators, we saw that a forced oscillator would oscillate at the forcing frequency, rather than the natural frequency. The closer the two frequencies were to each other, the greater the amplitude of the oscillation. This is true of electric oscillators, and the amplitude of the current of the oscillation is

$$I = \frac{\mathcal{E}_m}{\sqrt{R^2 + \left(\omega_d L - \frac{1}{\omega_d C}\right)^2}} = \frac{\mathcal{E}_m/R}{\sqrt{1 + \frac{L^2}{R^2}\left(\frac{\omega_d^2 - \omega^2}{\omega_d}\right)^2}}$$

where ω_d is the forcing frequency and $\mathcal{E}_m$ is the amplitude of the forcing voltage. If $R = 0$, then at resonance ($\omega_d = \omega$) this becomes infinite, but the resistance of the circuit won't ever be zero.

EXAMPLE

Choose approximate values for L and C for an FM radio so that the amplitude of the next station is $\frac{1}{100}$ the amplitude of the "tuned" station, given that the circuit has a resistance of 0.6 Ω.

Student: How am I supposed to do this? There are no numbers.
Tutor: They are merely disguised. FM stations have frequencies of about 100 MHz, and are 0.2 MHz apart.
Student: So I need $\omega = \frac{1}{\sqrt{LC}}$ to be 10^8.
Tutor: The frequency f is 10^8, and $\omega = 2\pi f$.
Student: Oh, that's right. So $\frac{1}{\sqrt{LC}} = 2\pi \times 10^8$. But if I don't know one, I can't get the other. Do I just make one up?
Tutor: There are other criteria to match. What's the amplitude of the current at resonance?
Student: It's $\mathcal{E}_m/R$.
Tutor: So if $\omega_d - \omega$ is 0.2 MHz, you need the amplitude to be $\frac{1}{100}$ of that.
Student: So the denominator must be 100.

$$\sqrt{1 + \frac{L^2}{R^2}\left(\frac{\omega_d^2 - \omega^2}{\omega_d}\right)^2} = 100$$

$$1 + \frac{L^2}{R^2}\left(\frac{\omega_d^2 - \omega^2}{\omega_d}\right)^2 = 10^4$$

$$\frac{L}{R}\left(\frac{\omega_d^2 - \omega^2}{\omega_d}\right) \approx 10^2$$

Student: Is 100 MHz the natural frequency ω or the driving frequency ω_d?
Tutor: The natural frequency ω is the frequency of the radio station that you want to hear. The driving frequency ω_d is the frequency of the next station over, 0.2 MHz either higher or lower.
Student: The difference is 0.2 mHz, but how do I get $\omega_d^2 - \omega^2$?
Tutor: Careful, mHz is millihertz instead of megahertz — a difference of 10^9. You can factor

$$(\omega_d^2 - \omega^2) = (\omega_d - \omega)(\omega_d + \omega) \approx (\omega_d - \omega)(2\omega) \approx (2\pi 0.2\text{ MHz})(2\pi 100\text{ MHz})$$

$$\frac{L}{R}\left(\frac{(2\pi 0.2\text{ MHz})(2\pi 100\text{ MHz})}{(2\pi 100\text{ MHz})}\right) \approx 10^2$$

$$\frac{L}{R}(2\pi 0.2\text{ MHz}) \approx 10^2$$

Tutor: So far, we haven't made any assumptions that weren't in the problem.
Student: The units don't seem right.
Tutor: A henry, the unit for L, is equal to Ω times a second. So L/R has units of seconds, and Hz is 1/s, so the left side is unitless.
Student: Okay. We're given the resistance R.

$$\frac{L}{R}\left(1.26 \times 10^6\text{ s}^{-1}\right) \approx 10^2$$

$$L \approx \frac{(10^2)(0.6\ \Omega)}{1.26 \times 10^6\text{ s}^{-1}} = 4.8 \times 10^{-5}\text{ H} = 48\ \mu\text{H}$$

Student: Now I can find the capacitance C.

$$\frac{1}{\sqrt{LC}} = 2\pi \times 10^8\text{ s}^{-1}$$

$$C = \frac{1}{(4.8 \times 10^{-5}\text{ H})(2\pi \times 10^8\text{ s}^{-1})^2} = 5.3 \times 10^{-14}\text{ F} = 0.053\text{ pF}$$

Student: How do you get farads out of that?
Tutor: A henry is an ohm times a second, divided by two factors of seconds is an ohm/second; take the inverse to get second/ohm, which is a farad. LC is (ohm second)(second/ohm) is second^2, then take the square root and invert is 1 over seconds.
Student: Okay. That seems like a small capacitance.
Tutor: It is a small capacitance. Every circuit has some capacitance, whether you put a capacitor in or not. You would need to control the "stray" capacitance at this level to tune the radio properly.
Student: Is this how a radio works?
Tutor: It's one way. All of the radio stations in the area broadcast their signal, and all of the signals arrive at your antenna and try to "force" your circuit. Only one of them has a frequency that matches the circuit's frequency, so all of the others produce much smaller amplitudes. If the current is 100 times less, then the power is 10^4 less and down by 40 dB.

Transformers change the voltage in a circuit. Imagine two coils wrapped together — that is, two wires wrapped around the same space. If the current in one changes, the magnetic field through each changes, so there is a voltage induced in the *other* wire. Because it is the *change* in the magnetic field that causes a voltage in the other coil, transformers only work with alternating current, and not with direct current.

The magnitude of the induced voltage increases if there are more turns in the second wire. We typically call the first wire the "primary" and the second, or output wire, the "secondary." The voltage out, or secondary voltage, is determined by the ratio of the number of turns.

$$\frac{V_s}{V_p} = \frac{N_s}{N_p}$$

Transformers cannot create energy, so the power out is equal to the power in. This is not true at every instant. The magnetic field stores energy and the power out will increase and decrease as the instantaneous current changes, taking energy stored in the magnetic field. But over a complete cycle, the power in and out are the same. Therefore, if the output (secondary) voltage is greater, the output current must be less.

$$V_p I_p = P_p = P_s = V_s I_s$$

Why do we use I here for current instead of i? Because I is not the instantaneous current but the *amplitude* of the oscillating current. The instantaneous current is $i(t) = i \sin \omega t$. Maximum current in the primary and secondary do not occur at the same time, but usually we are interested in the maximum current, or current amplitude I.

EXAMPLE

Electric power in the United States is 120 volts, but in Europe is 240 volts. When an American traveler goes to Europe, he may take a travel transformer, which allows him to use his American devices in Europe. How many turns must be on the primary and secondary sides of such a transformer?

Student: I need to use $V_s/V_p = N_s/N_p$.
Tutor: Yes, but which is the primary and which is the secondary?
Student: The higher voltage is the secondary.
Tutor: Always?
Student: Isn't it?
Tutor: No. "Step-down" transformers are used to decrease the voltage for neighborhoods, and in the home for dimmer switches and model trains.
Student: So the lower voltage is the secondary?
Tutor: "Step-up transformers" are used for long-distance transmission and high-voltage equipment, like

mosquito zappers.
Student: So which is which?
Tutor: The device is designed to work in America, so it must need a voltage of ...
Student: ... 120 volts.
Tutor: So the transformer has to output 120 volts.
Student: That's the secondary.
Tutor: And the voltage that the transformer has to work with is 240 volts. That's what it can get out of the wall.
Student: So that's the primary, the "in" to the transformer.
Tutor: Yes.

$$\frac{N_s}{N_p} = \frac{V_s}{V_p} = \frac{120\text{ V}}{240\text{ V}} = \frac{1}{2}$$

Student: But I still don't know how many turns are on either side.
Tutor: No, but you know that $N_p = 2N_s$, that there are twice as many turns on the primary as on the secondary.
Student: So it could be 200:100, or 60:30, or 2:1?
Tutor: Well, it's not likely to be 2:1. The magnetic field created in the transformer would be so weak that it wouldn't work as well.
Student: But it's not important to know how many turns there are?
Tutor: Only the *ratio* of turns.

Chapter 32

Maxwell's Equations; Magnetism of Matter

All of electromagnetism is summed up in Maxwell's equations.

$$\Phi_E = \oint \vec{E} \cdot d\vec{A} = \frac{q_{\text{net}}}{\epsilon_0}$$

$$\Phi_B = \oint \vec{B} \cdot d\vec{A} = 0$$

$$\oint \vec{E} \cdot d\vec{s} = -\frac{d}{dt}\Phi_B$$

$$\oint \vec{B} \cdot d\vec{s} = \mu_0 i_{\text{enc}} + \mu_0 \left(\epsilon_0 \frac{d}{dt} \Phi_E \right)$$

While not one of Maxwell's equations, a fifth is often included with the group:

$$\vec{F} = q\left(\vec{E} + \vec{v} \times \vec{B}\right)$$

The first equation is Gauss's law. The flux coming out of a closed surface is proportional to the enclosed charge. This is because electric field lines start at positive charges and end at negative charges.

The second equation, Gauss's law for magnetic fields, says that the magnetic flux coming out of a closed surface is zero. This is because magnetic field lines never end, so every magnetic field line that enters a surface has to leave it somewhere.

The third equation is Faraday's induced currents. The induced voltage is the sum of the electric field around the loop.

The fourth equation is a modified version of Ampere's law. The sum of the magnetic field around a loop is proportional to the current through the loop, plus the piece in parentheses called the "displacement current." When current passes through a capacitor, charging the capacitor, the current causes a magnetic field. There is also a magnetic field around the capacitor, even though no real current passes through the capacitor. But the changing electric field between the capacitor plates creates a magnetic field, just like a current would. It also leads to electromagnetic waves, seen in the next chapter.

EXAMPLE

A capacitor is made of two parallel circular disks, 6 cm in radius and 1 mm apart. A current of 4 A passes through the capacitor. Find the magnetic field magnitude 3 cm from the wire, and 3 cm from the center of the capacitor.

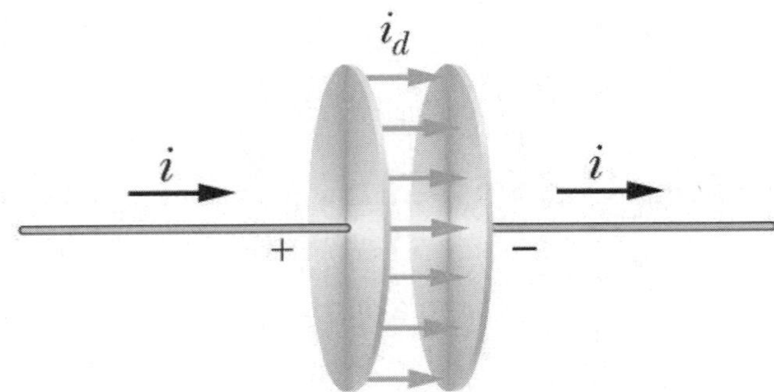

Student: Isn't this just like what we did with Ampere's law?
Tutor: Yes. Do you remember that?
Student: I think so. We draw a circle around the wire with a radius of 3 cm, so that the circle passes through our point. The magnetic field is the same everywhere on the circle, so $\oint \vec{B} \cdot d\vec{s} = BL = B(2\pi r)$. The enclosed current is 4 A, so

$$B(2\pi r) = \mu_0(4\text{ A})$$

$$B = \frac{\mu_0(4\text{ A})}{2\pi r} = \frac{(4\pi \times 10^{-7}\text{ T m/A})(4\text{ A})}{2\pi(0.03\text{ m})} = 26.7\ \mu\text{T}$$

Tutor: Excellent. Now for inside the capacitor.
Student: We need to find the electric field flux inside the capacitor.
Tutor: We need the change in the electric flux, but finding an equation for the flux is a step in that process.
Student: Can we use Gauss's law?
Tutor: What is the surface you are using? What is the loop you are using?
Student: I'm drawing a circle around the center of the capacitor with a radius of 3 cm.

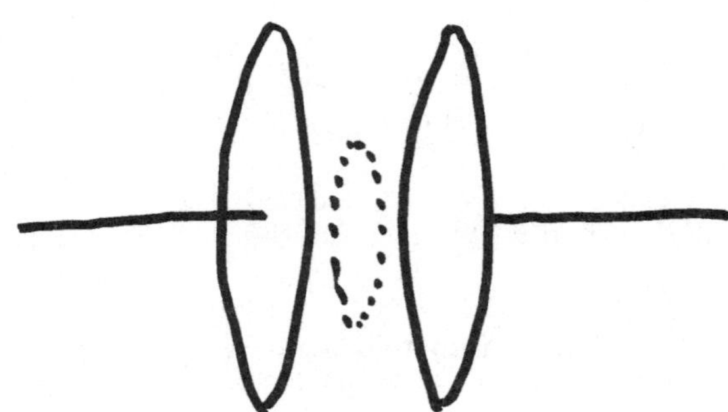

Tutor: The surface is the area surrounded by the loop. Is this a closed surface?
Student: No. That means no Gauss's law.
Tutor: Correct. What is the electric field inside the capacitor?
Student: There is a constant field between the two plates.
Tutor: Constant means that it doesn't change with time. Uniform means that the field is the same at all

points inside the region.
Student: I meant there is a uniform field between the two plates. The flux is $\Phi_E = \int E dA = EA$.
Tutor: Good. What is the area A?
Student: The area of my surface, right? It's $\pi(3\text{ cm})^2$.
Tutor: Yes. What's the electric field?
Student: Can we use $E = \frac{\sigma}{\epsilon_0}$? It's a uniform field near a plane of charge, with the field going only one way.
Tutor: Very good.

$$\Phi_E = EA = \frac{\sigma}{\epsilon_0}(\pi r^2)$$

Student: σ is the charge Q divided by the area πr^2.
Tutor: πr^2 is the area of your surface, but not the area of the capacitor.
Student: Oops. The area of the capacitor is πR^2, where R is 6 cm.

$$\Phi_E = EA = \frac{Q/(\pi R^2)}{\epsilon_0}(\pi r^2) = \frac{Q\pi r^2}{\pi R^2 \epsilon_0} = \frac{Q}{\epsilon_0}\frac{r^2}{R^2}$$

Tutor: Good. What we really want is the change in the flux.
Student: So I take the derivative.

$$\frac{d}{dt}\Phi_E = \frac{d}{dt}\frac{Q}{\epsilon_0}\frac{r^2}{R^2} = \frac{1}{\epsilon_0}\frac{r^2}{R^2}\frac{dQ}{dt} = \frac{1}{\epsilon_0}\frac{r^2}{R^2}i$$

Student: The change in the charge is the current. I don't need to know the charge, only the current.

$$\oint \vec{B}\cdot d\vec{s} = \mu_0 i_{\text{enc}} + \mu_0\left(\epsilon_0 \frac{d}{dt}\Phi_E\right)$$

$$B(2\pi r) = \mu_0(0) + \mu_0\left(\epsilon_0 \frac{1}{\epsilon_0}\frac{r^2}{R^2}i\right)$$

$$B(2\pi r) = \mu_0 \frac{(3\text{ cm})^2}{(6\text{ cm})^2}i = \mu_0\left(\frac{1}{4}\right)i$$

Student: It's just one-fourth of what we had before.
Tutor: That's because your loop enclosed one-fourth of the area, so one-fourth of the flux.
Student: If my loop had enclosed all of the flux, then we would have gotten the same answer.
Tutor: Yes. The increasing flux acts like a current, called the displacement current.

A magnet in a magnetic field tries to rotate so that its magnetic field is parallel to the existing field. This is how a compass works — the compass is a small magnet and it rotates so that its field lines up with the Earth's field.

Many atoms and molecules have magnetic moments, meaning that they act like tiny magnets. In some materials, the atoms are free to "rotate" so that their field becomes parallel to the external field. This makes the magnetic field even stronger. We call these materials paramagnetic. When the external magnetic field is removed, the atoms align randomly, so that they tend to cancel each other out and they no longer create a net magnetic field.

Ferromagnetic materials are similar to paramagnetic, except that the atoms can remain aligned after the external magnetic field is removed. Iron and some alloys of iron are ferromagnetic materials and are used to make permanent magnets. The atoms remain aligned because they have less energy aligned than when their magnetic fields are random. If such a magnet is heated then the energy put into the magnetic can demagnetize the material, overcoming the energy from being aligned. The Curie temperature is the point at which a magnetic becomes demagnetized.

Materials that are neither paramagnetic nor ferromagnetic are diamagnetic. Diamagnetic materials develop a small magnetic field *opposite* to the external field. The atoms are not free to rotate, but the external field changes the orbits of the electrons in the atoms. Diamagnetic materials are repelled by magnetic fields, rather than attracted.

EXAMPLE

A coin-shaped piece of iron has a thickness of 1 mm and a diameter of 1 cm. Iron has a density of 7900 kg/m^3 and an iron atom has a magnetic moment of 2×10^{-23} J/T. What is the maximum magnetic field strength that this magnet can create 3 mm above its surface?

Student: Each atom acts like a magnet, with its own magnetization. I find out how many atoms there are and multiply by the magnetic moment.
Tutor: Good. How will you find the number of atoms?
Student: Well, if I use the density and find the volume, I can find the mass of the magnet. Then I divide by the mass of a single atom.

$$N = \frac{M}{m} = \frac{V\rho}{m} = \frac{(\pi r^2 t)\rho}{m} = \frac{\pi(0.005 \text{ m})^2(0.001 \text{ m})(7900 \text{ kg/m}^3)}{(56)(1.67 \times 10^{-27} \text{ kg})} = 6.6 \times 10^{21}$$

Tutor: Nicely done.
Student: Now I multiply by the magnetic moment of each atom.

$$\mu = N\mu_{\text{Fe}} = (6.6 \times 10^{21})(2 \times 10^{-23} \text{ J/T}) = 0.13 \text{ J/T}$$

Tutor: But that's not the magnetic field. That's the magnetic moment.
Student: Oh. How do I find the magnetic field?
Tutor: The magnetic field from a magnetic moment is

$$\vec{B} = \frac{\mu_0}{2\pi}\frac{\vec{\mu}}{z^3}$$

Tutor: This is the field along the axis of the magnetic moment.
Student: So

$$B = \frac{4\pi \times 10^{-7} \text{ T m/A}}{2\pi}\frac{0.13 \text{ J/T}}{(0.003 \text{ m})^3} = 0.96 \text{ J/A}\cdot\text{m}^2$$

Student: Shouldn't the units be teslas?
Tutor: We typically measure magnetic field in tesla. Is this the same as teslas?
Student: A joule is a newton meter, so it's newtons per amp·meter.
Tutor: I remember teslas by using $F = iLB$, so a newton is an amp·meter·tesla.
Student: Then a newton per amp·meter *is* a tesla. We get a magnetic field of 1 tesla.

Chapter 33

Electromagnetic Waves

Maxwell's equations show that a changing electric field creates a magnetic field, and a changing magnetic field creates an electric field. In we were to create an electric field that varied like sine, then the change in the electric field would create a magnetic field. Because the rate at which the electric field is changing is changing (it has a second derivative too), the magnetic field would be changing, so it would create an electric field, and so on.

The result is a wave equation involving E and B. The solutions to the wave equation are waves, and these electromagnetic waves are radio wave, visible light, and X-rays. The solution also gives the speed for the wave.

$$c = \frac{1}{\sqrt{\mu_0 \epsilon_0}} = \frac{E}{B} = 3 \times 10^8 \text{ m/s}$$

Because the speed of light shows up so much in physics, it is given its own symbol c. $\vec{E}$ and $\vec{B}$ are always in phase with each other and perpendicular to each other.

The instantaneous intensity is given by the Poynting vector

$$\vec{S} = \frac{1}{\mu_0} \vec{E} \times \vec{B}$$

The magnitude of the Poynting vector is equal to the intensity, and the direction of the vector gives the direction of travel of the wave. We can replace B with E/c from above. Remember that the intensity of any wave is proportional to the amplitude squared, and electromagnetic waves are no exception. To find the average or RMS intensity, we need a factor of two.

$$I = \frac{\text{power}}{\text{area}} = \frac{1}{c\mu_0} E_{\text{rms}}^2$$

Remember that the RMS is maximum divided by $\sqrt{2}$.

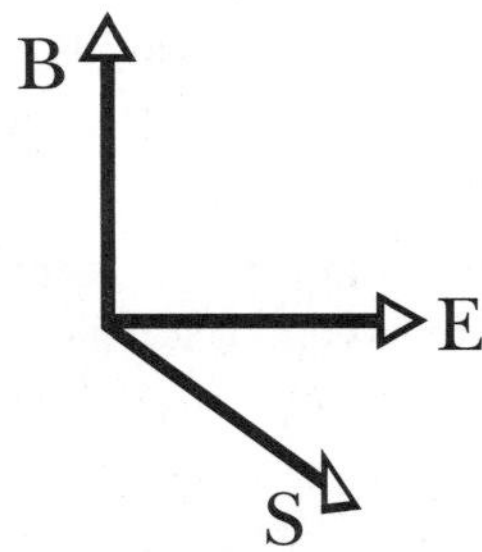

EXAMPLE

A 100 watt light bulb is 1.4 m away. The light bulb is 5% efficient. What is the amplitude of the electric field oscillation?

Student: Is 100 watts the maximum intensity or the average intensity?
Tutor: It's the average electrical *power*. The intensity is the power per square meter.
Student: The 5% means that the light power is not the same as the electrical power.
Tutor: Correct. For a typical incandescent light bulb, only 5% of the electrical energy is turned into light. The rest is turned into heat. You can save on your lighting bills by using compact fluorescent bulbs, which are about 20% efficient.
Student: So you get four times the light.
Tutor: Or the same amount of light with one-fourth the electricity. What is the light power from the bulb?
Student: It's 0.05×100 W or 5 watts.
Tutor: Over how much area is that power spread?
Student: If it radiates uniformly, then over a sphere of radius 1.4 m.

$$A = 4\pi r^2 = 4\pi(1.4 \text{ m})^2 = 24.6 \text{ m}^2$$

Tutor: What's the average intensity?
Student: Intensity is power per area.

$$I = \frac{5 \text{ W}}{24.6 \text{ m}^2} = \frac{1}{c\mu_0} E_{\text{rms}}^2$$

$$0.203 \text{ W/m}^2 = \frac{1}{(3 \times 10^8 \text{ m/s})(4\pi \times 10^{-7} \text{ T m/A})} E_{\text{rms}}^2$$

$$E_{\text{rms}}^2 = 76.5 \text{ W}\cdot\text{T/s}\cdot\text{A}$$

$$E_{\text{rms}} = 8.75 \sqrt{\text{W}\cdot\text{T/s}\cdot\text{A}}$$

Student: I'm not sure about the units. They ought to be N/C or V/m.
Tutor: True. The way I remember teslas is with $F = iLB$, so N = A·m·T, and 1 T = 1 N/A·m.

$$E_{\text{rms}} = 8.75 \sqrt{\text{W}\cdot\text{N/s}\cdot\text{m}\cdot\text{A}^2} = 8.75 \sqrt{(\text{N}\cdot\text{m})\cdot\text{N/s}^2\cdot\text{m}\cdot\text{A}^2} = 8.75 \sqrt{\text{N}^2\ /\text{s}^2\cdot\text{A}^2} = 8.75 \text{ N/C}$$

Student: It worked.
Tutor: What we have now is the RMS value of the electric field.
Student: And we want the amplitude.

$$E_{\text{RMS}} = \frac{E_m}{\sqrt{2}}$$

$$E_m = E_{\text{RMS}}\sqrt{2} = 8.75 \text{ N/C} = 12.4 \text{ N/C}$$

Electromagnetic waves are transverse waves, meaning that the electric field is perpendicular to the direction that the wave travels. This is clear mathematically from the Poynting vector, because the cross product has to be perpendicular to vectors that are cross products. To describe the direction in which the electric and magnetic field oscillate, we use polarization.

Most light that you see is not polarized. One "photon" of light may have its electric field vertical, and the next horizontal, and the next somewhere in between. Light is polarized when all of the light is oriented the same way. We can polarize light by putting it through a polarizer. When we put unpolarized light through a polarizer, half of it comes through:

$$I = \frac{1}{2} I_{\text{unpolarized}}$$

but the light that does come through is now polarized parallel to the orientation of the polarizer. When we put polarized light through a polarizer, the fraction that comes through is

$$I = I_{\text{polarized}} \cos^2 \theta$$

where θ is the angle between the polarizer and the polarization of the light.

Light can also be polarized when it is reflected. When light reflects from a surface at an angle (not normal incidence), one polarization is reflected better than the other. At Brewster's angle, one of the polarizations is not reflected at all, so only the other polarization remains. This is why polarizing sunglasses work — they remove half of unpolarized sunlight, or light that reflects off of irregular surfaces like trees, but they can remove much more than half of glare from reflective surfaces. They are especially effective at glare from water.

EXAMPLE

700 W/m^2 of sunlight goes through three polarizers. The three polarizers are oriented at angles of 15°, 40°, and 105° from the common reference axis. How much light comes out of the third polarizer?

Tutor: How much light intensity goes into the first polarizer?
Student: Wouldn't that just be 700 W/m^2?
Tutor: It would. Is this light polarized?
Student: Natural sunlight isn't polarized — no.
Tutor: How much light comes through the first polarizer?
Student: Because the light isn't polarized, half of what went in: $\frac{1}{2}(700 \text{ W/m}^2) = 350 \text{ W/m}^2$.
Tutor: How much light goes into the second polarizer?
Student: What comes out of the first one: 350 W/m^2.
Tutor: How much comes out of the second?
Student: The light going into the second one is polarized, isn't it?
Tutor: Yes, because it has gone through a polarizer.
Student: Is there a way to de-polarize light?
Tutor: Sure. If you shine polarized light on a wall or anything that isn't mirror-like, the reflected light will be unpolarized.
Student: Since it is polarized, the amount that comes through is

$$I = I_{\text{polarized}} \cos^2 \theta$$

$$I_2 = I_1 \cos^2 40° \quad \textbf{?}$$

Tutor: Why 40°?
Student: Because that's the angle of the polarizer.
Tutor: θ is the angle between the light and the polarizer, not the polarizer and the reference axis. What is the direction of polarization of the light that reaches the second polarizer?
Student: It's not 0°, right? It matches the first polarizer. So the angle θ is 25°.
Tutor: Correct.

$$I_2 = I_1 \cos^2 25° = (350 \text{ W/m}^2) \cos^2 25° = 287 \text{ W/m}^2$$

Student: And this light is parallel to the second polarizer.
Tutor: Correct.
Student: So the angle to use when it goes through the third polarizer is 65°.

$$I_3 = I_2 \cos^2 25° = (287 \text{ W/m}^2) \cos^2 65° = 51 \text{ W/m}^2$$

Student: Now if the angle was 90°, then $\cos 90° = 0$, so no light would get through.
Tutor: Correct. If you try to put horizontal light through a vertical polarizer, none gets through. The

component of the electric field in the vertical direction is zero.
Student: But if you stick another polarizer in between them, some light gets through. Our first and last polarizers were 90° apart.
Tutor: True. Because some gets through the second one, and now that light has a nonzero component in the direction of the third polarizer.
Student: Strange.

It is a general principle of waves that when a wave encounters a change of medium, in which the wave would go a different speed, some of the wave is reflected and some is transmitted. This is true of light as well.

We characterize the speed of light in a material with the index of refraction of the material n.

$$v = \frac{c}{n}$$

n is one for vacuum, and increases from there — light only goes slower as it goes through material. Because n is the ratio of two speeds, it has no units.

Some Indexes of Refraction	
vacuum	1 (exact)
air	1.0003
water	1.33
glass	1.5

One effect of the change in speed is that the transmitted light is deflected. We can this refraction, and the equation that we use is Snell's law.

$$n_1 \sin\theta_1 = n_2 \sin\theta_2$$

This important thing to remember is that we **always measure angles from the normal to the surface, never the surface itself**.

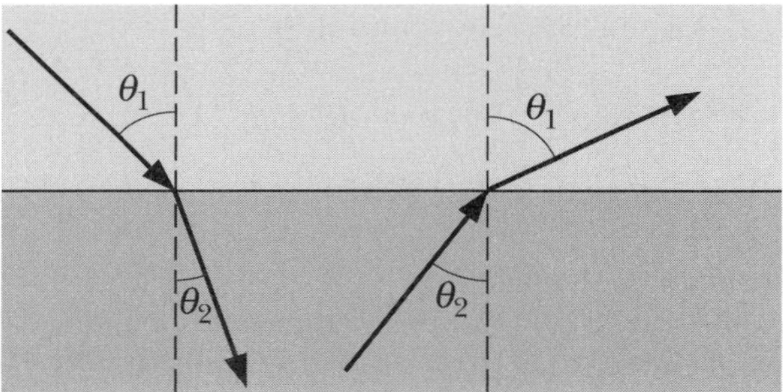

Whichever material has the larger n has to have the smaller $\sin\theta$, and thus the smaller angle θ. In the figure, material 2 has the smaller angle (measured from the normal), so it must have the larger index of refraction. This is true regardless of the material that the light starts in. This can lead to an interesting situation, as shown on the right side of the figure. If the angle θ_2 was to increase, then θ_1 would reach 90°. If θ_2 increased still more, then θ_1 would be unsolvable — the sine of θ_1 would have to be more than one. Since the light can't exit, it must all be reflected back into material 2, called total internal reflection. This can only happen when the refracted (outgoing) angle is larger than the incident (incoming) angle, or when the light is going into a material with lower n. Note that there would be some light reflected anyway, but when the incident angle is too large, all of the light is reflected. Reflected light is always reflected at the same angle as the incident angle.

EXAMPLE

Light enters the glass block ($n = 1.57$) from air as shown. Find the marked angle.

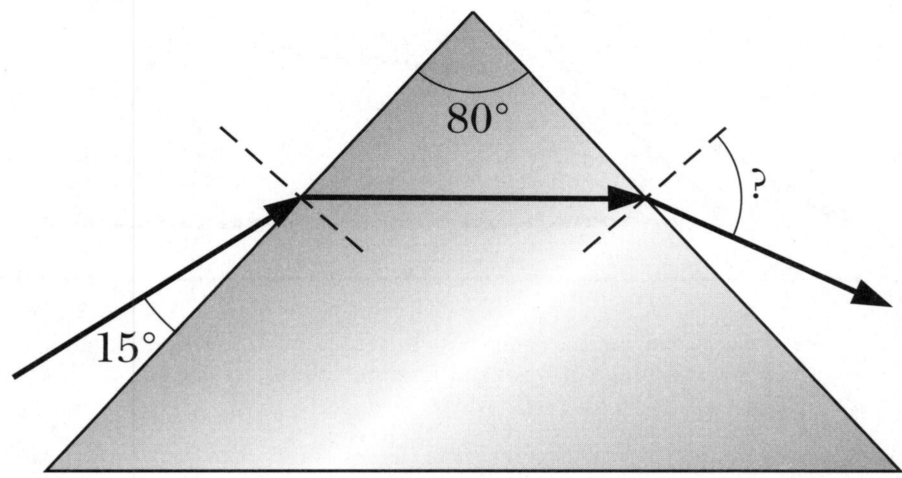

Student: The angle isn't measured compared to the normal.
Tutor: Then it isn't the angle that you need to use.
Student: So I need to use $\theta_1 = 90° - 15° = 75°$.

$$n_1 \sin\theta_1 = n_2 \sin\theta_2$$

$$(1.0003)\sin 75° = (1.57)\sin\theta_2$$

Tutor: Even though the index of refraction of air isn't exactly 1, we often use 1 anyway.
Student: Whatever.

$$\theta_2 = \arcsin\left(\frac{(1.0003)\sin 75°}{(1.57)}\right) = 38.0°$$

Tutor: Now you need to do the second surface.
Student: But what's the angle?
Tutor: Use the geometry of the triangle.
Student: The bottom-left angle of the triangle is $90° - 38.0° = 52.0°$. The bottom-right angle of the triangle is $180° - 80° - 52.0° = 48.0°$. Then the incident angle at the second surface is $90° - 48.0° = 42.0°$.
Tutor: Very good.

$$n_1 \sin\theta_1 = n_2 \sin\theta_2$$

$$(1.57)\sin 42.0° = (1.0003)\sin\theta_2$$

$$\theta_2 = \arcsin\left(\frac{(1.57)\sin 42.0°}{(1.003)}\right) = \arcsin(1.048)$$

Student: How do you take the arcsine of a number greater than 1?
Tutor: You don't.
Student: So, does that mean there is total internal reflection?
Tutor: Yes.

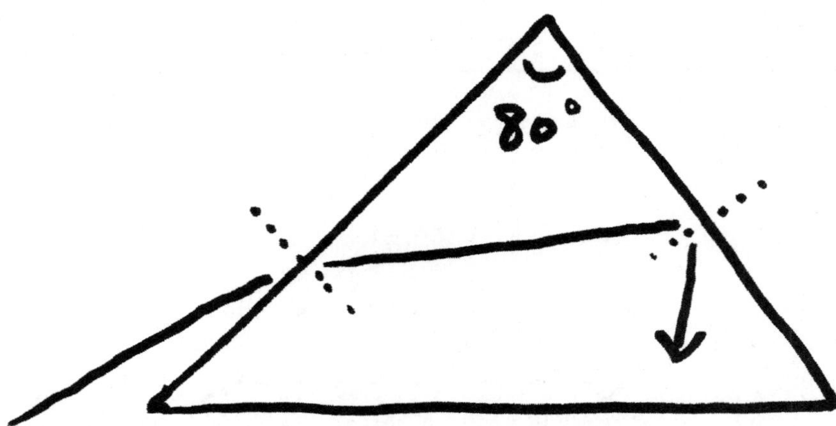

We use names for ranges of electromagnetic frequencies. Infrared, for example, is about 10^{11} Hz to 10^{15} Hz. Visible light is the relatively narrow range of wavelengths from 400 nm to 700 nm. (We use wavelengths for visible light and higher frequency electromagnetic waves.)

The reason for the different names is that electromagnetic waves of different frequencies interact with matter in different ways. Therefore you have to use different techniques to generate and detect infrared waves than you use for ultraviolet waves. There is not a distinct cutoff where a technique stops working — it is not true that your eye can see 699 nm but not 701 nm, but as you go further into the infrared your eye works less well, and we use 700 because it's about right.

Chapter 34

Images

When light reflects off of a mirror, the reflected angle is equal to the incident angle. This results in the creation of an image. Examine the figure below, and you find that all of the light that leaves the nose and then reflects off of the mirror appears to come from one spot, located behind the mirror. Someone looking at nose in the mirror will really be looking at a spot behind the mirror. You can try this yourself — put a sticker or tape a paper onto a mirror and then look at yourself in the mirror; when your reflection (your image) is in focus, the sticker on the mirror will not be, and vice versa. Your reflection is not *at* the mirror, but *behind* it.

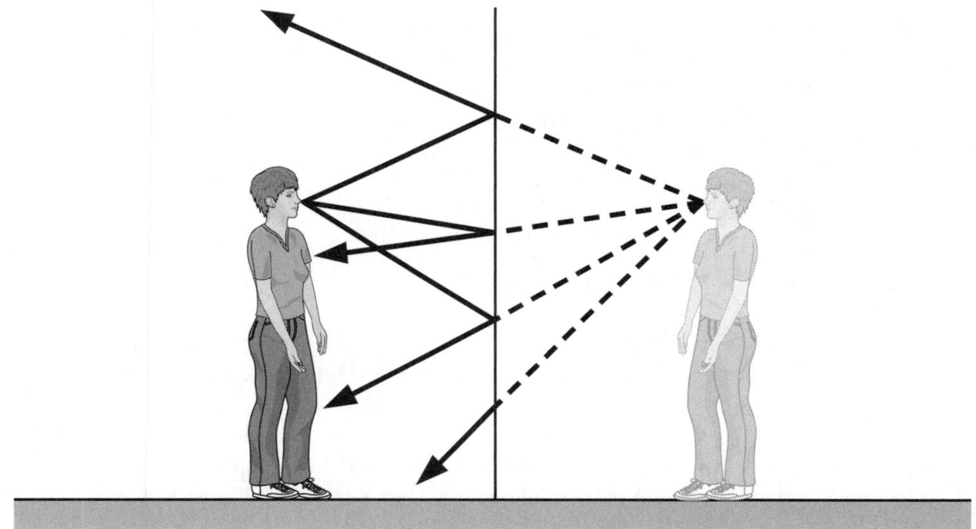

This allows us to deal with mirrors without working out all of the reflected angles. Instead, we locate the image, and pretend that the object of our attention is indeed there. The object is not there, of course, but all of the light that we see from the mirror is the same as if it were there. The location of this image is behind the mirror, the same distance from the mirror as the object but on the other side.

One myth that persists about mirrors is that they reverse left and right. This is not true. Imagine that you are reading this book and decide that you want to read its reflection in a mirror. To view the book in the mirror, you turn the book around — you reverse left and right by turning the book. The mirror reverses front and back, so to see the text of the book you have to turn it around. You could flip it upside down, in which case you also could see the text, and in doing so left and right would not be switched, because you didn't switch them.

EXAMPLE

Joanne (at the bottom) wants to take a picture of Betty's reflection. How far away does Betty (at the top) appear to be from Joanne (what focal length must she use for the camera lens)?

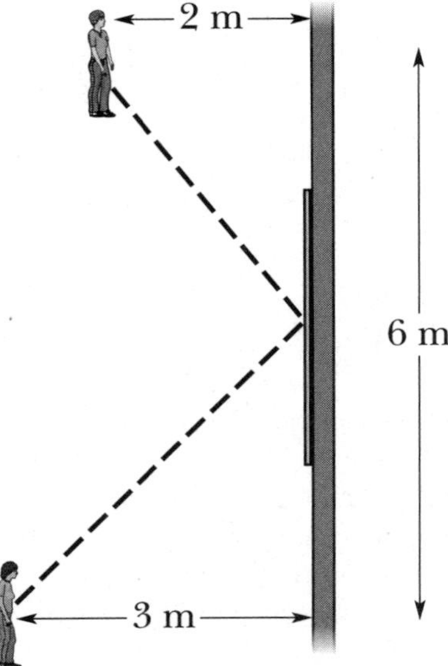

Student: We could draw the path of the light, and then set up similar triangles.
Tutor: We could, and it would work, but there's an easier way. Where is the *image* of Betty?
Student: Doesn't that depend on where Joanne is?
Tutor: No. Anyone looking at Betty's reflection will see it at the same place, the location of the image.
Student: Is that 2 m behind the mirror?
Tutor: Yes it is. How far is it from Betty's image to Joanne?
Student: In a straight line?

$$\sqrt{(6\text{ m})^2 + (5\text{ m})^2} = 7.8\text{ m}$$

Student: Is that it?
Tutor: Yes. Once we know where the image is, it's the same as if Betty was there. All of the light from Betty appears to come from there, just as if she was there and the mirror wasn't.
Student: It feels like cheating, somehow, to not do an extensive calculation.
Tutor: I felt that way for years, because I first saw this "solution" in a book of brain teasers. But I've learned that it works *because* images work that way — the light behaves as if it comes from the image.

What if the mirror is curved? There is still an image, but the location of the image is different than if the mirror was flat. The same is also true of lenses. The math is nearly the same for mirrors and lenses, so we'll deal with them together.

Every curved mirror has a focal point, and lenses have one on each side. When parallel beams of light reach the mirror or lens, they are deflected so that they all hit the focal point.

The equation that tells us where to find the image is

$$\frac{1}{f} = \frac{1}{p} + \frac{1}{i} \quad \text{or} \quad \frac{1}{f} = \frac{1}{d_o} + \frac{1}{d_i}$$

where f is the focal length, or distance from the mirror/lens to the focal point, p is the distance from the mirror/lens to the object (many books use d_o), and i is the distance from the mirror/lens to the image (many books use d_o). The magnification is the ratio of the size of the image to that of the object

$$m = \frac{h'}{h} = -\frac{i}{p} \qquad \text{or} \qquad m = \frac{h_i}{h_o} = -\frac{d_i}{d_o}$$

If we know any two of f, p, i, and m, we can find the other two.

A common stumbling block in dealing with mirrors/lenses is in handling the minus signs. Each of f, p, i, m, h', and h can be positive or negative, and there are meanings attached to each.

	+	−
f	converging	diverging
p or i	real	virtual
m	not flipped	flipped
h or h'	right side up	upside down

A mirror/lens can be either converging or diverging. A converging mirror/lens causes the light beams to converge more rapidly, or diverge less rapidly. A diverging mirror/lens causes the light beams to diverge more rapidly, or converge less rapidly. Concave mirrors and convex lenses are converging, and have positive focal lengths. Convex mirrors and concave lenses are diverging, and have negative focal lengths.

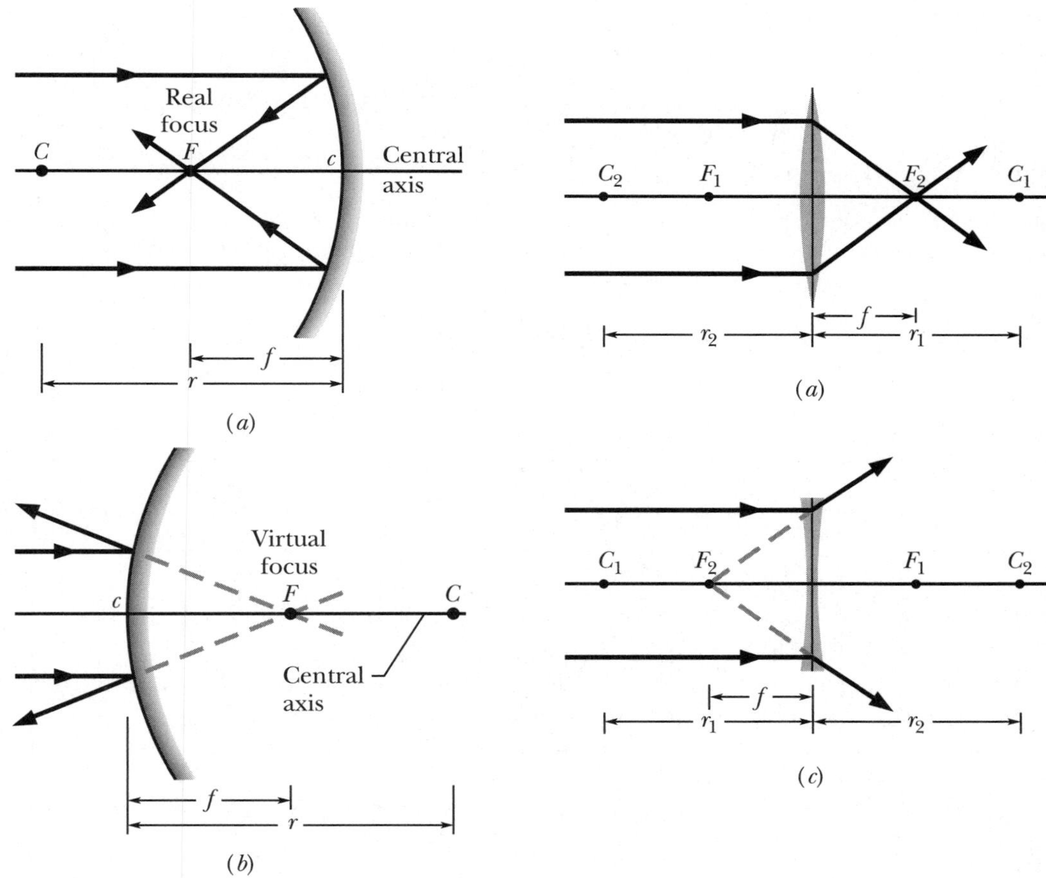

If you can hold a sheet of paper up where the image should be, and see the image on the piece of paper, then it is a real image with a positive image distance. If you can't ,then it is a virtual image with a negative image distance. Key to this is knowing which way is the positive direction for measuring the image distance. **Whatever direction the light goes when it leaves the mirror/lens is the positive direction for measuring the image distance**. Therefore, if the image is where the light goes, it is a real image. If the image is where the light doesn't go, it is a virtual image. You can still see a virtual image — you have to be looking at the mirror/lens, seeing the light that comes from it.

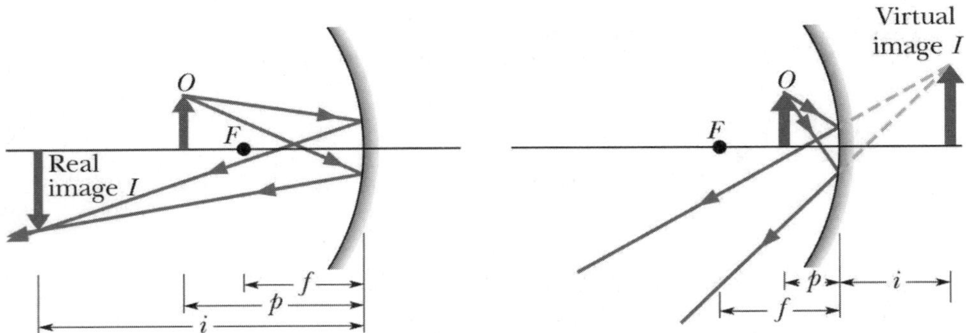

It is also possible to have a negative object distance. Whatever direction the light comes from as it reaches the mirror/lens is the positive direction for measuring the object distance. You can only have a negative object distance, meaning a "virtual object," if you have more than one mirror/lens.

Sometimes the image is flipped — slide projectors flip the image, so the slide must be inserted upside down so that the image is right side up. A negative magnification corresponds to a flipped image. A negative magnification does not mean an upside down image — the object might be upside down and the image right side up. If the absolute value of the magnification is greater than one, then the image is larger than the object, and if it is less than one, then the image is smaller than the object. Rear-view mirrors with the warning "objects in mirror are closer than they appear" create smaller images, so that you can see more, but may distort your mental perception if you do not account for the demagnification.

The last piece before we start some examples is how to make a mirror. If a mirror has a radius of curvature r, then the focal length is

$$f = \frac{1}{2}r$$

There is a similar but more difficult equation for lenses, which we won't deal with here.

EXAMPLE

The side mirror on a car is designed so that an upright truck 100 m behind the car will appear upright and half as big when viewed in the mirror. What is the radius of the mirror, and is it convex or concave?

Student: The image is half as big, so the magnification $m = 1/2$.
Tutor: Is it positive 1/2 or negative 1/2?
Student: The image needs to be upright, so the magnification is positive.
Tutor: For the magnification to be positive, the height h' needs to be positive. A positive magnification does not mean that the image is upright, or right side up.
Student: Aren't you being a little picky? If the image isn't flipped, then it's right side up, right?
Tutor: The object really could be upside down.
Student: Okay. The object car is right side up, and so is the image, so the image isn't flipped. That means that the magnification is positive, so it's $+1/2$.
Tutor: Good.

Student: I need another one. The object is the truck, and the truck is 100 m from the mirror, so the object distance is 100 m.
Tutor: Positive or ...
Student: The light comes from behind the car, and the truck is behind the car, so it's +100 m. Also, there is only one mirror or lens, so the object distance p has to be positive.
Tutor: Good. Now you should be able to solve.
Student: I can find the image distance, then find the focal length, and then the radius.

$$m = -\frac{i}{p}$$

$$+\frac{1}{2} = -\frac{i}{+100 \text{ m}} \quad \rightarrow \quad i = -50 \text{ m}$$

Tutor: What does the negative image distance mean?
Student: The table says that it's a virtual image.
Tutor: Where is the image?
Student: 50 meters from the mirror.
Tutor: But on which side? Which way is the positive direction for measuring the image?
Student: The mirror faces the back of the car, and the light that bounces off of the mirror goes toward the back of the car, so that's the positive direction for the image. The image distance is positive, so the image is 50 m toward the front of the car.
Tutor: Yes. The driver looks forward to see the image of the truck.
Student: If the image distance was +50 m, would he look behind him to see the image?
Tutor: The image would be 50 m behind the car, but he couldn't see it. If he looks backward, he wouldn't see the light that reflected off of the mirror.
Student: Oh.
Tutor: If there was a screen or piece of paper 50 m in front of the car, would there be an image on the screen?
Student: No, the light bounces off of the mirror and doesn't go there.
Tutor: Correct. That's why it's a virtual image.

$$\frac{1}{f} = \frac{1}{p} + \frac{1}{i}$$

$$\frac{1}{f} = \frac{1}{+100 \text{ m}} + \frac{1}{-50 \text{ m}} = \frac{1-2}{100 \text{ m}} = \frac{-1}{100 \text{ m}}$$

$$f = -100 \text{ m}$$

Student: Is there a connection between the 100 m to the truck and the 100 m focal length?
Tutor: No, it's coincidence; besides, it's −100 m.
Student: Negative focal length means diverging mirror.
Tutor: Yes.
Student: Is it true that a negative focal length causes a negative image distance so that a diverging mirror always creates a virtual image?
Tutor: Not necessarily. If p is positive and f is negative, then i has to be negative, so your statement is true if the object is a real object. If you have a virtual object, then your statement might not be true. We'll do one of those later.
Student: How do you remember convex and concave?
Tutor: A concave mirror/lens is one in which the bear can hide. Seriously, this is how I learned to remember the shapes. To remember that a concave mirror is converging, think about where parallel incoming beams of light would go — would they converge or diverge?
Student: So a diverging mirror is a convex mirror.
Tutor: Yes, but the opposite for lenses. A diverging lens is a concave lens.

$$f = \frac{1}{2}r$$

$$r = 2f = 2(-100 \text{ m}) = -200 \text{ m}$$

Student: That seems like a big radius.
Tutor: Pick a point 2 football fields in front of the car, and that's the point around which the mirror is centered. It's not a lot of curvature.

EXAMPLE

Find the object distance necessary for a +36–cm–focal–length lens to form a real image three times as large as the object.

Student: I have to know two of the four. The focal length f is +36 cm and the magnification is 3.
Tutor: Is the magnification positive or negative?
Student: It doesn't say, so it must be positive.
Tutor: It may not say explicitly, but that doesn't mean that the magnification is positive. What do you know about the object and image distances?
Student: That I can solve for them.
Tutor: But you don't know what the magnification is, so you could only determine two possible sets of values.
Student: There is only one lens, so it's a real object with a positive object distance p.
Tutor: Good.
Student: The problem says that the image is real, so it's positive.
Tutor: Now look at the equation for magnification $m = -i/p$, with what you've just said.
Student: If p and i are both positive, then m must be negative. m is -3.
Tutor: Good. Now you can do the problem.

$$m = -\frac{i}{p} \qquad \text{and} \qquad \frac{1}{f} = \frac{1}{p} + \frac{1}{i}$$

$$-3 = -\frac{i}{p} \qquad \rightarrow i = 3p$$

$$\frac{1}{(+36 \text{ cm})} = \frac{1}{p} + \frac{1}{3p} = \frac{3+1}{3p} = \frac{4}{3p}$$

Student: Do I need to convert the 36 cm to meters?
Tutor: No. It does help if you have common units for f, p, and i.

$$p = \frac{4}{3}(+36 \text{ cm}) = +48 \text{ cm}$$

Tutor: Signs are important, but sometimes they aren't explicit. If you wanted a virtual image, then you'd need a negative i, even if the problem didn't come right out and say so.

We can explain a plane mirror using the same equations. The radius of a plane mirror is infinite — the mirror gets flatter as the radius increases, and we increase the radius until the mirror is completely flat.

$$\overset{0}{\cancel{\frac{1}{\infty}}} = \frac{1}{p} + \frac{1}{i} \qquad \rightarrow \qquad i = -p$$

The light bounces off of the mirror, and the virtual image is behind the mirror, the same distance behind the mirror that the object is in front of it. The magnification is

$$m = -\frac{i}{p} = -\frac{-p}{p} = +1$$

so the image is the same size and unflipped.

We can likewise examine a flat piece of glass. The focal length of a flat piece of glass is infinite. Again, $i = -p$ and $m = +1$. The virtual image is on the opposite side of the glass that the light goes, but the same side that it comes from, and the same distance away as the object — it is at the object.

EXAMPLE

A 2.4–mm–high object is 15 cm in front of a +10 cm convex lens. 24 cm behind the convex lens is a −10 cm concave lens. What is the size of the final image?

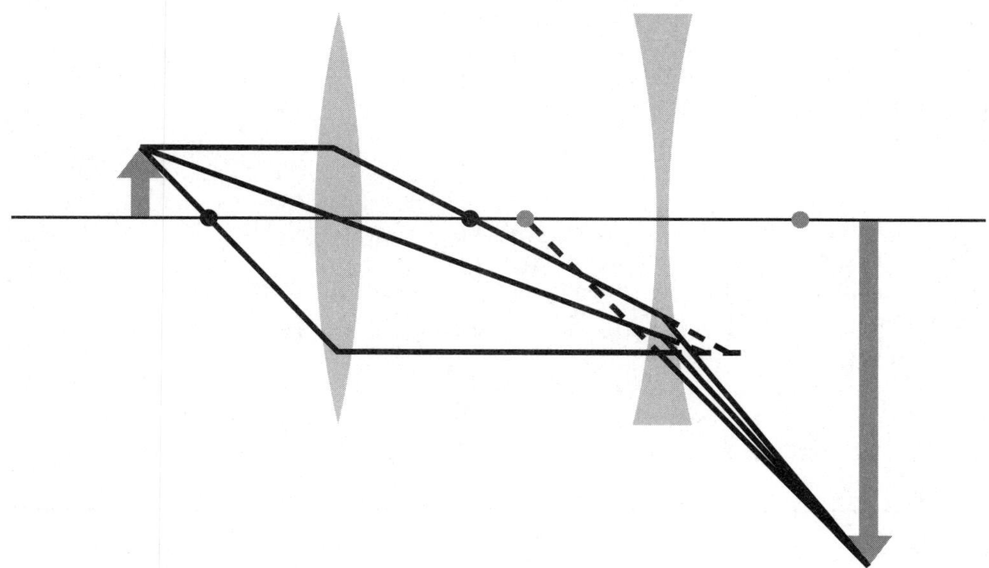

Student: We have two lenses, so we might have a negative object distance.
Tutor: Correct. Take the lenses one at a time.
Student: Okay, the first lens. The light comes from the left, so the left is the positive direction for measuring the object. The object is on the left, so the object distance p is positive 15 cm. The focal length is +10 cm. I have two of the four, so I can solve for the other two.

$$m = -\frac{i}{p} \qquad \text{and} \qquad \frac{1}{f} = \frac{1}{p} + \frac{1}{i}$$

$$\frac{1}{+10 \text{ cm}} = \frac{1}{+15 \text{ cm}} + \frac{1}{i}$$

$$\frac{1}{i} = \frac{1}{+10 \text{ cm}} - \frac{1}{+15 \text{ cm}} = \frac{3-2}{30 \text{ cm}} = \frac{1}{30 \text{ cm}}$$

$$i = +30 \text{ cm}$$

Student: The light that leaves the first lens goes to the right of the first lens, so the image from the first lens is 30 cm to the right of the first lens.

$$m_1 = -\frac{i}{p} = -\frac{+30 \cancel{\text{cm}}}{+15 \cancel{\text{cm}}} = -2$$

Tutor: So far so good. Now we do the second lens.
Student: Is the object distance for the second lens +39 cm?
Tutor: No. The object for the second lens is the image created by the first lens. That's where the light

into the second lens appears to come from.
Student: But the image from the first lens is on the right side of the second lens.
Tutor: And the light into the second lens comes from the left, so it's a negative object distance. The light into the second lens doesn't appear to be coming from a spot, but going to a spot 6 cm to the right of the second lens.
Student: So the object distance for the second lens is −6 cm?
Tutor: Yes.

$$\frac{1}{-10\text{ cm}} = \frac{1}{-6\text{ cm}} + \frac{1}{i}$$
$$\frac{1}{i} = \frac{1}{-10\text{ cm}} - \frac{1}{-6\text{ cm}} = \frac{3-5}{-30\text{ cm}} = \frac{-2}{-30\text{ cm}}$$
$$i = +15\text{ cm}$$

Student: The final image is a real image, 15 cm to the right of the second lens — right because that's where the light goes when it leaves the second lens.
Tutor: Very good.

$$m_2 = -\frac{i}{p} = -\frac{+15\,\cancel{\text{cm}}}{-6\,\cancel{\text{cm}}} = +2.5$$

Student: How do I combine the magnifications?
Tutor: The first image is $-2\times$ the first object, then it becomes the second object, and the second image is $2.5\times$ the second object.
Student: So I multiply.

$$(2.4\text{ mm}) \times (-2) \times (+2.5) = -12\text{ mm}$$

Student: And the negative height means that it's upside down.

Chapter 35

Interference

When we looked waves and sound, we encountered interference. Interference takes place when two waves add. For light, this usually involves two beams of light that are parallel, or nearly so.

If you have two identical waves travelling side-by-side, they add and you get one wave that's twice the size. If you have two waves, identical except that one is half a wavelength behind the other, or half a period in time, then they add to zero. This is like adding 3 and -3. If two waves are identical except that one is one wavelength behind the other, then it looks the same as the first wave and they again add to give a wave twice as large. The general idea is that if one wave is an integer number of waves behind the first, we have constructive interference, and the sum is twice the amplitude. If the second wave is a half-integer (0.5, 1.5, 2.5, ...) behind the first, then we have destructive interference, and the sum is zero.

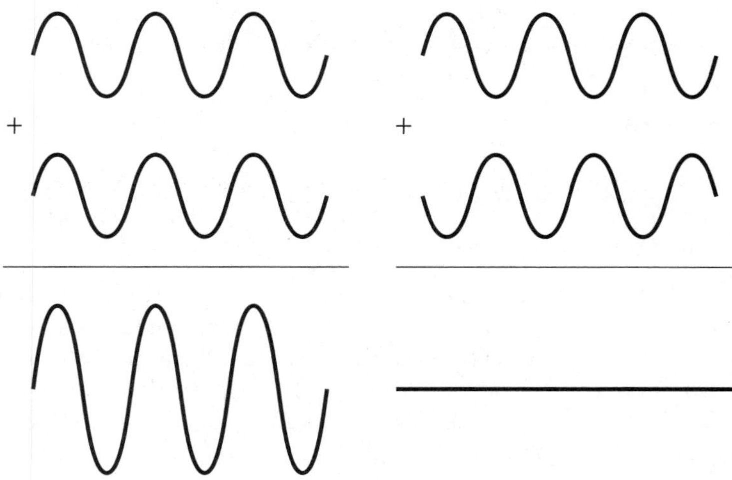

Typically the two light waves start at the same time, so one is behind the other because it has a greater distance to travel. We don't care what the two distances are, but we do care what the difference in the two distances ΔL is. There are many geometries we can use to create the difference in the two path lengths, but regardless of geometry,

$$\Delta L = \begin{cases} m\lambda & \text{constructive interference} \\ (m+\frac{1}{2})\lambda & \text{destructive interference} \end{cases}$$

where m is an integer, any integer.

EXAMPLE

A Michelson interferometer is shown below. When one of the mirrors is moved by 1.407 μm, the pattern shifts by 6 fringes. What is the wavelength of the light?

Student: Shouldn't the waves either cancel or not? The viewer should see bright or dark.
Tutor: In reality, there are fringes due to slight differences in the paths to the telescope. When the difference in the path length changes by one wavelength, the fringes shift by one.
Student: So the question is: What do I have to do to change the path length difference by a wavelength?
Tutor: Yes.
Student: Do I just move one of the mirrors or both?
Tutor: Typically one mirror is fixed and we move the other.
Student: Do I move it a wavelength?
Tutor: The light has to go out and back. If you move one mirror out by a wavelength, then that path increases by two wavelengths and the other doesn't change.
Student: So the difference ΔL would increase by 2λ. So one fringe corresponds to moving one mirror by $\frac{1}{2}\lambda$.
Tutor: Yes.
Student: We saw six fringes, so we moved one mirror by $6 \times \frac{1}{2}\lambda = 3\lambda$.

$$3\lambda = 1.407\ \mu\text{m}$$

$$\lambda = \frac{1.407\ \mu\text{m}}{3} \times \frac{10^{-6}}{1\mu} \times \frac{1\text{ n}}{10^{-9}} = 469\text{ nm}$$

Tutor: It's handy to remember the wavelengths of light. Visible light goes from 400 nm to 700 nm. If you're using visible light, then you should expect the result to be in that range.
Student: And it is.

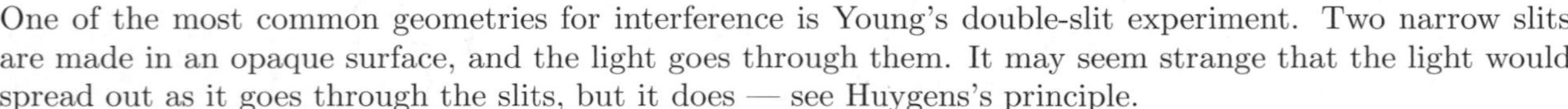

One of the most common geometries for interference is Young's double-slit experiment. Two narrow slits are made in an opaque surface, and the light goes through them. It may seem strange that the light would spread out as it goes through the slits, but it does — see Huygens's principle.

When the light goes straight through, the light from the two slits goes the same distance, the difference is zero, and the light is bright. We call this a bright fringe or a maximum. As you go a little ways away from the center, light from one of the slits has to go further than light from the other. When this difference reaches $\Delta L = \frac{1}{2}\lambda$, the two lights cancel and there is no light. We call this a dark fringe or a minimum. If we go further, so that $\Delta L = \lambda$, we get to another maximum. Maxima (plural of maximum) occur when $\Delta L = m\lambda$ and minima at $\Delta L = (m + \frac{1}{2})\lambda$.

The difference in the two path lengths for this geometry is

$$\Delta L = d \sin\theta$$

where d is the distance or separation between the two slits. So

$$d\sin\theta = \Delta L = \begin{cases} m\lambda & \text{constructive interference} \\ (m+\frac{1}{2})\lambda & \text{destructive interference} \end{cases}$$

To find the position y on the screen of the maxima and minima, we use trigonometry:

$$y = D\tan\theta$$

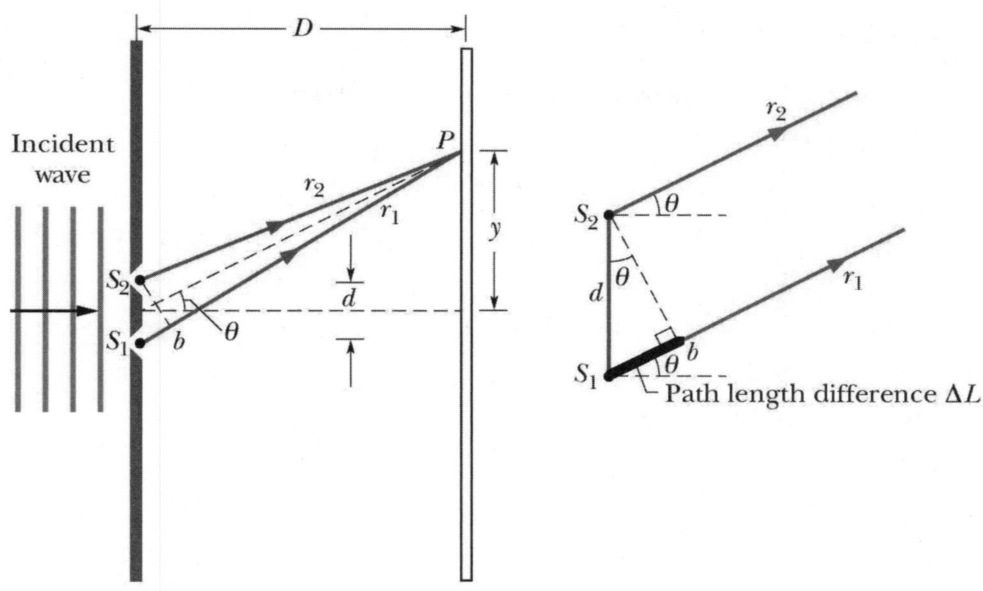

where D is the distance to the screen. Often in double-slit experiments the angle θ is small, so we can use

$$\theta(\text{in radians}) \approx \sin\theta \approx \tan\theta \qquad \text{when } \theta \text{ is small}$$

To check whether the angle is small, check for either $\lambda \ll d$ or $y \ll D$, where $\ll$ typically means 100 times less than.

EXAMPLE

Light of wavelength 562 nm is incident on two slits with slit spacing 80 μm. On a screen 2.6 m away, how far from the central maximum is the third minimum?

Student: I use $d \sin\theta = \Delta L = (m + \frac{1}{2})\lambda$ for a minimum, with $m = 3$ for the third one.
Tutor: What is m for the central maximum?
Student: At the center, $\Delta L = 0$, $\theta = 0°$, and $m = 0$.
Tutor: Good. What's m for the first maximum?
Student: The central maximum isn't the first one?
Tutor: No. The central maximum is the "zeroth" one, with $m = 0$.
Student: So the first maximum is at $m = 1$, and the second maximum is at $m = 2$.
Tutor: Yes. What is at $m = \frac{1}{2}$, between the zeroth and first maxima?
Student: m has to be an integer.
Tutor: Okay. The central maximum is at $\Delta L = 0$ and the first maximum is at $\Delta L = \lambda$. What is at $\Delta L = \frac{1}{2}\lambda$?
Student: So that's what you mean by $m = \frac{1}{2}$. There is a minimum there.
Tutor: Yes. What would you call it — the first minimum?
Student: Sure.
Tutor: So the first minimum occurs at $m = 0$ and $\Delta L = (0 + \frac{1}{2})\lambda$.
Student: And the second minimum is at $\Delta L = (1 + \frac{1}{2})\lambda$, and the third at $\Delta L = (2 + \frac{1}{2})\lambda$.

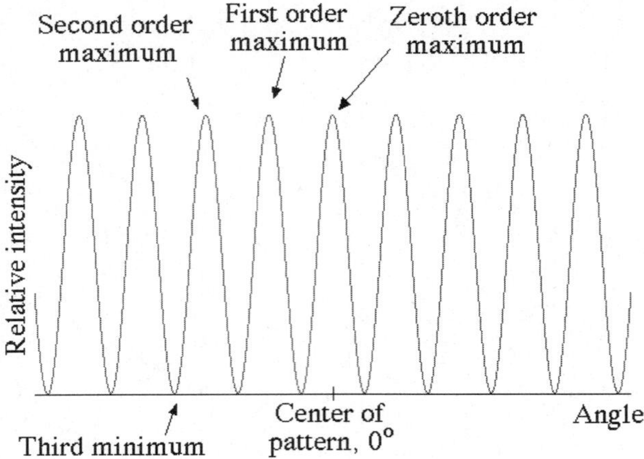

Student: Then the angle is

$$d \sin\theta = \Delta L = \left(2 + \tfrac{1}{2}\right)\lambda$$

Tutor: Before doing the arcsine, is the angle small?
Student: How small?
Tutor: Small enough to use $\sin\theta \approx \tan\theta$, as in $\lambda \ll d$.
Student: λ is 562 nm, or 0.562 μm. d is 80 μm, so λ is less than a hundredth of d.
Tutor: So the angle is small.

$$\frac{\frac{5}{2}\lambda}{d} = \sin\theta \approx \tan\theta = \frac{y}{D}$$

$$y \approx \frac{\frac{5}{2}\lambda D}{d} = \frac{\frac{5}{2}(0.562\ \mu\text{m})(2.6\ \text{m})}{(80\ \mu\text{m})} = 0.0457\ \text{m} = 45.7\ \text{mm}$$

EXAMPLE

Light of wavelength 562 nm is incident on two slits with slit spacing 19 μm. How many interference maxima are there in the whole pattern?

Student: Why should there be a limit?
Tutor: As you go to higher values of m, you get larger angles. Eventually you get to an angle that you can't use.
Student: You mean past 90°?
Tutor: Yes. How high would m have to be so that the angle θ is 90°?

$$d \sin\theta = \Delta L = m\lambda$$

$$m = \frac{d \sin 90^\circ}{\lambda} = \frac{(19\ \mu\text{m})(1)}{(0.562\ \mu\text{m})} = 33.8$$

Student: Of course, m has to be an integer, so we round.
Tutor: m does have to be an integer. If $m = 33.8$, the angle θ is 90°. What would the angle be if we used $m = 34$?
Student: It would be more than 90°.
Tutor: No. You would need to find an angle such that the sine of the angle is greater than 1.
Student: You can't have a sine greater than 1.

Tutor: So you can't use $m = 34$.
Student: So we truncate rather than rounding.
Tutor: Yes. We get the 1st through 33rd interference maxima on each side, plus the central maxima for $m = 0$.
Student: So the total is

$$33 + 1 + 33 = 67$$

Student: Is there some reason?
Tutor: To get the 12th interference maximum, you need light from one slit to be 12 wavelengths behind the other. The slits here are only 33.8 wavelengths apart, so the greatest path length difference we can get is 33.8λ. You can't get a path length difference of 34 wavelengths, so you can't get the 34th interference maximum.

Remember that a general principle of waves is that when a wave reaches a change of medium, some of the wave is transmitted and some is reflected. If a beam of light strikes a piece of glass, some of the light is reflected from both the front and the back surfaces of the glass. We will assume that the light hits at normal incidence, despite the figure.

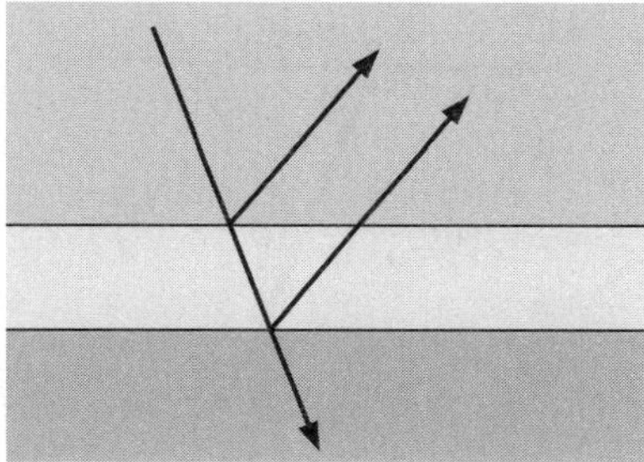

One beam goes $2t$ further than the other, where t is the thickness of the glass. (Most glass is too thick and doesn't have a uniform enough thickness — we usually observe this with much thinner films.) Because the extra distance occurs in the film, we need the extra distance to be equal to an integer number of wavelengths, using the wavelength in the film.

$$\lambda_n = \frac{\lambda}{n}$$

So

$$2t = \Delta L = \begin{cases} m\frac{\lambda}{n} & \text{constructive interference} \\ (m + \frac{1}{2})\frac{\lambda}{n} & \text{destructive interference} \end{cases}$$

There is one more change in thin films. When a wave reflects off of a surface in which it would go slower than it would in the medium it is in, the wave is flipped. So if the material that the wave bounces off of has a greater n, the wave is flipped. If both waves or neither wave is flipped, then nothing changes. But if one is flipped and the other isn't, then what would have been constructive interference becomes destructive, and vice versa.

EXAMPLE

A thin film of soap is stretched across a coat hanger. The film is 575 nm thick and has an index of refraction of 1.24. Light reflects normally off of the film. Find all wavelengths of visible light for which the reflection is especially bright.

Student: I start with the formula.

$$2t = \Delta L = \begin{cases} m\frac{\lambda}{n} & \text{constructive interference} \\ (m+\frac{1}{2})\frac{\lambda}{n} & \text{destructive interference} \end{cases}$$

Tutor: Okay, but make a quick drawing.
Student: Really?
Tutor: You need to determine whether the waves flip on reflection, so a quick drawing is important.
Student: Okay. The light is in air, with $n = 1.00$. It bounces off of soap, with $n = 1.24$. The thing it bounces off of has a greater n than what it's in, so it is flipped.
Tutor: Very good.
Student: Is the transmitted light also flipped?
Tutor: No. The transmitted light isn't flipped as it moves into the second medium, regardless of whether the reflected light is flipped.
Student: Okay. At the second surface, the light is travelling in soap, with $n = 1.24$. It bounces off of air, with $n = 1.00$. The thing it bounces off of has a lesser n than what it's in, so it is not flipped.
Tutor: Good.

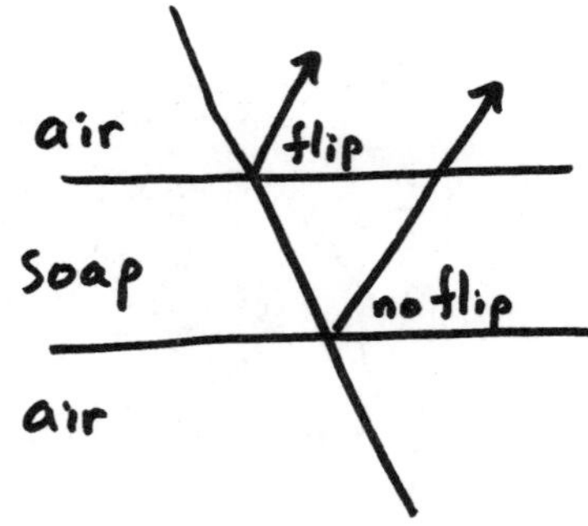

Student: One of the waves is flipped and the other isn't, so the equations switch.

$$2t = \Delta L = \begin{cases} m\frac{\lambda}{n} & \text{\sout{constructive} destructive interference} \\ (m+\frac{1}{2})\frac{\lambda}{n} & \text{\sout{destructive} constructive interference} \end{cases}$$

Tutor: Good.
Student: I want brighter light, which means constructive interference.

$$2t = (m+\tfrac{1}{2})\frac{\lambda}{n}$$

$$\lambda = \frac{2nt}{(m+\frac{1}{2})}$$

Student: What value do I use for m?
Tutor: All of them.
Student: All? That could take a while.
Tutor: Try it.

$$\lambda = \frac{2nt}{(m+\frac{1}{2})} = \frac{2(1.24)(575\text{ nm})}{(m+\frac{1}{2})} = \frac{1426\text{ nm}}{(m+\frac{1}{2})}$$

$$m = 0 \quad \rightarrow \quad \lambda = \frac{1426\text{ nm}}{0.5} = 2852\text{ nm}$$

$$m = 1 \quad \rightarrow \quad \lambda = \frac{1426\text{ nm}}{1.5} = 951\text{ nm}$$

$$m = 2 \quad \rightarrow \quad \lambda = \frac{1426 \text{ nm}}{2.5} = 570 \text{ nm}$$

$$m = 3 \quad \rightarrow \quad \lambda = \frac{1426 \text{ nm}}{3.5} = 407 \text{ nm}$$

$$m = 4 \quad \rightarrow \quad \lambda = \frac{1426 \text{ nm}}{4.5} = 317 \text{ nm}$$

Tutor: Which of these correspond to visible light?
Student: The first two are infrared, and the last one is ultraviolet. Only 570 nm and 407 nm are visible light.
Tutor: Do you need to keep going?
Student: No. As m gets even bigger, the wavelength will get smaller, and it's already too small for visible light.

EXAMPLE

Two pieces of glass are placed with a thin wire separating them at one end. There are 114 bright fringes between the left end and the wire, when using 632 nm light. Find the diameter of the wire.

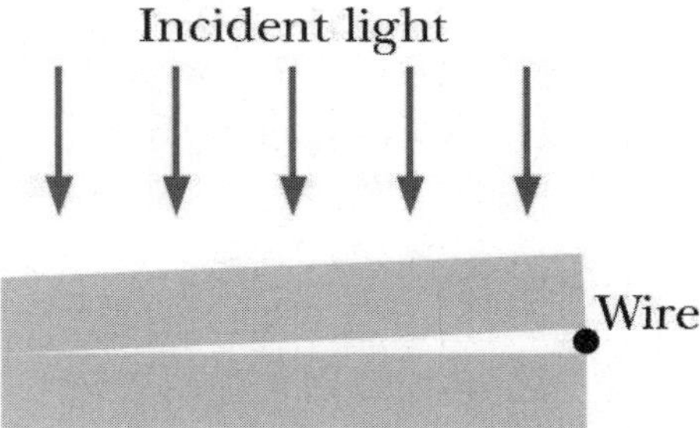

Student: What does the diameter of the wire have to do with fringes?
Tutor: The light reflects off of the top and bottom of each piece of glass, a total of four surfaces. The light from the top and bottom of the top surface interfere constructively or destructively, or something in between, but whatever it is it doesn't change. The light that reflects off of the bottom of the top piece of glass, and the light that reflects off the top of the bottom piece of glass, also interfere. This interference will change from constructive to destructive to constructive and so on as the thickness of the "air gap" increases.
Student: So the *changing* interference causes fringes.
Tutor: Yes. Like the two-slit experiment, when we move away from the center of the pattern, the path length difference increases and we see fringes.

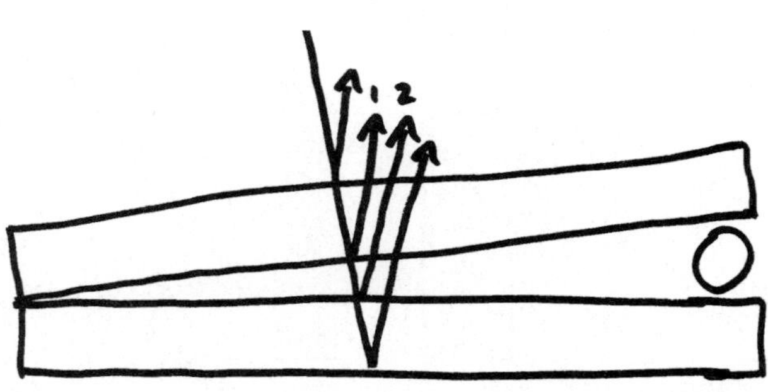

Student: So the first thing I do is write down the equation.

$$2t = \Delta L = \begin{cases} m\frac{\lambda}{n} & \text{constructive interference} \\ (m+\frac{1}{2})\frac{\lambda}{n} & \text{destructive interference} \end{cases}$$

Student: Then I make a little drawing.

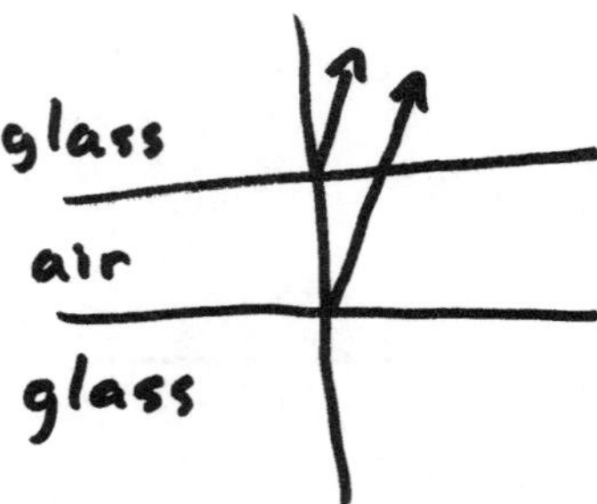

Student: The top beam is traveling in glass ($n \approx 1.5$) and bounces off of air ($n \approx 1.0$), so it is not flipped. The bottom beam is traveling in air and bounces off of glass, so it is flipped. One is flipped and the other is not, so the equations are flipped.

$$2t = \Delta L = \begin{cases} m\frac{\lambda}{n} & \cancel{\text{constructive}}\ \text{destructive interference} \\ (m+\frac{1}{2})\frac{\lambda}{n} & \cancel{\text{destructive}}\ \text{constructive interference} \end{cases}$$

Tutor: Doing well. At the far left end, is it dark or bright?
Student: That would be the top equation, so it's dark. I'm not sure that makes sense.
Tutor: The second beam goes no extra distance, so there is no change from that. One of the beams is flipped, so the two beams are opposites and add to zero.
Student: Now I need to relate the fringes to the thickness t, which is equal to the diameter.
Tutor: The first dark fringe is at $m = 0$ and $t = 0$. Where is the first bright fringe?
Student: At $m = 0$ in the bottom equation. So the 114th bright fringe is at $m = 113$ in the bottom equation.

$$2t = (113 + \tfrac{1}{2})\frac{\lambda}{n}$$

Student: What value should I use for n?
Tutor: What material is there where the extra distance takes place?
Student: Air. So I use $n = 1.0$.

$$t = \tfrac{1}{2}(113.5)\lambda = \tfrac{1}{2}(113.5)(632\text{ nm}) = 35866\text{ nm} = 35.9\ \mu\text{m}$$

Chapter 36

Diffraction

In the previous chapter we examined interference from two sources. It is possible to have interference from more than two sources.

A diffraction grating (or interference grating) is a series of slits, so that there is interference from 100 or more sources. If the slits are evenly spaced, then the math is the same as with two slits. If the light from the second slit is one wavelength behind that from the first, then light from the third is one wavelength behind that from the second, and so on. The maxima are in the same place, but are narrower because, if you go a little ways off from the maximum, there are more sources and they cancel each other out sooner.

A more interesting question is what to do with a single slit. It may seem that with only one source, there should be no interference. But light from the far edge of the slit has to go a little further than light from the near edge of the slit. To deal with one slit properly, we need to divide it into an infinite number of slits. When the interference is from different points in the same slit, we call it diffraction.

Consider the point in the middle of the slit. If the light from here goes half a wavelength further than light from the near edge, then they will cancel. Working across the slit, light from each point in the near half is canceled by light from a point in the far half of the slit. All of the light cancels, and we get a minimum. This occurs at $\frac{a}{2}\sin\theta = \frac{\lambda}{2}$, where a is the slit width. In general, the diffraction minima occur at

$$a\sin\theta = m\lambda$$

This equation is similar to the one used in two-slit interference, except then it was for interference (two or more slit) maxima, and here it is for diffraction (one slit) minima.

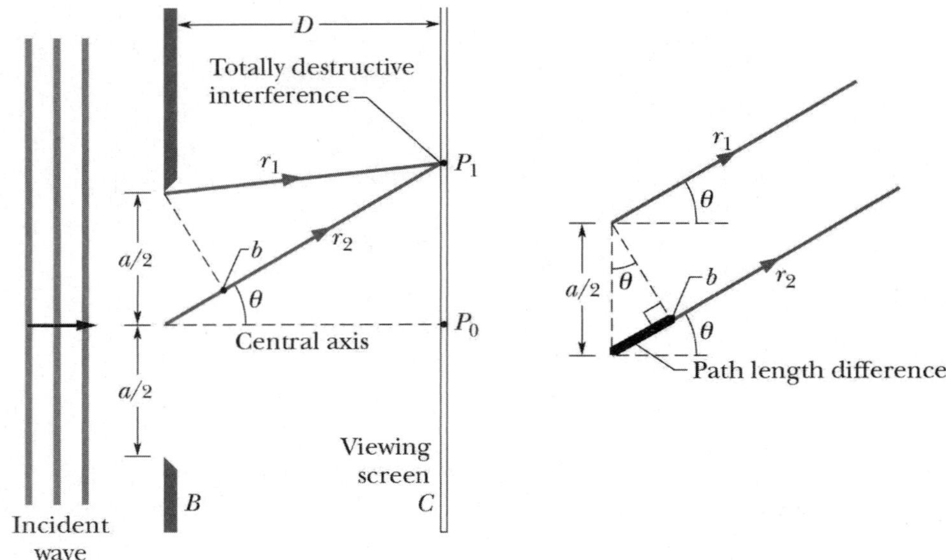

EXAMPLE

Light of wavelength 570 nm goes through a slit of width 30 μm. What is the width of the central diffraction maximum on a screen 1.5 m away?

Student: What's the "central diffraction maximum?"
Tutor: At the middle of the pattern on the screen, light from all points in the slit travels the same distance, so they all interfere constructively, and the light is the brightest. If you move away from the center to the spot where $a \sin\theta = \lambda$, then the light from the bottom half of the slit cancels the light from the top half and there is a minimum.
Student: Wait a minute. In single-slit diffraction, the minima occur at $a \sin\theta = m\lambda$. If $m = 0$ then $\theta = 0$, so shouldn't there be a minimum in the center of the pattern?
Tutor: A good question. In diffraction the minima occur at every integer value of m *except* $m = 0$. There has to be a maximum there because all of the light travels the same distance. Because there is no minimum at $m = 0$, the central maximum is twice as wide as the others.
Student: So there is a maximum at $\theta = 0$, which is $m = 0$. How do I determine the width of this central diffraction maximum?
Tutor: The edge of the maximum is where the minima are.
Student: So I need to find the minimum corresponding to $m = 1$?
Tutor: Yes.
Student: And they are at

$$a \sin\theta = \cancelto{1}{m}\lambda$$

Tutor: Yes. There is one on each side, the same distance away from the center.
Student: So I can find one and double it. To find the distance on the screen I use

$$y = D \tan\theta$$

Student: Is the angle small enough to use θ(in radians) $\approx \sin\theta \approx \tan\theta$?
Tutor: How do you check?
Student: With two-slit interference, we compared the wavelength to the slit separation d.
Tutor: So with single-slit diffraction, compare the wavelength to the slit width a. If $\lambda \ll a$, then $\sin\theta$ is small.

Student: So the ratio λ/a is

$$\frac{\lambda}{a} = \frac{0.57\ \mu\text{m}}{30\ \mu\text{m}} = \frac{1}{53} = 0.019$$

Student: It needs to be 1/100 or less, doesn't it?
Tutor: Try it both ways and see.
Student: If I make the approximation, then

$$\frac{y}{D} = \tan\theta \approx \sin\theta = \frac{\lambda}{a}$$

$$\text{width} = 2y = 2\frac{\lambda D}{a} = 2\frac{(0.57\ \cancel{\mu\text{m}})(1.5\ \text{m})}{(30\ \cancel{\mu\text{m}})} = 0.057\ \text{m} = 57\ \text{mm}$$

Tutor: What if you don't do the approximation?
Student: Then I have to use arcsine to find the angle.

$$\sin\theta = \frac{\lambda}{a} = \frac{0.57\ \mu\text{m}}{30\ \mu\text{m}} = 0.019$$

$$\theta = \arcsin(0.019) = 1.0887^\circ$$

$$\text{width} = 2y = 2D\tan\theta = 2(1.5\ \text{m})(0.0190034) = 0.0570103\ \text{m} = 57.0103\ \text{mm}$$

Student: So the difference is that with the approximation, the width is 57.0 mm instead of 57.0103 mm.
Tutor: Correct. 57.0103 mm is the true width of the central diffraction minimum.
Student: So I didn't need to be worried that the angle was too big?
Tutor: It depends on how much precision you want. The difference between sine and tangent reaches 1% at about 8° or 0.14 radians. If two significant figures is enough, then you need $\frac{\lambda}{a} < 0.14$. To keep the error below 0.1%, you need an angle less than 0.04 radians. Your angle was less than 0.02 radians.
Student: So if the slit width is $1/0.04 = 25$ times as large as the wavelength, then I can use the approximation and my error will be less than 0.1%?
Tutor: Yes, using 1/100 is more restrictive than you need to be.

Having determined the location of the minima, we now ask what the intensity is at any spot on the screen. For two slits, each of width a and separated by d, the intensity is given by

$$I(\theta) = I_m \left(\frac{\sin\alpha}{\alpha}\right)^2 (\cos^2\beta)$$

where

$$\alpha = \frac{\pi a}{\lambda}\sin\theta \qquad \text{and} \qquad \beta = \frac{\pi d}{\lambda}\sin\theta$$

I_m is the maximum intensity, found at the center of the pattern ($\theta = 0$). This may seem mysterious, but is not that bad.

The $\cos^2\beta$ factor is the effect of the interference of the light from the two slits. At the center of the pattern, $\beta = 0$ and $\cos^2\beta = 1$. When $\beta = \pi/2$, $\cos^2\beta = 0$, so the intensity is zero, corresponding to the first interference minimum. This happens at

$$\frac{\pi}{2} = \beta = \frac{\pi d}{\lambda}\sin\theta \quad \rightarrow \quad \frac{1}{2}\lambda = d\sin\theta$$

That is, $\cos^2\beta$ goes to zero at the angle where we know the first interference minimum is. The first interference minimum always occurs when $\cos^2\beta = \pi/2$, regardless of the wavelength and slit separation, because β uses the ratio of these two. Interference maxima occur whenever $\cos^2\beta = 1$, or at $\beta = 0, \pi, 2\pi, \ldots$, which is at

$d\sin\theta = 0, \lambda, 2\lambda, \ldots$ Interference minima occur whenever $\cos^2\beta = 0$, or at $\beta = \frac{\pi}{2}, \frac{3\pi}{2}, \frac{5\pi}{2}, \ldots$, which is at $d\sin\theta = \frac{1}{2}\lambda, \frac{3}{2}\lambda, \ldots$

The $\left(\frac{\sin\alpha}{\alpha}\right)^2$ factor is the effect of diffraction of each of the slits. At the center of the pattern, $\alpha = 0$ and $\left(\frac{\sin\alpha}{\alpha}\right)^2 = 1$ (as a limit). When $\alpha = \pi$, $\sin\alpha = 0$, so the intensity is zero, corresponding to the first diffraction minimum. This happens at

$$\pi = \alpha = \frac{\pi a}{\lambda}\sin\theta \quad \rightarrow \quad \lambda = a\sin\theta$$

That is, $\sin\alpha$ goes to zero at the angle where we know the first diffraction minimum is. The first diffraction minimum always occurs when $\sin\alpha = \pi$, regardless of the wavelength and slit widths, because α uses the ratio of these two. Diffraction minima occur whenever $\sin\alpha = 0$, or at $\alpha = 0, \pi, 2\pi, \ldots$, which is at $a\sin\theta = \lambda, 2\lambda, \ldots$ The diffraction maxima are close to $\alpha = \frac{\pi}{2}, \frac{3\pi}{2}, \frac{5\pi}{2}, \ldots$, but not at exactly these spots.

Consider a point in the pattern where $\cos^2\beta = 1$ and $\sin\alpha = 0$. The $\cos^2\beta$ part says that the point is an interference maximum — that the light from the two slits is in phase, because light from one slit goes an integer number of wavelengths further than light from the other. The $\sin\alpha$ part says that the point is a diffraction minimum — that for each slit the light from part of the slit cancels the light from the other part, so that there is no light from the slit. The light from the two slits would interfere constructively, except that there is no light from either slit. The intensity is $I = I_m(0/\alpha)(1) = 0$.

The α in the denominator causes the pattern to lose intensity as we move away from the center of the pattern. In the first diffraction maximum, light from the middle third of the slit cancels light from the top third, and only light from the bottom third remains. In the second diffraction maximum, light from the second fifth of the slit cancels light from the top fifth, light from the fourth fifth cancels light from the third fifth, and only light from the bottom fifth remains. As we continue, each diffraction maximum is less intense than the one before.

In the last chapter, we treated the slits as being infinitesimally narrow. This is a practical impossibility, since no light would go through such a narrow slit, but it works well mathematically. When $a \rightarrow 0$, then α becomes zero regardless of the angle, and $\left(\frac{\sin\alpha}{\alpha}\right)^2 = 1$ regardless of the angle. Light diffracts out evenly in all directions. We had to take only interference into account.

EXAMPLE

Light of wavelength 614 nm goes through two slits with slit width 73 μm and slit separation 214 μm. What is the intensity of the second interference maximum, compared to the intensity at the center of the pattern?

Student: So I find the angle θ, then calculate α and β.
Tutor: There's an easier way. What's β?
Student: Don't I need to know the angle?
Tutor: β is zero at the center and π at the first interference maximum, regardless of d, λ, or θ.
Student: So $\beta = 2\pi$.
Tutor: Yes. What's α?
Student: Presumably you'll say I don't need the angle for that one either.
Tutor: Yes, you don't. What's the ratio of β/α?
Student: Most of the stuff cancels.

$$\frac{\beta}{\alpha} = \frac{\frac{\pi d}{\lambda}\sin\theta}{\frac{\pi a}{\lambda}\sin\theta} = \frac{d}{a}$$

Tutor: You can use the ratio to find α.

$$\alpha = \frac{a}{d}\beta = \frac{73\ \mu\text{m}}{214\ \mu\text{m}}(2\pi) = 2.143$$

Student: That *was* easy. Does it always work like that?
Tutor: If you know one of α and β, then you can use the ratio to find the other.
Student: That must mean something.
Tutor: It does. The ratio d/a tells us what the last interference maximum inside the central diffraction maximum is. The ratio here is $214/73 = 2.93$, so the second interference maximum is inside the edge of the central diffraction maximum, but the third interference maximum just misses.. This is true regardless of wavelength. If we increased the wavelength, the pattern would get narrower but would keep the same shape and intensity ratios, the whole pattern getting narrower together.
Student: Interesting. Now I can calculate the intensity.

$$I(\theta) = I_m \left(\frac{\sin\alpha}{\alpha}\right)^2 \left(\cos^2\beta\right)$$

$$\frac{I}{I_m} = \left(\frac{\sin(2.143)}{2.143}\right)^2 \left(\cos^2(2\pi)\right)$$

Tutor: When you take the sine of α, remember to have your calculator in radians mode.
Student: Is that important?
Tutor: Yes. The equation was derived with α and β in radians.
Student: Does θ have to be in radians, if we had found θ?
Tutor: It can be in whatever you want, and have your calculator in the correct mode. A hint that α and β are in radians is the π in the equation for them.

$$\frac{I}{I_m} = (0.392)^2\,(1.0) = 0.154 = 15.4\%$$

Student: So the second interference maximum has 15% of the intensity of the central interference maximum, regardless of the wavelength of the light.
Tutor: Yes. It is determined by the ratio d/a.

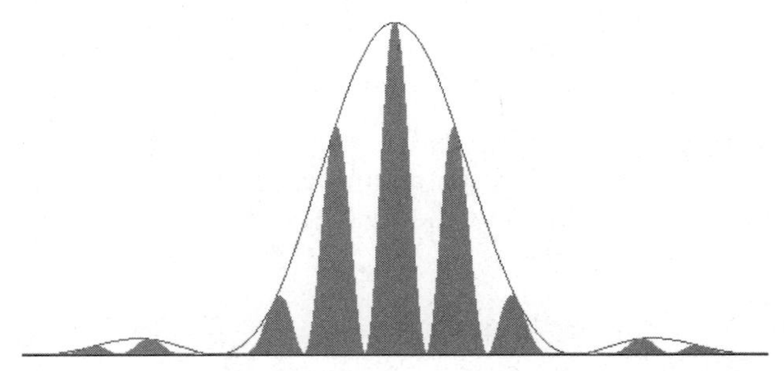

EXAMPLE

The figure shows intensity as a function of angle (0° in the center) for a number of slit arrangements. Which figure corresponds to

- a single narrow ($a \ll \lambda$) slit?
- two slits with slit separation three times the slit width ($d = 3a$)?
- many narrow ($a \ll \lambda$) slits?

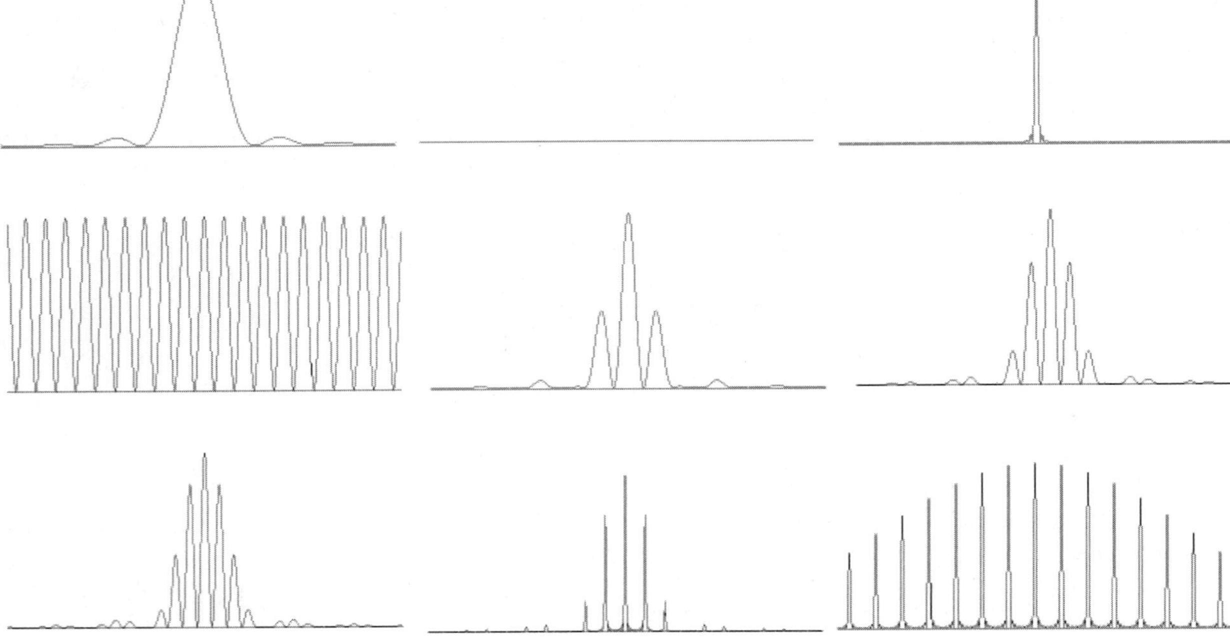

Student: What kind of evil bonehead question is this?
Tutor: One meant to see if you can distinguish between interference and diffraction.
Student: There's nothing to calculate.
Tutor: You might find a little. If $a < \lambda$, where is the first diffraction minimum?
Student: The diffraction minima are at

$$a \sin\theta = m\lambda$$

$$\sin\theta = (1)\frac{\lambda}{a}$$

Student: If $a < \lambda$, then the right side is greater than one. We can't solve for the first diffraction minimum.
Tutor: There isn't one. The first diffraction minimum occurs when light from one side of the slit goes a wavelength further than light from the near side. Because the slit is less than a wavelength wide, there isn't room for this to happen.
Student: So is that the top-middle graph?
Tutor: Yes. When the slit is very narrow, then light spreads way out.
Student: And when the slit is wide, the diffraction minima move way in. If $a \gg \lambda$, then do we get the top-right graph?
Tutor: Yes.
Student: And when they're about the same, we get the top-left figure.
Tutor: Yes. You can tell that the minima are caused by diffraction rather than interference because each minimum is so much smaller than the previous one, and because the central maximum is twice as wide as the others.
Student: The left-middle graph looks like interference. The central maximum is the same size as the others and they don't get much smaller as we go out.
Tutor: Good. Is it two slits or many slits?
Student: How can I tell?
Tutor: Compare the left-middle graph with the bottom-right graph.
Student: They're similar, but the maxima in the bottom-right graph are much narrower.
Tutor: That's what you get when you have many slits — narrower bright spots.
Student: So the left-middle is two slits and the bottom-right is many slits.

Tutor: Yes. How wide are the slits?
Student: There isn't much diffraction, and there's no diffraction minimum, so they're like the top-middle graph, where $a < \lambda$.
Tutor: Yes. The slits in the bottom-right graph are wider than those in the left-middle, but still smaller than a wavelength.
Student: Is that the difference between the right-middle and the bottom-middle graphs? They look the same but with narrower slits.
Tutor: Very good. Notice that the third interference maximum in each is missing.
Student: Is that because there is a diffraction minimum there?
Tutor: Yes. There is a third interference maximum, but the intensity is zero there. The light from one slit travels three wavelengths further than light from the other, but there is no light from either slit due to diffraction.
Student: The middle graph has the second interference maximum missing. So $d = 2a$ for the middle graph?
Tutor: Yes, and $d = 3a$ for the right-middle and bottom-middle graphs. What about the bottom-left graph?
Student: The fourth interference maximum is missing, so $d = 4a$.

When light travels through a circular hole rather than a slit, it again forms a diffraction pattern. The diffraction minima and maxima are circular in shape, like the hole, and get weaker in intensity as you go further out. The angle to the first diffraction minimum is

$$\theta = 1.22\frac{\lambda}{d}$$

where θ is in radians and the small angle approximation applies ($\theta \approx \sin\theta \approx \tan\theta$), and d is the diameter of the hole.

If we look at two objects that are close together but far from us, they may appear to be a single object. Rayleigh's criterion for being able to tell the two objects apart is that the center of the second object has to be outside the first diffraction minimum of the first object. This diffraction minimum depends on the size of the hole through which we view the object.

$$\theta_{\mathrm{R}} = 1.22\frac{\lambda}{d}$$

EXAMPLE

An athlete is at the far end of a sports field, 100 m away. How large a telescope would be needed to distinguish the athlete's nose, 3 cm long?

Student: Why do I want to see his nose?
Tutor: To see if his opponent broke it.
Student: Okay. The small angle approximation applies, so

$$\theta \approx \tan\theta = \frac{y}{D}$$

Student: where y is the length of the nose and D is the distance to the nose.
Tutor: Excellent.
Student: Is this the same angle θ as in the previous equation?
Tutor: Yes.
Student: What wavelength should I use? The problem doesn't say.
Tutor: Pick one in the visible light range.

Student: Like 500 nm?
Tutor: Sure.

$$\frac{y}{D} \approx \theta = 1.22\frac{\lambda}{d}$$

$$d = 1.22\frac{\lambda D}{y} = \frac{1.22(500 \times 10^{-9}\ \text{m})(100\ \text{m})}{0.03\ \text{m}} = 0.0020\ \text{m} = 2.0\ \text{mm}$$

Tutor: That's about the size of the pupil of the human eye.
Student: So I should be able to see the nose.
Tutor: So you should just be able to see that there is a nose, but not tell the shape of the nose. This is assuming that diffraction in the eye is what limits your eyesight — there are other things that could keep you from seeing the nose, but diffraction limits you to just barely seeing a nose. If you use 25–mm–diameter binoculars, then you can see details 12 times smaller.
Student: Let's see. If d goes up, then y goes down. But what if the binoculars are 7× binoculars?
Tutor: If the limiting factor is diffraction, then it's the ratio of the diameters that matter.

Chapter 37

Relativity

When two people are moving compared to each other, they measure the world differently. This is not because one of them is less competent, nor is one right and the other wrong. The world is just different to two people who are moving compared to each other.

Note that we don't say that one is standing still and the other is moving. One of the points of relativity is that there is no "standing still" frame of reference, compared to which all measurements are made. If you drive by me at 60 miles per hour, I believe that I am standing still and that you are moving. But you see yourself as stationary, you see the car not moving, and you see me — by the side of the road — as moving toward the rear of your car at 60 mph. Both of us are right, in our own frame of reference.

The important thing to remember is that nothing in relativity changes what we've covered so far. As long as you stay in one frame of reference, all of physics works just fine. The issues begin when you need to change from one frame of reference to another.

There are two basic postulates that we start with in relativity, and these lead to some interesting results.

The first postulate is that **the laws of physics are the same in any inertial frame of reference**. An inertial frame of reference is one in which Newton's laws apply. Note that any frame of reference that moves with a constant velocity compared to an inertial frame of reference is also an inertial frame of reference.

What is a noninertial frame of reference? Imagine that you are in a car, holding out your hand, with a Ping-Pong ball on your hand. If the driver of the car steps on the brakes, the car will slow, but the Ping-Pong ball will continue moving forward. From your standpoint, you were just sitting there when, all of a sudden, the Ping-Pong ball accelerated forward without any horizontal forces on it. Noninertial frames of reference typically involve acceleration that causes objects to accelerate without forces.

The second postulate is that the speed of light is the same for all observers. In particular, if I shine a laser beam and measure the speed of the light, I get c. If you are moving compared to me, and measure the speed of the same light beam, you get the same speed c, rather than $c + v$ or $c - v$.

The most immediate result of these two postulates is that **two frames moving compared to each other will not agree on whether two events are simultaneous or not**. You believe that two events happened simultaneously. As I move compared to you and look at the same events, I believe that they are not simultaneous.

This is not because of the delay in my seeing the events. If an event happens somewhere in my frame of reference other than in front of me, I subtract off the travel time of the signal to determine when the event happened. For example, If I see something at $t = 0$ but 3×10^6 m away from me, I realize that it took 0.01 seconds for the light of the event to travel to me, so that the event really occurred at $t = -0.01$ s. All observers are considered smart enough to compensate for the travel time of the signals, but this does not explain the effects of relativity.

A second result from these two postulates is that two observers, moving compared to each other, can measure the distance between two points, or the time between two events, and will get different values. For each, it is still true that $d = vt$, but they have different values of d, v, and t.

To convert space-time coordinates between frames of reference, we use the Lorentz transform.

$$\begin{cases} \Delta x' = \gamma(\Delta x - v\,\Delta t) \\ \Delta t' = \gamma(\Delta t - \frac{v}{c^2}\,\Delta x) \end{cases}$$

where γ is the Lorentz factor

$$\gamma = \frac{1}{\sqrt{1-(\frac{v}{c})^2}}$$

and v is the velocity between the two frames of reference.

Many of the most important effects of relativity can be understood in terms of proper time Δt_0 and proper length L_0. One frame of reference measures the "proper time" Δt_0, where proper is a name and not a description, and the proper time is not the correct time. Similarly, one frame of reference measures the "proper length" L_0, where proper is a name and not a description, and the proper length is not the correct length. The person who sees both events (start and end) happen directly in front of them measures the proper time. The person who sees the two ends stationary measures the proper length. These are never the same person.

$$L = L_0/\gamma \quad \text{and} \quad \Delta t = \gamma \Delta t_0$$

EXAMPLE

Captain Krank leaves Earth to go to Vapid to visit Smerk while McClay stays behind on Earth. Captain Krank travels at $0.866c$ and Vapid is 26 lt yr away, both as measured by McClay. How long does Krank's trip take, as measured by Krank?

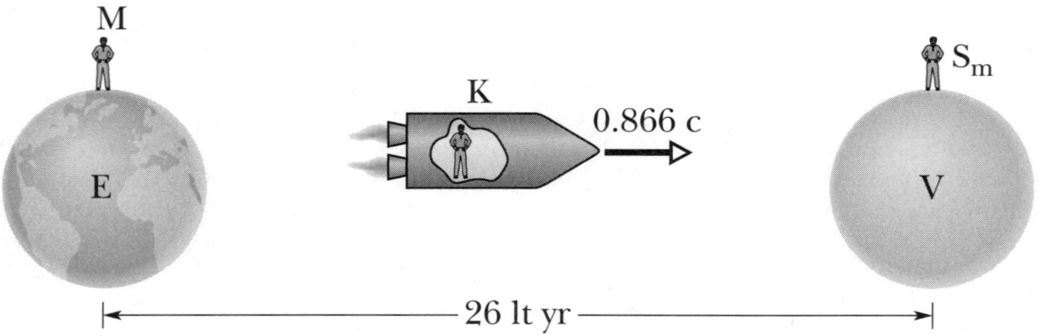

Student: If we can use proper time and proper length, then that will be easiest?
Tutor: Yes. Who measures the proper length for Krank's trip?
Student: McClay is not moving, so he is the one measuring proper time and length.
Tutor: Being on Earth does not mean that he isn't moving. From Krank's point of view, Krank isn't moving and McClay is moving to the left at $0.866c$.
Student: How do we tell who is moving and who is standing still?
Tutor: There is no "standing still." Each person is standing still in their own frame of reference, and everyone else is moving. Everyone is correct for their own frame of reference.

Student: But I can use $d = vt$, right?
Tutor: For each person and each frame of reference, you can do that.
Student: So ...

$$t = \frac{d}{v} = \frac{26 \text{ lt yr}}{0.866c} = \frac{26 \text{ lt yr}}{0.866 \text{ lt yr/yr}} = 30 \text{ yr}$$

Tutor: That is the distance measured by McClay and the velocity measured by McClay, so it's the time measured by McClay.
Student: And if I take the distance measured by Krank and the velocity measured by Krank, I get the time measured by Krank, right?
Tutor: Right. What is the distance measured by Krank?
Student: It isn't 26 lt yr.
Tutor: Correct. McClay sees the proper length because the endpoints aren't moving in his frame of reference. To him, Earth and Vapid are stationary.
Student: So McClay measures the proper length for the trip *because* the planets are not moving in his frame of reference.
Tutor: Correct. And what is this "proper length"?
Student: 26 lt yr.
Tutor: What is the length as measured by Krank?
Student: He measures a different length.

$$L = L_0/\gamma \quad \text{so} \qquad \rightarrow \qquad L_{\text{Krank}} = L_{\text{McClay}}/\gamma$$

Student: What is γ ?
Tutor: γ comes from the difference in speeds between the two frames of reference.

$$\gamma = \frac{1}{\sqrt{1-(\frac{v}{c})^2}} = \frac{1}{\sqrt{1-(\frac{0.866c}{c})^2}} = \frac{1}{\sqrt{1-(0.866)^2}} = 2$$

Tutor: How far is the trip according to Krank?
Student: Since McClay measures the proper length, we can use the Krank-McClay γ to find the distance according to Krank.

$$L_{\text{Krank}} = L_{\text{McClay}}/\gamma = (26 \text{ lt yr})/(2) = 13 \text{ lt yr}$$

Tutor: Good. How fast is Krank going according to Krank?
Student: Trick question — Krank isn't moving according to Krank.
Tutor: Very good. How fast is Krank going according to McClay?
Student: Is it just $0.866c$?
Tutor: Yes. The $\vec{v}_{\text{AB}} = -\vec{v}_{\text{BA}}$ from relative motion still works.
Student: So I need a minus sign.
Tutor: Gamma is not negative. It is always a positive number greater than 1. When we square the negative velocity the sign disappears. There are some times when the sign of the velocity matters, but not yet.
Student: Okay. Now we calculate how long the trip takes according to Krank.

$$t_{\text{Krank}} = \frac{d_{\text{Krank}}}{v_{\text{Krank}}} = \frac{13 \text{ lt yr}}{0.866c} = \frac{13 \text{ lt yr}}{0.866 \text{ lt yr/yr}} = 15 \text{ yr}$$

Tutor: We could also have found the time according to McClay and converted the time into Krank's frame.

$$t_{\text{McClay}} = \frac{d_{\text{McClay}}}{v_{\text{McClay}}} = \frac{26 \text{ lt yr}}{0.866c} = \frac{26 \text{ lt yr}}{0.866 \text{ lt yr/yr}} = 30 \text{ yr}$$

Tutor: Who measures the proper time?
Student: Wouldn't that be McClay, since he measures the proper length?
Tutor: No, the proper time is measured by the person who is at Earth when Krank leaves Earth, and is

also at Vapid when Krank arrives at Vapid.
Student: That would be Krank.
Tutor: Correct.
Student: But if McClay measures proper length and Krank measures proper time, which one is the correct frame of reference?
Tutor: Neither. There is no "correct" frame of reference. Each frame of reference measures things and gets different results, but their measurements are correct for their frame of reference. We use relativity to change measurements from one frame of reference to another.
Student: So I can use $d = vt$ as long as all three are measured in the same frame of reference.

$$\Delta t = \gamma\, \Delta t_0 \quad \rightarrow \quad \Delta t_{\text{McClay}} = \gamma\, \Delta t_{\text{Krank}}$$
$$\Delta t_{\text{Krank}} = \Delta t_{\text{McClay}}/\gamma = (30\text{ yr})/(2) = 15\text{ yr}$$

Student: Now I've heard of this so-called "twin paradox," where one twin stays on Earth and the other goes away and comes back. Can't we say that the one in the rocket stays put while the one on Earth goes away and comes back?
Tutor: Good question. When the twin in the rocket turns around, everything in the rocket ship that is not bolted down picks itself up and flies against the front of the ship. This never happens to the twin on Earth. This moment of noninertial frame of reference is the difference. The equations we use in this chapter assume a constant velocity between two frames of reference. When the rocket ship turns around, the velocity between the twins changes, if only from positive to negative. So we have to do the problem in two pieces, and in each piece the twin in the rocket ship experiences the proper time, which is shorter, so he is younger when they get back together.

Just as two observers will measure lengths and times differently, so they will also measure velocities differently. To transform a velocity from one equation into another, we use

$$u = \frac{u' + v}{1 + u'v/c^2}$$

where u is the velocity measured in one frame and u' is the velocity measured in the other.

EXAMPLE

In the previous example, Scootie leaves Earth toward Vapid at $0.943c$. How long does Krank's trip take according to Scootie?

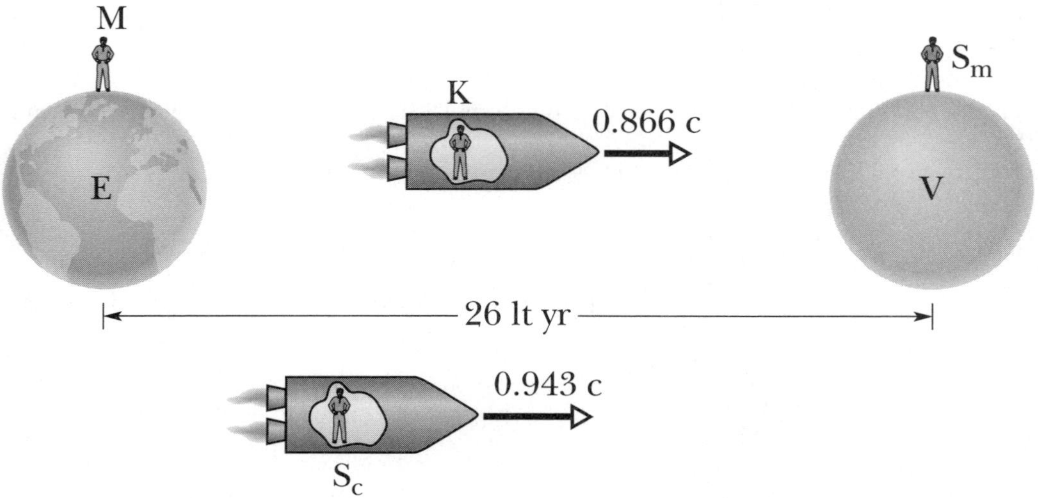

Student: Can we use $\Delta t = \gamma\, \Delta t_0$?
Tutor: We can. We would find the proper time, and then find the gamma factor γ between the frame with the proper time and Scootie's frame.
Student: Krank measures the proper time, so $\Delta t = 15$ yr.
Tutor: Good. Now we need γ between Krank and Scootie.
Student: So I take the speed between Krank and Scootie ...
Tutor: Which is?
Student: $0.943c$.
Tutor: That's the speed between McClay and Scootie.
Student: Can I subtract?

$$0.943c - 0.866c = 0.077c \quad \textbf{?}$$

Tutor: That is the speed at which McClay sees Scootie gaining on Krank. It is the difference of two velocities, but it is still in McClay's frame of reference.
Student: So we need to use the complicated formula. How do I keep u and u' and v straight?
Tutor: It's possible, but not necessary. In the numerator, do what you already did. If you saw two cars on the highway, one going 86.6 mph and the other chasing it at 94.3 mph, you would say that the second was gaining on the first at 7.7 mph. Then, whatever values you used in the numerator, just use them in the denominator.
Student: Like this?

$$u = \frac{u' + v}{1 + u'v/c^2} = \frac{0.943c - 0.866c}{1 + (0.943c)(0.866c)/c^2} \quad \textbf{?}$$

Tutor: More like this.

$$u = \frac{u' + v}{1 + u'v/c^2} = \frac{(0.943c) + (-0.866c)}{1 + (0.943\cancel{c})(-0.866\cancel{c})/\cancel{c^2}} = 0.42c$$

Student: So if one of them is negative in the numerator, it is negative in the denominator?
Tutor: Precisely.
Student: Then I find γ from this velocity.

$$\gamma = \frac{1}{\sqrt{1 - (\frac{v}{c})^2}} = \frac{1}{\sqrt{1 - (\frac{0.42c}{c})^2}} = \frac{1}{\sqrt{1 - (0.42)^2}} = 1.10$$

Student: Now if I used $0.943c$ and found γ, I'd get 3. With $0.866c$, I got 2. Is there some way to go from 2 and 3 to 1.10?
Tutor: Not really. The equations in relativity are highly nonlinear.
Student: Okay. Now I can find the time.

$$\Delta t = \gamma\, \Delta t_0 \quad \rightarrow \quad \Delta t_{\text{Scootie}} = \gamma\, \Delta t_{\text{Krank}} = (1.10)(15 \text{ yr}) = 16.5 \text{ yr}$$

Student: Since Scootie is going faster, shouldn't he see a shorter time?
Tutor: γ is always more than 1, so any other time is longer than the proper time. Whoever measures the proper time measures the shortest time, and everyone else measures a longer time.
Student: Could we have done the calculation in Scootie's frame of reference?
Tutor: Yes. According to Scootie, the world looks like this.

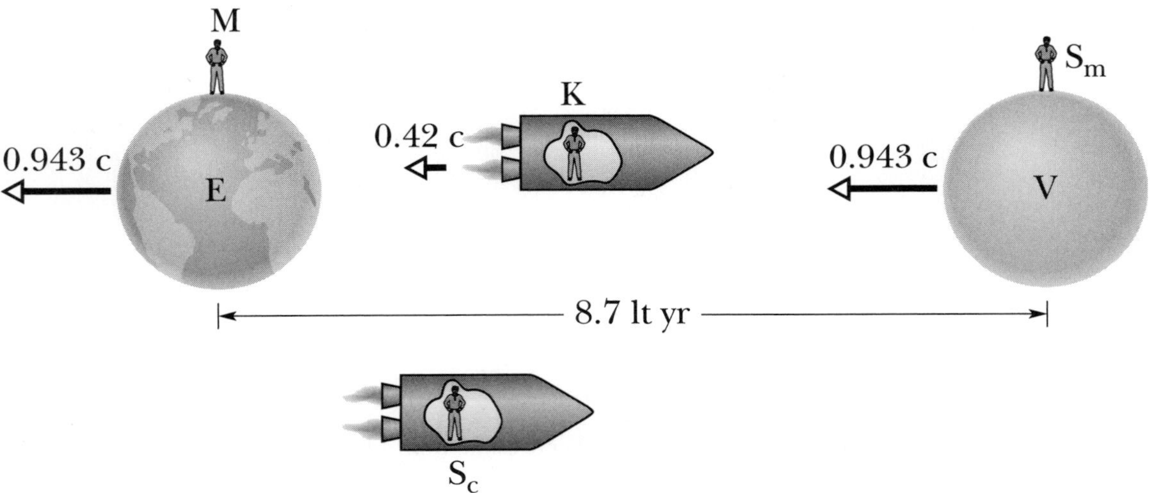

Tutor: The question for Scootie is: how fast is Vapid gaining on Krank?
Student: So Krank is facing toward the right but moving toward the left?
Tutor: Everyone sees Krank facing toward the right, but Scootie sees Krank moving toward the left — in the picture. That is, Scootie is gaining on Krank, but Scootie is gaining on Vapid faster than he is gaining on Krank.

$$\Delta t_{\text{Scootie}} = \frac{\Delta d_{\text{Scootie}}}{v_{\text{Scootie}}}$$

Student: What is $\Delta d_{\text{Scootie}}$?
Tutor: How much distance Vapid must gain on Krank, according to Scootie.
Student: According to McClay, that's 26 lt yr.
Tutor: Very good. What is that distance according to Scootie?
Student: McClay measures the proper length, and γ between Scootie and McClay is 3.

$$L_{\text{Scootie}} = L_{\text{McClay}}/\gamma = (26 \text{ lt yr})/(3) = 8.7 \text{ lt yr}$$

Tutor: v_{Scootie} is the speed at which Vapid gains on Krank, according to Scootie.
Student: Is that $v_{\text{Scootie}} = 0.42c - 0.943c = -0.523c$?
Tutor: Yes.

$$\Delta t_{\text{Scootie}} = \frac{\Delta d_{\text{Scootie}}}{v_{\text{Scootie}}} = \frac{8.7 \text{ lt yr}}{0.523 \text{ lt yr/yr}} = 16.6 \text{ yr}$$

Student: It's not quite the same.
Tutor: Because of rounding in intermediate steps.

EXAMPLE

Ashley *observes* two flashes, one red and one blue, both at $t = 0$. Also at $t = 0$, Helen races by Ashley at $v = 0.5c$. Which flash does Helen observe first, and by how much?

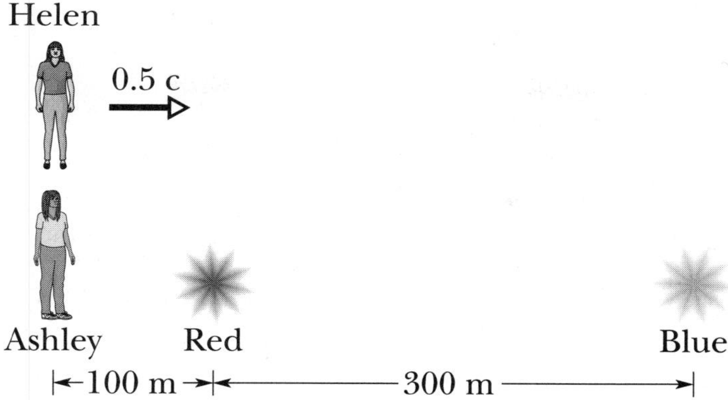

Student: What do we mean by "observes?"
Tutor: Ashley does not see the light of the flashes at the same time. She sees the red one first, and then the blue one 1 μs later. But she concludes that, because the blue light had to travel further, they really happened at the same time. Her "observation" is that they happened at the same time, once she corrects for the travel time of the signal.
Student: And we want to know which one Helen "observes" first, which is different from which one she "sees" first.
Tutor: They might be the same, but yes, after she subtracts off the travel time of the signals, she might conclude that the one she "saw" first really happened second.
Student: So which of Helen and Ashley measures the proper time?
Tutor: Which one is at the red light when it goes off and at the blue light when it goes off?
Student: Neither.
Tutor: So neither of them measures the proper time.
Student: Does that mean that we need to use the Lorentz transform?
Tutor: Yes.

$$\begin{cases} \Delta x' = \gamma(\Delta x - v\,\Delta t) \\ \Delta t' = \gamma(\Delta t - \frac{v}{c^2}\,\Delta x) \end{cases}$$

Tutor: Take Ashley as the "unprimed" frame, so that t, x, and v are measured by Ashley, and t_B is the time of the red flash as measured by Ashley.
Student: Then $t_B = 0$.
Tutor: Yes. Take Helen as the "primed" frame, so that t', x', and v' are measured by Helen, and t'_B is the time of the red flash as measured by Helen.
Student: We don't know that.
Tutor: Correct. We do know t_R, t_B, x_R, x_B, and v.
Student: v is the velocity of Helen as measured by Ashley, measured by Ashley because it is unprimed. Is v positive or negative?
Tutor: According to Ashley, is Helen moving in the positive or negative direction?
Student: Positive. So v is positive. Does that mean v' is negative?
Tutor: It does. To use the Lorentz transform, both people must have there x-axes parallel, not antiparallel. For both of them, $t = t' = x = x' = 0$ at the moment that the one passes the other.
Student: Does it have to be that way?
Tutor: No, but if we didn't then the equations would be messier. Like when we chose the potential of a point charge to be zero at infinity — we didn't have to, but not doing so would lead to messier equations.
Student: Okay. Now I can find the times according to Helen.

$$t' = \gamma\left(t - \frac{v}{c^2}\,x\right)$$

$$\gamma = \frac{1}{\sqrt{1-(\frac{v}{c})^2}} = \frac{1}{\sqrt{1-(\frac{0.50c}{c})^2}} = \frac{1}{\sqrt{1-(0.50)^2}} = 1.155$$

$$t'_R = \gamma\left(t_R - \frac{v}{c^2}\,x_R\right) = (1.155)\left((0) - \frac{0.5c}{c^2}\,(+100\text{ m})\right) = -\,(1.155)\left(\frac{1}{2}\frac{1}{c}\,(+100\text{ m})\right)$$

$$t'_R = -\frac{(1.155)(+100\text{ m})}{2(3\times10^8\text{ m/s})} = -1.92\times10^{-7}\text{ s} = -0.192\ \mu\text{s}$$

$$t'_B = \gamma\left(t_B - \frac{v}{c^2}\,x_B\right) = (1.155)\left((0) - \frac{0.5c}{c^2}\,(+400\text{ m})\right) = -\frac{(1.155)(+400\text{ m})}{2(3\times10^8\text{ m/s})} = -7.70\times10^{-7}\text{ s} = -0.770\ \mu\text{s}$$

Tutor: The negative times mean that Helen "observes" both of the flashes before she passes Ashley, which is at $t = 0$.
Student: And because the blue time is more negative, it happens first.
Tutor: It happens first in Helen's frame of reference. They happened at the same time in Ashley's frame of reference. For a third person that Ashley sees going in the opposite direction as Helen, that person would observe the red flash first.
Student: So one observer might say that "the bomb went off before the fuse was lit."
Tutor: An interesting point. In Ashley's frame of reference, there would not be time for the red flash to "cause" the blue one, or vice versa. The second one already happened before the light or signal from the first one arrived. It is possible to show that, because one can't cause the other in Ashley's frame of reference, it can't cause the other in any inertial frame of reference. Therefore two observers can observe different orders for the events. If the red flash happened after the blue one in Ashley's frame, long enough after ($\geq 1\ \mu$s) that the blue one could have caused the red one, then there is no frame in which the red one happened first.
Student: So if I find the distance between the flashes in Helen's frame, it will be greater than the time difference between the flashes in Helen's frame times the speed of light?
Tutor: Yes, so that Helen concludes that the blue one could not have caused the red one.
Student: Let's check.

$$\Delta x' = \gamma(\Delta x - v\,\Delta t) = (1.155)\Big((300\text{ m}) - (0.5c)(0)\Big) = 346\text{ m}$$

$$d = ct = (3\times10^8\text{ m/s})\left|\Big((-0.770\ \mu\text{s}) - (-0.192\ \mu\text{s})\Big)\right| = 173\text{ m}$$

Student: Helen sees the flashes 346 m apart, but in the time between them light could only travel 173 m. So she concludes that one couldn't have caused the other.

Energy and momentum also have relativistic values. The momentum is

$$\vec{p} = \gamma m\vec{v}$$

For objects moving much slower than the speed of light, $\gamma \approx 1$ and we get the same equation for momentum that we have always had.

One of the most well known equations from relativity is $E = mc^2$. This is the rest energy, the energy that something has even when it isn't moving. The total energy is

$$E = \gamma mc^2$$

The difference between the total energy and the rest energy is the kinetic energy K.

$$E = \gamma mc^2 = mc^2 + K$$

For slow speeds, $K \approx \frac{1}{2}mv^2$.

Another useful equation involving energy is

$$E^2 = (pc)^2 + (mc^2)^2$$

Here E is the total energy. Often nuclear physicists will describe particles with their kinetic energy. For example, a 100 keV electron is one with a kinetic energy of 100,000 electron volts. Note that E, pc, and mc^2 all have units of energy.

Chapter 38

Photons and Matter Waves

In the last few chapters, we have examined how light acts like a wave. Interference and diffraction are wave behaviors, and anything that demonstrates interference and diffraction is acting like a wave. Now we ask the question, what kind of behavior is a characteristic behavior of particles? If we could identify such a behavior, and then found something that showed this behavior, we would know that this something was a particle.

The characteristic behavior of particles is that they can't be arbitrarily divided. If I want to reduce the power of a wave, I can always reduce the amplitude. I cannot continue to reduce baseballs — eventually I get to one baseball and I can't go any further.

The stunning observation is that light shows this indivisibility feature, and small particles like electrons show interference behavior. Therefore **light sometimes acts like particles and electrons sometimes act like waves**.

Light behaves like little particles called photons. The energy in each photon is

$$E = hf = \frac{hc}{\lambda}$$

where $h = 6.63 \times 10^{-34}$ J s is the Planck constant, or you can use the convenient $hc = 1240$ nm·eV. The amount of energy that can be taken from light has to be an integer multiple of the photon energy. You can't take half of a photon's worth of energy from a beam of light. In this way light is indivisible and behaves like a stream of particles.

When does light behave like a wave and when does it behave like particles? In general, when light interacts with light, it behaves like a wave, showing interference. When light interacts with matter like atoms, it behaves like particles, showing indivisibility. In particular, an atom can absorb only one photon and therefore only the energy of one photon.

Consider what happens when light hits a metal. Given sufficient energy, electrons can escape from the atoms in the metal. The energy needed, called the work energy Φ, is typically a few electron volts. But an atom can absorb only one photon. So if a single photon has enough energy, or $E_\gamma \geq \Phi$, then an electron can escape from the metal. Any excess energy that the photon has above the work function shows up as kinetic energy of the electron after it escapes. If the photons do not have enough energy, then the electrons can't escape at all, no matter how intense the light. More intense light is more photons, but not greater energy photons. This is called the photoelectric effect. People knew about the photoelectric effect, but it took Einstein to explain it using photons.

It is possible in these next few chapters to do the calculations in joules, meters, and the other units that you've become familiar with. It is easier to use units of about the right size, such as electron volts (eV) and nanometers. Remember that 1 eV = $(e)(1\text{ V}) = 1.6 \times 10^{-19}$ J and that 1 nm = 10^{-9} m. In particular, look for the item in the left column, try to turn it into the combination in the middle column, and substitute the value in the right column.

Look for	Turn into	Value
h	hc	1240 nm·eV
m_e	m_ec^2	511 keV = 511,000 eV
m_p	m_pc^2	938 MeV = 938×10^6 eV
p	pc	momentum in energy units
v	v/c	velocity parameter
k	ke^2	1.44 nm·eV

EXAMPLE

When light of 470 nm strikes a metal, electrons are ejected with a maximum kinetic energy of 0.16 eV. What is the stopping voltage and the speed of the ejected electrons when 410 nm light is used?

Student: I need to know the energy of the photons.
Tutor: Yes, that would help.
Student: And I use hc/λ to find that.

$$E = hf = \frac{hc}{\lambda} = \frac{1240\text{ nm}\cdot\text{eV}}{470\text{ nm}} = 2.64\text{ eV}$$

Student: And for the 410 nm photons.

$$E = \frac{hc}{\lambda} = \frac{1240\text{ nm}\cdot\text{eV}}{410\text{ nm}} = 3.02\text{ eV}$$

Student: Is the kinetic energy of the electrons proportional to the photon energy?
Tutor: No, because the work energy does not change.
Student: Then what can I do?
Tutor: Can you find the work energy?
Student: How do I do that? Don't I need a formula?
Tutor: How much energy is in the 470 nm photon?
Student: 2.64 eV.
Tutor: And after the atom absorbs the energy from the photon, 0.16 eV is left over and goes into kinetic energy of the electron.
Student: So 2.64 eV − 0.16 eV = 2.48 eV must be the work function.
Tutor: Yes. Now the 410 nm light hits the metal.
Student: But the work function is still the same. So 3.02 eV − 2.48 eV = 0.54 eV must be energy left over to become the kinetic energy of the electron.
Tutor: Yes.
Student: Why do they talk about the "maximum kinetic energy?"
Tutor: Some electrons are easier to remove from the atom than others. The work function is the energy needed to remove the easiest electrons. When a harder-to-remove electron is ejected, there is less energy left over for kinetic energy.
Student: I see. And what's this "stopping voltage" about?
Tutor: It comes from the way in which the experiment is done. It's easy to detect the presence of an electron, but hard to detect it's velocity. So we use an electric field to slow it down until it stops. We create a potential difference and gradually increase the potential difference until the electrons can't make it any more. So the stopping voltage is the kinetic energy divided by the charge of the electron. If the kinetic energy is 0.54 eV, then the stopping voltage is 0.54 V.
Student: That seems like a strange way to do things.

Tutor: Imagine that you want to find the kinetic energy of popcorn as it is popped. One way to do that would be to take the lid off of the popper and see how high the popcorn goes.
Student: Because the kinetic energy turns into potential energy, yes?
Tutor: Yes. But what if you couldn't see popcorn. Then you might take the lid off and hold out your hand and keep moving it up until you don't feel any more popcorn hitting it. The potential energy of a popcorn where your hand was when you quit feeling them tells you about the maximum energy of the popcorn.
Student: So when we look at electrons we're blindfolded.
Tutor: In the sense that we can't just know everything about them.
Student: Okay. I have the kinetic energy of the electron in electron-volts. I can convert that to joules, put in the mass of the electron, and solve for the speed.
Tutor: You can, but there's an easier way.

$$KE = \frac{1}{2}mv^2 = \frac{1}{2}(mc^2)\left(\frac{v}{c}\right)^2$$

Student: Is this the "combinations" thing?
Tutor: Yes.

$$0.54\cancel{eV} = \frac{1}{2}(511000\cancel{eV})\left(\frac{v}{c}\right)^2$$

$$\left(\frac{v}{c}\right)^2 = \frac{2(0.54)}{511,000}$$

$$\frac{v}{c} = \sqrt{\frac{2(0.54)}{511,000}} = 0.00145$$

$$v = 0.00145c = 0.00145(3 \times 10^8 \text{ m/s}) = 4.36 \times 10^5 \text{ m/s}$$

Tutor: The advantage to this is that you aren't dragging big exponents all over the equations.

Just as light can act like a particle, matter like electrons can act like waves. When they do so, their wavelength is the de Broglie wavelength,

$$\lambda = \frac{h}{p}$$

This equation is also true regarding the momentum of a photon. A common mistake is to use $\lambda = \frac{hc}{E}$ for electron. This equation does not work for anything that has mass — it only works for photons.

EXAMPLE

What is the de Broglie wavelength of a 14 eV electron?

Student: I can't use $\lambda = \frac{hc}{E}$ for the electron because the electron has mass.
Tutor: Good.
Student: I need to find the momentum. So I use the kinetic energy to find the speed, just like in the last example. Then I use the speed and $p = \gamma mv$ to find the momentum.
Tutor: You could do that. Because the kinetic energy of our electron is so much less than the rest energy, the electron is going much slower than the speed of light, so you could use $p = mv$.
Student: So the way to tell is to compare the energy to the rest energy.
Tutor: That's one way. If $K \ll mc^2$, then you're safe with $p = mv$. Also, you don't really need the speed, do you? That's just an intermediate step.
Student: Correct. But I need to find the momentum.

Tutor: If we find the speed from the energy, and put the speed into the momentum, we get

$$\frac{1}{2}mv^2 \quad \text{and} \quad p = mv \quad \rightarrow \quad p = \sqrt{2mK}$$

Student: Is there a reason that we're using K for kinetic energy now instead of KE?
Tutor: Because we have a total energy E that we didn't have before, so someone might get confused and think KE was the product of two values.
Student: Okay, so I have $p = \sqrt{2mK}$. I want to do this combinations thing.
Tutor: It will be much easier.
Student: I put two factors of c inside the square root to get mc^2, and add one factor of c to the left, and I get pc. It works out perfectly.

$$p = \sqrt{2mK} \quad \rightarrow \quad pc = \sqrt{2(mc^2)K}$$

Tutor: Then you can find the wavelength.
Student: Yes. I need to use

$$\lambda = \frac{h}{p} = \frac{hc}{pc} = \frac{hc}{\sqrt{2(mc^2)K}} = \frac{(1240 \text{ nm}\cdot\text{eV})}{\sqrt{2(511,000 \text{ eV})(14 \text{ eV})}} = 0.328 \text{ nm} = 328 \text{ pm}$$

Student: What's a pm?
Tutor: A picometer, or 10^{-12} m. Just remember that our earlier step that led to $\sqrt{2mK}$ required that the speed be slow.
Student: What if the speed isn't slow?
Tutor: Then we find the total energy $E = K + mc^2$, use $E^2 = (pc)^2 + (mc^2)^2$ to find the combination pc, and use $\lambda = h/p = hc/pc$ to find the de Broglie wavelength. There's more algebra involved, but the idea is the same.

Now that we know that an electron acts like a wave, we can show you a picture of an electron. Remember that the de Broglie wavelength of the electron is connected to the momentum of the electron, with shorter wavelengths corresponding to greater momentum.

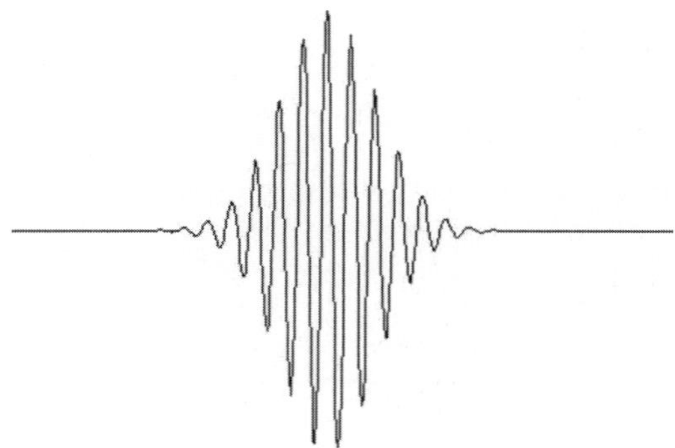

Two interesting questions to ask are: where is the electron and how fast is it going? Because of the width of the "wave packet," it is hard to say exactly where the electron is. To improve the determination of the position of the electron, we need to shrink the width of the wave packet (shown horizontally). But then the wave packet will consist of fewer waves, so we can't determine the wavelength as well, and the wavelength tells us the momentum.

This effect is the **Heisenberg uncertainty principle**, which states that the uncertainty with which the position and momentum are measured must have a minimum value.

$$\Delta x \, \Delta p_x \geq \hbar = \frac{h}{2\pi}$$

We cannot determine the position of an electron exactly without losing all information on its velocity.

EXAMPLE

An electron has a speed $10\ 1450 \pm 25$ m/s. What is the minimum uncertainty in its position?

Student: I need to use Heisenberg's uncertainty principle.
Tutor: How did you decide that?
Student: Because it asks for minimum uncertainty. Nothing else we've done has required a minimum uncertainty.

$$\Delta x\ \Delta p_x \geq \hbar = \frac{h}{2\pi}$$

$$\Delta x\ m\ \Delta v_x \geq \frac{h}{2\pi}$$

Tutor: Can you spot the combinations in the equation?
Student: You mean like $h \rightarrow hc$?

$$\Delta x\ (mc^2)\ \frac{\Delta v}{c} \geq \frac{hc}{2\pi}$$

$$\Delta x \geq \frac{hc}{2\pi(mc^2)(\Delta v/c)} = \frac{1240\ \text{nm}{\cdot}\text{eV}}{2\pi(511,000\ \text{ev})(25/(3\times 10^8))} = 4634\ \text{nm} = 4.6\ \mu\text{m}$$

Student: Where does the 1450 m/s go?
Tutor: The minimum uncertainty in the position depends on the uncertainty in the velocity, not the velocity itself. To know the momentum better, we need the wave packet to contain more "waves" so that we can better determine the wavelength. That makes the wave packet longer, spread out over more distance.
Student: So there's no way to measure the position of an electron to better than 4.6 μm.
Tutor: Sure there is, but not while measuring its velocity to a precision of 25 m/s. In a previous example, less than 1 eV of kinetic energy gave an electron a speed of half a million meters per second. So this is a very slow moving electron. For an electron moving with a speed of 5×10^5 m/s, this is an error of 0.005%.
Student: I would think that a slow-moving electron would be easier to pin down than a fast one.
Tutor: The way you make a wave packet is by adding waves of difference frequency or wavelength. The narrower the wave packet, the more wavelengths you need. If you use fewer different wavelengths, so that you know the wavelength better, then the wave packet will be wider.
Student: Why don't we see the Heisenberg uncertainty principle in larger objects? When a policeman pulls me over for speeding, he says he knows my speed to 1 mph. Then he must not know where I am.
Tutor: 25 m/s is about 50 mph, so he claims to know your speed 50 times better than our electron, and the uncertainty in the position is 50 times worse, or 0.25 mm. But the mass of the car is much bigger, so that the minimum uncertainty in the position of the car is about 10^{-34} m while knowing the velocity to the nearest 1 mph.
Student: So the Heisenberg uncertainty principle is not going to keep me from getting a ticket.

Chapter 39

More About Matter Waves

In the previous chapter, we treated a moving electron as a traveling wave. What happens when an electron acts like a standing wave? In a standing wave, we forced nodes at particular places, and the result was that only specific frequencies would work. For an electron in a similar situation, only specific energies are allowed.

Imagine an electron in a long skinny tube. The tube is so skinny that we worry about the motion of the electron in only one direction. Because the electron behaves like a wave, it must form a standing wave with a node at each end. The allowed wavelengths of the electron are $2L/1$, $2L/2$, $2L/3$, $2L/3$, ... From these we can find the allowed momenta.

$$p = \frac{h}{\lambda} = \frac{h}{2L/n} = \frac{hn}{2L}$$

The potential energy doesn't change throughout the box, so the energy is the kinetic energy.

$$E = \frac{1}{2}mv^2 = \frac{1}{2}m\left(\frac{p}{m}\right)^2 = \frac{p^2}{2m} = \frac{1}{2m}\left(\frac{hn}{2L}\right)^2 = \frac{h^2}{2mL^2}n^2 = \frac{(hc)^2}{2(mc^2)L^2}n^2$$

There are specific values of the energy of the electron that work, and all other values don't work. We call this a one-dimensional infinite square well. The electron can move in only one dimension, it takes an infinite energy to get past the ends of the tube, and the potential energy is "flat" within the well, or square.

A similar thing happens inside an atom, where the electrons orbit the positively charged nucleus. A hydrogen atom consists of an electron orbiting a proton. When the electron makes one complete orbit, the distance it travels must be an integer number of wavelengths of the electron $2\pi r_n = n\lambda_n$. This is the Bohr model of the atom. Because only specific values of the wavelength are allowed, only specific values of the energy are allowed. From this, it is possible to determine the allowed energies.

$$E_n = -\frac{1}{2}m\left(\frac{2\pi ke^2}{hn}\right)^2 = -\frac{13.6 \text{ eV}}{n^2}$$

Why is the energy negative? Remember that the potential energy of the proton and electron when they are far apart is zero. As the electron approaches the proton, it gains kinetic energy and loses potential energy, so that the potential energy is negative. This means that we would have to do work to move the negative charge away from the positive charge and back to an infinite distance apart. While the electron is close to the proton, we take away some of its kinetic energy. The kinetic energy is positive and the potential energy is negative, but the potential energy is more negative than the kinetic energy is positive, so the total energy is negative.

It is a general principle that the energy of anything that is stable is less than the energy that the pieces have when they are apart. Things tend to move toward lower energy, so if the energy together was positive, then it would tend to move apart, and be unstable. The energies of the one-dimensional infinite square well are positive, but the energy when the electron is outside the "box" is infinite rather than zero.

A system that can have only specific values is said to be quantized, and the study of quantized systems is called quantum mechanics. Not all quantized things are microscopic. When you buy eggs you can buy a dozen, or a half dozen, or perhaps a single egg, buy you can't buy 1.39 eggs, so eggs are quantized. You can go to the butcher's store and buy 1.39 pounds of ground beef, or any value you like, so ground beef is not quantized, but is a continuum. **The smallest possible value that a quantized system can have is called a quantum**. Therefore a "quantum leap" is the smallest possible change in a quantized system.

The different allowed values in a quantized system are called states. The lowest energy state is called the ground state, and the next lowest state is the first excited state. To make a quantum leap, or change from one state to another, in a hydrogen atom involves moving to a different energy. To move to a higher n value, or higher energy, requires that the atom gain energy. To move to a lower n value, or lower energy, requires that the atom lose energy.

Atoms gain and lose energy by absorbing or emitting a photon of light. The energy of the photon must be equal to the difference of the energies of the initial and final states.

$$\Delta E = |E_{\text{high}} - E_{\text{low}}| = E_\gamma$$

Because both the initial and final energies are quantized, the photon must have exactly the right amount of energy, to about 8 significant figures. If the energy of the photon is not exactly right, then the atom can't absorb the photon and can't make a transition. This is the principle behind spectroscopy.

EXAMPLE

What happens to a hydrogen atom in the ground state when it is illuminated with light of wavelength 99.255 nm, with light of wavelength 97.255 nm, and with light of wavelength 90.255 nm?

Student: The atom absorbs the energy of the photon.
Tutor: Not necessarily. If the energy from the photon puts the atom between allowed energy states, or energy levels, then the atom can't go there and won't absorb the photon.
Student: So I need to see whether the photon has the right amount of energy.

$$E = \frac{hc}{\lambda} = \frac{1240 \text{ nm}\cdot\text{eV}}{99.255 \text{ nm}} = 12.493 \text{ eV}$$
$$E = \frac{hc}{\lambda} = \frac{1240 \text{ nm}\cdot\text{eV}}{97.255 \text{ nm}} = 12.750 \text{ eV}$$
$$E = \frac{hc}{\lambda} = \frac{1240 \text{ nm}\cdot\text{eV}}{90.255 \text{ nm}} = 13.739 \text{ eV}$$
$$E_n = -\frac{13.6 \text{ eV}}{n^2} = 12.493 \text{ eV} \quad ?$$

Tutor: What are you trying to do?
Student: I need to find out if the energy of the photon matches one of the allowed energy levels. So I'll solve for n and see if it's an integer.
Tutor: Your idea is partially correct. The question isn't whether the photon energy matches 13.6 eV divided by n^2, but whether the energy of the atom after absorbing the photon matches an allowed energy.
Student: Ah, so I need to add the photon energy to the starting energy to find the after energy.
Tutor: Yes.
Student: What's the starting energy?
Tutor: The hydrogen atom is in the ground state. What value of n gives the lowest energy?
Student: $n = 1$. So the initial energy is 13.6 eV.
Tutor: Negative 13.6 eV.

Student: Oops. With the first photon, the energy afterward is

$$-13.6 \text{ eV} + 12.493 \text{ eV} = -1.107 \text{ eV}$$
$$E_n = -\frac{13.6 \text{ eV}}{n^2} = -1.107 \text{ eV}$$
$$n = \sqrt{\frac{-13.6 \text{ eV}}{-1.107 \text{ eV}}} = 3.51$$

Student: 3.51 is not an integer, so we can't shine 99.255 nm light on a hydrogen atom.
Tutor: Oh, we can shine the light on a hydrogen atom, but the atom doesn't absorb any of the photons.
Student: Could the hydrogen atom absorb a photon and go up to $n = 3$?
Tutor: The photon has too much energy. $E_3 = -1.511$ eV.
Student: But the atom could emit a photon with the excess energy, and then it would be okay, right?
Tutor: That doesn't happen. It's possible to cause effects similar to that to happen, but that's for advanced quantum mechanics and we won't get into that stuff.
Student: Okay, so nothing happens. With 97.255 nm light,

$$-13.6 \text{ eV} + 12.750 \text{ eV} = -0.850 \text{ eV}$$
$$E_n = -\frac{13.6 \text{ eV}}{n^2} = -0.850 \text{ eV}$$
$$n = \sqrt{\frac{-13.6 \text{ eV}}{-0.850 \text{ eV}}} = 4.00$$

Student: n is an integer, so it works. The atom absorbs the photon and jumps to the $n = 4$ state.
Tutor: Good. Then what happens?
Student: I don't know.
Tutor: The atom wants to be in the lowest possible state, remember.
Student: So it jumps back down to the ground state?
Tutor: After a while, like 10 or 20 ns, it jumps to a lower state, either $n = 3$ or $n = 2$ or $n = 1$. It has to lose energy in the process so it emits a photon with the appropriate energy.
Student: Not the energy of the $n = 3$ or $n = 2$ or $n = 1$ state, but the *difference* in the energies of the two states, right?
Tutor: Right. This photon goes off in a random direction, not necessarily the same direction as the incident photons. So if we see photons going off in other directions, that is a sign that the energy of the incident light matches a transition energy.
Student: Is that how scientists do spectroscopy, by changing the wavelength of the incident light until they see light coming off?
Tutor: A simplistic but substantially correct explanation.
Student: Cool. Now with 90.255 nm light,

$$-13.6 \text{ eV} + 13.739 \text{ eV} = +0.139 \text{ eV}$$
$$E_n = -\frac{13.6 \text{ eV}}{n^2} = +0.139 \text{ eV}$$
$$n = \sqrt{\frac{-13.6 \text{ eV}}{+0.139 \text{ eV}}} = \sqrt{-97.8}$$

Student: How am I supposed to take the square root of a negative number?
Tutor: You aren't. There are no values of n for which the energy of the hydrogen atom is positive.
Student: So nothing happens.
Tutor: Something does happen. What happens at energy equals zero?
Student: The hydrogen atom has zero energy when the electron and proton are separated.
Tutor: So a positive energy is more than enough for the electron to leave the proton.
Student: Like the photoelectric effect.
Tutor: Yes, but we call it ionization. The ionization energy is the energy needed to extract an electron.

Student: How is the ionization energy different from the work function?
Tutor: It isn't really, except that we use the work function only for metals.
Student: Okay. What are the allowed positive energies?
Tutor: Any positive energy is allowed. Once the electron is free of the atom, it can have any kinetic energy it wants. Above zero energy, the energy of the hydrogen atom is a continuum. Also, remember to not use $-13.6\text{ eV}/n^2$ for a particle in a one-dimensional box; it only applies to the hydrogen atom.
Student: What about other atoms, like helium?
Tutor: When there are two electrons in an atom, the problem becomes extremely difficult to solve. A helium ion He^+ is similar to a hydrogen atom, because it has one electron. For one-electron atoms with greater charges, like the helium atom, the allowed energy levels are

$$E_n = -\frac{(13.6\text{ eV})Z^2}{n^2}$$

Tutor: where Z is the charge of the nucleus. For atoms with more than one electron, someone has already measured them so we look them up in a book.

EXAMPLE

The shortest wavelength in a series of Li^{++} is 40.5 nm. What is the longest wavelength in the series?

Student: Lithium has a charge of 3, so $Z = 3$ and the energies are $-13.6\text{ eV} \cdot Z^2/n^2$. But I don't know the initial or final levels, and what's a "series"?
Tutor: A series is *emission lines* from transitions that end with the same n. An electron in the $n = 4$ state can drop to $n = 3$, but an electron in the $n = 5$ state can also drop to $n = 3$, and so on. When we get to high values of n, the states have energies that are close together, so a series is transitions with photon wavelengths that are close together. For example, in hydrogen all of the transitions that end at $n = 1$ have energies > 10 eV and have wavelengths in the ultraviolet. All of the hydrogen transitions that end at $n = 3$ have photon wavelengths in the infrared.
Student: So the names Balmer and Paschen mean the level at which the transition ends.
Tutor: They mean the lower-energy level of the two, which is the ending level when photon emission takes place. The Paschen series in hydrogen is all transitions down to $n = 3$.
Student: What about up to $n = 3$, do we have a name for that?
Tutor: No. It's not easy to go up from $n = 2$ to $n = 3$. When we put the hydrogen atom in $n = 2$, it stays there for only a short time, like 10 or 20 ns, before going back down to $n = 1$. For this reason, the emission lines are all of the ones available but the absorption lines are only the ones that start at $n = 1$ (or the ground state).
Student: What's a "line?"
Tutor: When we excite an atom into higher energy states and it decays, it gives off photons.
Student: And the photons have energies equal to the difference between the energies of the initial and final states, right?
Tutor: Yes. When we view this light through a spectrometer, like a diffraction grating, it appears as lines. So sometimes we call the photons "lines."
Student: Okay. I have a 40.5 nm photon, with an energy of

$$E = \frac{hc}{\lambda} = \frac{1240\text{ nm}\cdot\text{eV}}{40.5\text{ nm}} = 30.6\text{ eV}$$

Student: But I don't know what level the transition finishes at.
Tutor: Look at the other end. At what level does it start?
Student: The problem doesn't say that either.
Tutor: The problem does say that you have the shortest wavelength in the series. Shorter wavelengths correspond to what?

Student: Greater energies.
Tutor: Yes. If this is the shortest wavelength and greatest energy, is the initial energy higher or lower?
Student: It would be the highest starting energy, or the biggest n. How big can n get?
Tutor: As big as you can make it. Experimentally, it is hard to hold atoms together when n gets to be 100 or so. The energy gets too close to zero.
Student: So I can take zero as the starting energy, and the energy of the final state is -30.6 eV.

$$E_n = -\frac{(13.6 \text{ eV})(3^2)}{n^2} = -30.6 \text{ eV} \quad \rightarrow \quad n = 2$$

Student: The series is all transitions down to $n = 2$.
Tutor: Good. What's the longest wavelength in the series?
Student: Longer wavelengths are lower energies, so I want the smallest transition energy. That would be when the electron goes to $n = 3$.

$$E_3 = -\frac{(13.6 \text{ eV})(3^2)}{3^2} = -13.6 \text{ eV}$$

$$E_\gamma = \Delta E = |E_3 - E_2| = |(-13.6 \text{ eV}) - (-30.6 \text{ eV})| = 17 \text{ eV}$$

$$\lambda = \frac{hc}{E_\gamma} = \frac{1240 \text{ nm·eV}}{17 \text{ eV}} = 72.9 \text{ nm}$$

The solutions that we've presented are overly simplistic. To solve these and other problems properly we use the wavefunction. A wavefunction Ψ is a function of both position and time $\Psi(x, y, z, t) = \psi(x, y, z)e^{-i\omega t}$, and often the value of the wavefunction is a complex number. To solve a quantum system we have to find the wavefunctions. Only specific wavefunctions work, where working means to satisfy the Schrödinger equation

$$\frac{d^2\psi}{dx^2} + \frac{8\pi^2 m}{h^2}\left[E - U(x)\right]\psi = 0$$

The significance of the wavefunction is two-fold. Using the wavefunction we can calculate energy levels, momentum, angular momentum, and transition probability between levels, but the math gets nasty. Also, the probability of finding the electron at a given spot is proportional to the square of the wavefunction ψ^2 at that spot. Therefore, the integral of the square of the wavefunction is the sum of all the probabilities of finding the electron everywhere, so it must be equal to 1.

For the electron in the one-dimensional infinite square well, the wavefunctions that work are

$$\psi_n(x) = \sqrt{\frac{2}{L}}\sin\left(\frac{n\pi}{L}x\right)$$

These resemble the amplitude of a standing wave. Because the wavefunction is zero at each end, the chances of finding the electron very close to the end of the one-dimensional box are nil. For the $n = 2$ state, ψ is zero at the middle, so the electron is never there, but for the $n = 1$ state that is the most likely place for it to be.

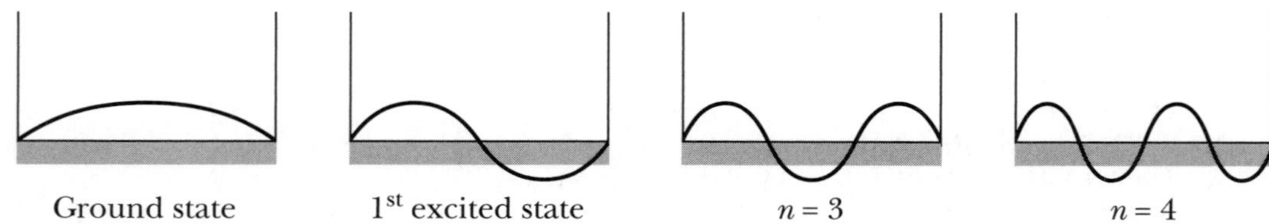

Consider the one-dimensional finite well shown. The wider lines represent the potential well — it is possible to get out of the well without an infinite amount of energy, and it takes energy to get from the left side to the right. The horizontal line represents the energy of one of the wavefunctions. The electron has enough energy to get to both the left and right parts of the energy well but not enough to get out. The sinusoidal line is one of the wavefunctions.

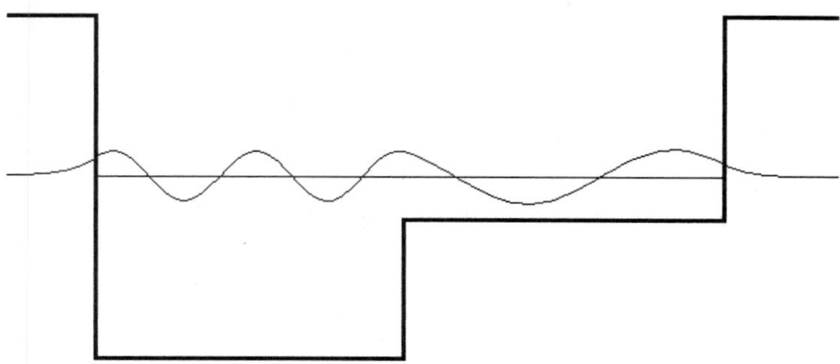

The potential energy on the right side is greater, so if the electron is there it must have less kinetic energy. Less kinetic energy means less momentum, and a longer wavelength. Also, the wavefunction extends past the walls of the well. There is a nonzero chance of finding the electron outside of the well, even though the electron doesn't have enough energy to get out. This is a purely quantum mechanical effect that doesn't happen in the old "classical" physics.

EXAMPLE

An electron is in a one-dimensional infinite square well of width 0.14 nm. The electron has an energy of 76.76 eV. What is the probability of finding the electron in the leftmost 0.02 nm of the infinite well?

Student: So I need to find the wavefunction and integrate it from 0 to 0.02 nm, yes?
Tutor: Mostly correct. You want to find the probability of the electron being at each spot in that region. Because the probabilities are not uniform, we divide the region into pieces so small that we can treat the probability as being the same throughout the region.
Student: Infinitesimally small pieces.
Tutor: Yes. The probability is the *square* of the wavefunction.
Student: Okay. Which state n is the electron in?
Tutor: You need to determine that. How can you tell the states apart?
Student: They have different energies. I know the energy, so I can find the state that has the right energy.

$$E_n = -\frac{13.6 \text{ eV}}{n^2} = 76.76 \text{ eV} \quad \textbf{?}$$

Tutor: The $-13.6/n^2$ applies only for the hydrogen atom. What are the energies for the one-dimensional square well?
Student: Oh, that's right.

$$E = \frac{(hc)^2}{2(mc^2)L^2}\, n^2$$

$$76.76 \text{ eV} = \frac{(1240 \text{ nm}\cdot\text{eV})^2}{2(511,000 \text{ eV})(0.14 \text{ nm})^2}\, n^2$$

$$76.76 \text{ eV} = (19.19 \text{ eV})n^2 \quad \rightarrow \quad n = 2$$

Student: The electron is in the $n = 2$ state.
Tutor: Otherwise known as the first excited state.

Student: What if I hadn't gotten an integer for n?
Tutor: n has to be an integer, so either you or the problem author made an error.
Student: Now that I know that $n = 2$, I can write down the wavefunction.

$$\psi_2(x) = \sqrt{\frac{2}{L}} \sin\left(\frac{2\pi}{L}x\right)$$

Tutor: What is the probability that the electron is in an infinitesimal piece of width dx?
Student: That's the square of the wavefunction.

$$\text{probability} = (\psi_2(x))^2 = \frac{2}{L}\sin^2\left(\frac{2\pi}{L}x\right) \quad \textbf{?}$$

Tutor: The probability of finding the electron in any infinitesimally small piece has to be infinitesimally small, but nothing in your equation is infinitesimally small. The square of the wavefunction gives you the relative probability, but to get the probability for a given piece you need to multiply by the width of the piece. For an infinitesimally small piece this is infinitesimally small.

$$\text{probability} = (\psi_2(x))^2\, dx = \frac{2}{L}\sin^2\left(\frac{2\pi}{L}x\right) dx$$

Student: Then the probability of finding the electron between 0 and 0.02 nm is the sum of all the probabilities, or an integral.
Tutor: Yes.

$$\int_0^{0.02\text{ nm}} \frac{2}{L}\sin^2\left(\frac{2\pi}{L}x\right) dx$$

Student: To find the integral, I replace the $\sin^2$ with the trigonometric identity.
Tutor: Or just look it up in a book.

$$\int \sin^2 ax = \frac{x}{2} - \frac{\sin 2ax}{4a}$$

$$\begin{aligned}\int_0^{0.02\text{ nm}} \frac{2}{L}\sin^2\left(\frac{2\pi}{L}x\right) dx &= \frac{2}{L}\left[\frac{x}{2} - \frac{L}{8\pi}\sin\left(\frac{4\pi}{L}x\right)\right]_0^{0.02\text{ nm}} \\ &= \frac{2}{L}\left[\frac{(0.02\text{ nm})}{2} - \frac{L}{8\pi}\sin\left(\frac{4\pi(0.02\text{ nm})}{(0.14\text{ nm})}x\right)\right] \\ &= \frac{(0.02\text{ nm})}{(0.14\text{ nm})} - \frac{1}{4\pi}\sin\left(\frac{4\pi}{7}x\right) = 0.0653\end{aligned}$$

Student: The chances are 6.53%.

Chapter 40

All About Atoms

The Pauli exclusion principle says that no two electrons in an atom can have the same quantum numbers. The idea is that once one electron is in the $n = 1$ state, the next electron must go into the $n = 2$ state, and so on.

But there are more quantum numbers than just n. An electron in an atom is specified by the quantum numbers $\{ n, l, m_l, m_s \}$. l is the orbital angular momentum of the electron, m_l is the z component of this angular momentum, and m_s is the z component of the electron's spin.

The preferred way to distinguish between states of the same n is by the angular momentum. As the electron orbits the nucleus it has orbital angular momentum. The quantum number l has to be an integer between 0 and $n - 1$, so for $n = 3$ l can be 0 or 1 or 2. We attach names to the values of l, and call them by the first letter, so $l = 0$ is s, $l = 1$ is p, $l = 2$ is d, $l = 3$ is f, and continues with g and h. So the state $n = 2$, $l = 0$ is called $2s$, and the state $n = 4$, $l = 2$ is called $4d$.

The next quantum number, m_l, is the z component of l. (In quantum mechanics, we use z as the first axis chosen, rather than x.) The z component could be all of the angular momentum, or none of it, or even in the $-z$ direction. So m_l has to be an integer between $-l$ and l. For the $3p$ state, m_l could be -1 or 0 or $+1$.

As the electron orbits the nucleus, it also spins on its own axis. The angular momentum of the spinning is always $s = \frac{1}{2}$, so we don't bother writing it. But the z component can be $+\frac{1}{2}$ or $-\frac{1}{2}$ (from the minimum to the maximum in steps of 1).

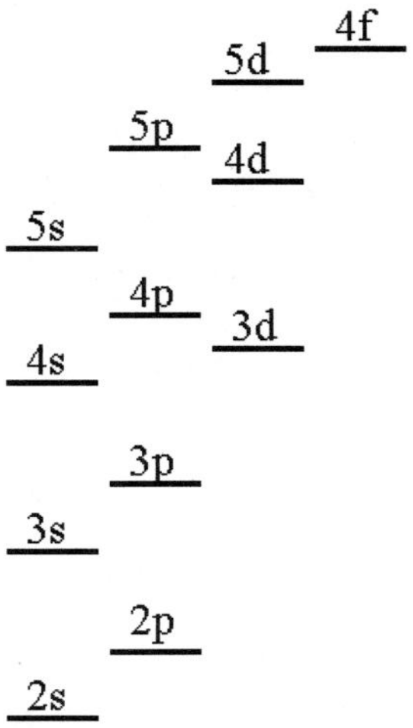

The Pauli exclusion principle still says that no two electrons in an atom can have the same quantum numbers. The first electron is in the state $\{ 1, 0, 0, +\frac{1}{2} \}$, the second in $\{ 1, 0, 0, -\frac{1}{2} \}$, and the third in $\{ 2, 0, 0, +\frac{1}{2} \}$, because the $n = 1$ level is filled. Previously, the energies of the states depended only on n, so that $2s$ and $2p$ have the same energy. This is not entirely true. As the number of electrons in an atom increases, they have more effect on each other, and the energy of $2p$ is slightly greater than $2s$, so $2s$ is filled first. The energy of $3d$ is greater than the energy of $3p$ is greater than the energy of $3s$, so much so that the energy of $3d$ is greater than the energy of $4s$ (in some atoms). By filling up the levels from the lowest energy, we create the periodic table (located at the end of the book).

One way to write the electron configuration of an electron is $nl^{\#}$. The # is the number of electrons in the nl level. So after two electrons, we could write $1s^2$. The electron configuration of aluminum ($Z = 13$) is $1s^2\ 2s^2\ 2p^6\ 3s^2\ 3p^1$. A configuration of $1s^2\ 2s^2\ 2p^6\ 3s^2\ 4s^1$ would be aluminum in an excited state.

EXAMPLE

What is the electron configuration of arsenic (As, $Z = 33$) in the first excited state?

Student: So there can be only two electrons in the $1s$ state, right?
Tutor: Correct, $\{ 1, 0, 0, +\frac{1}{2} \}$ and $\{ 1, 0, 0, -\frac{1}{2} \}$.
Student: And there can't be one in the $1p$ state, because p means $l = 1$ and l has to be less than n, right?
Tutor: Correct.
Student: So how can there be six electrons in the $2p$ state?
Tutor: Because $l = 1$, so m_l can be -1 or 0 or 1. There are three possibilities for m_l, and for each of them there are two possibilities for m_s, and $3 \times 2 = 6$.
Student: Does that mean that there can be 6 electrons in any p level?
Tutor: Yes, and 10 in any d level.
Student: There can be 2 electrons in 1, and $2 + 6 = 8$ in $n = 2$, and $2 + 6 + 10 = 18$ in $n = 3$. Is there some easy way to determine how many can fit in an n level?
Tutor: 2, 8, 18 is twice 1, 4, 9, so $2n^2$.
Student: $2(4)^2 = 32$, so As is one electron past 32.
Tutor: There can be 32 electrons in the $n = 4$ level, but you need to fill the $n = 1$, 2, and 3 levels first.
Student: Oops. 2 in $n = 1$ and 8 in $n = 2$ and 18 in $n = 3$ is 28. Arsenic is 5 electrons past $n = 3$.
Tutor: Careful. $4s$ has lower energy than $3d$, so you start $n = 4$ before finishing $n = 3$.
Student: Okay.

$$1s^2\ 2s^2\ 2p^6\ 3s^2\ 3p^6$$

Student: That's 18, and I need 15 more. Then comes $4s^2$, and then $3d^10$, so I fill the $3d$ level.
Tutor: Yes. Sometimes you'll see a level called a "shell."
Student: And the last three go into $4p$.
Tutor: That's the lowest energy place for them to go. For the first excited state, move one of them up to the next highest level.
Student: $5s$ is lower than $4d$.

$$1s^2\ 2s^2\ 2p^6\ 3s^2\ 3p^6\ 4s^2\ 3d^{10}\ 4p^2\ 5s^1$$

We use an x-ray spectrum to identify the nucleus of an atom. We take an atom and bombard it with high-energy electrons. Sometimes the incoming electron knocks out one of the electrons. An electron from a higher level comes down to fill the lower energy level, giving off the extra energy as an emitted photon. This photon typically has a high energy and is in the x-ray spectrum.

When an electron drops from a higher level to a lower one, it emits a photon with a characteristic energy. The energy of the photon must equal the difference in energy of the two levels. When an electron drops from $n = 2$ to $n = 1$ we call it a K_α photon. When an electron drops from $n = 3$ to $n = 1$ we call it a K_β photon. We also get photons of other energies, up to a maximum equal to the energy of the incoming electron.

Previously we used -13.6 eV $\times\ Z^2/n^2$ for the energy of the electron. This was applied to a single electron in an atom — a "one electron atom." With more than one electron the energy becomes very difficult to calculate, but we can get an estimate. Because there is one electron in the $n = 1$ level (the other has been knocked out), we use the combined charge of the nucleus plus the $1s$ electron.

EXAMPLE

What are the minimum wavelength and K_β wavelength for cesium (Cs, $Z = 55$) when bombarded with 42 keV photons?

Student: K_β means from $n = 3$ to $n = 1$. I find the energy of the two states, then the difference in the energies is the energy of the K_β photon.
Tutor: Very good.
Student: But I use $Z - 1$ to find the energies of the states.

$$E_n = -\frac{(13.6\text{ eV})(Z-1)^2}{n^2}$$

$$E_1 = -\frac{(13.6\text{ eV})(55-1)^2}{1^2} = -39.66\text{ keV}$$

$$E_3 = -\frac{(13.6\text{ eV})(55-1)^2}{3^2} = -4.41\text{ keV}$$

$$E_\gamma = \Delta E = |(-39.66\text{ keV}) - (-4.41\text{ keV})| = 35.25\text{ keV}$$

Student: What if the incoming electron doesn't have 39.66 keV of energy.
Tutor: Then it wouldn't be able to knock the electron out of the $n = 1$ level.
Student: So the wavelength of the K_β photon is

$$\lambda_\beta = \frac{hc}{E} = \frac{1240\text{ nm}\cdot\text{eV}}{35,251\text{ eV}} = 0.0352\text{ nm} = 35.2\text{ pm}$$

Tutor: Good.
Student: How do I get the minimum wavelength?
Tutor: What's the most energy that you could possibly get from a 42 keV photon?
Student: 42 keV.
Tutor: And the largest energy is the shortest wavelength.
Student: So the minimum wavelength photon is the one with an energy of 42 keV.

$$\lambda_{\min} = \frac{hc}{E} = \frac{1240\text{ nm}\cdot\text{eV}}{42,000\text{ eV}} = 0.0295\text{ nm} = 29.5\text{ pm}$$

Chapter 41

Conduction of Electricity in Solids

When atoms form a solid, they are very close to each other, about 0.1 nm away from each of their nearest neighbors. This is about the same size as an individual atom. (For comparison, in a gas at room temperature they are about 2 nm away from each other, leading to a density about 3000 times smaller.) When atoms are this close, the electron wavefunctions begin to overlap. The Pauli exclusion principle says that no two electrons can be in the same state. To avoid overlapping electrons being in the same state, they change their energies just a little. The result is that instead of discrete energy levels, the energies become energy bands, with many energies very close together.

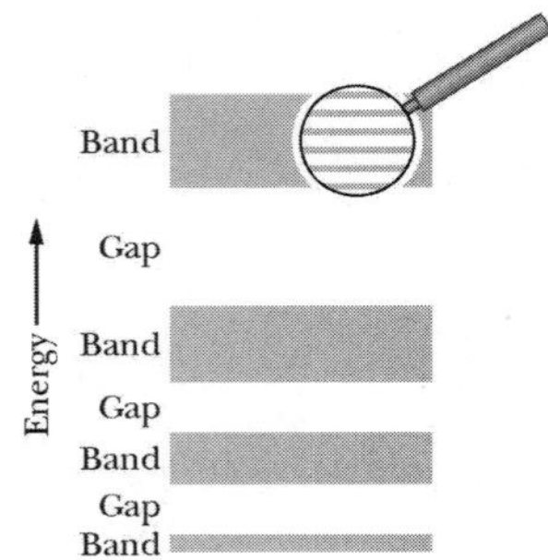

How close together are these energy levels? The density of energy levels in a 3-D material is

$$N(E) = \frac{8\sqrt{2}\pi m^{3/2}}{h^3}\sqrt{E} = \frac{8\sqrt{2}\pi(mc^2)^{3/2}}{(hc)^3}\sqrt{E} = (6.81/\text{eV}\cdot\text{nm}^3)\sqrt{E/(1\text{ eV})}$$

(For a 2-D material, such as a very thin film, one does a difference calculation and gets a difference result.) While this number may look small, the units conceal a large density, because nanometers are much smaller than the typical solid. For a (1 mm)^3 solid, the energy levels are about 10^{-19} eV apart.

The conduction of electric current in a solid depends on whether the energy states in these bands are filled. To move rapidly, electrons need room to work, and without it they can move only slowly. In the camp game "shuffle your buns," a couple dozen people sit in a circle of chairs, with one empty chair and one person in the middle of the circle. The "it" person in the middle tries to sit down, but one of the people next to the empty chair "shuffles" over into the chair, and the next person shuffles into the chair the first one vacated. In this way the people shuffle one way and the empty chair shuffles the other way. None of the chairs move, of course, but the location of the empty chair moves. Because there is but a single empty chair, the movement is relatively slow. Imagine a single empty chair among 10^{23} electrons, and you have virtually no current at all, as each electron waits for its turn to move. But if half of the chairs were empty, then each electron could constantly be in motion.

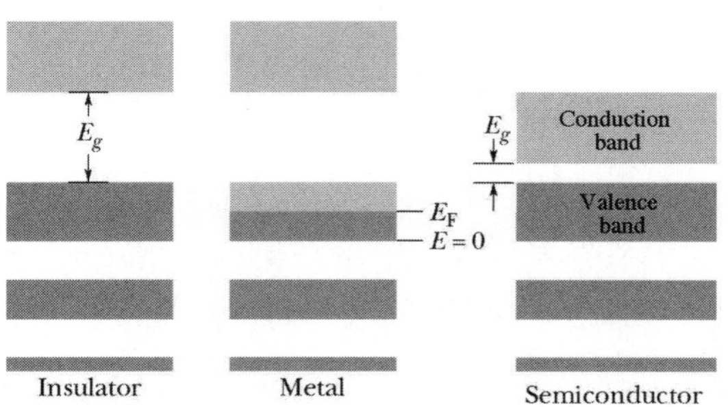

In the insulator shown to the left of the figure, the energy band is completely full. There is no room for electrons to move around, and very little current can flow. Electrons would have plenty of room to move around if they could jump up to the next band, but that requires a considerable energy, perhaps 3 eV or more. This corresponds to atoms where the outer shell of electrons is completely filled.

In the conductor shown in the middle, the energy band is only half full, and there is plenty of room to move around. Metals are typically conductors. Each metal atom has one or more electrons in a partly filled shell, so there are available states at nearly the same energy, and plenty of room to move about. The Fermi energy E_{F} is the energy of the highest occupied state when all lower states are filled.

$$E_{\mathrm{F}} = \left(\frac{3}{16\sqrt{2}\pi}\right)^{2/3} \frac{h^2}{m} n^{2/3} = \left(\frac{3}{16\sqrt{2}\pi}\right)^{2/3} \frac{(hc)^2}{(mc^2)} n^{2/3} \approx \left(\frac{4}{11}\ \mathrm{nm}^2 \cdot \mathrm{eV}\right) n^{2/3}$$

To the right is a semiconductor. Semiconductors have a completely filled band, but the jump to the next band is relatively small, perhaps 1 eV or less. Some of the electrons might get enough energy to move to the next band, where they would have plenty of room to move around, and a current could flow.

Remember from thermodynamics that temperature is a measure of internal energy. At higher temperatures, more electrons have the energy to move to higher levels. The probability of an energy level being filled by an electron is

$$P(E) = \frac{1}{e^{(E-E_{\mathrm{F}})/kT} + 1}$$

kT is the typical energy available to electrons, where k is the Boltzmann constant of 1.38×10^{-23} J/K, or about 1 eV/11,600 K. To get a value of kT of even 0.5 eV requires a temperature equal to the temperature of the Sun.

How could an electron hope to jump 1 eV to the next energy band at only room temperature? Though the probability is small, the number of electrons is very large. For a conductor like copper the density of "valence electrons" $n = 9 \times 10^{28}\ /\mathrm{m}^3$. For a semiconductor like silicon, the density of charges in the higher energy band is about 10^{16}, so that only a trillionth of the electrons have the energy to make it, but the number of electrons trying to make it is much more than a trillion.

One of the interesting result of this temperature dependence is that the resistance of semiconductors decreases with temperature. For metals, as the temperature increases, the rate of collisions for the moving electrons increases. More energy is lost in the collisions, and the power turned to heat increases, corresponding to increased resistance. For semiconductors, higher temperature means more available energy, and the density of electrons in the higher conduction band increases, so the current can increase.

EXAMPLE

For a 1 cm copper cube, how much more energy than the Fermi energy is the most energetic electron likely to have?

Student: That's it? I'm supposed to be able to figure this out from only that information?
Tutor: Let's see how far we can get. You could start by finding the Fermi energy.
Student: I need the density n. Is that something that I need to look up?
Tutor: You can find it from the density. Each copper atom contributes 1 electron.
Student: How do you know that it's one?
Tutor: The electrons in the ground state are in $1s^2\ 2s^2\ 2p^6\ 3s^2\ 3p^6\ 3d^{10}\ 4s^1$. Only the last electron is available to move around.
Student: I thought that $4s$ had lower energy than $3d$. Shouldn't it be $4s^2\ 3d^9$?
Tutor: Multiple electrons in an atom cause the energy levels to change. It's better, as in lower energy, to fill the $3d$ shell than the $4s$ shell.
Student: So if I know the density of copper atoms, the density of valence electrons is the same.

Tutor: You look up the density (8.96 g/cm^3) and the atomic mass (63.54 g/mol). You already have the electrons per atom, and you know the number of atoms per mol.
Student: Oh, Avogadro's number.

$$\frac{\text{electrons}}{\text{cm}^3} = \frac{\text{electrons}}{\text{atom}} \times \frac{\text{atom}}{\text{mol}} \times \frac{\text{mol}}{\text{g}} \times \frac{\text{g}}{\text{cm}^3}$$

$$n = (1)\left(6.02 \times 10^{23}\right)\left(\frac{1}{63.54}\right)(8.96) = 8.5 \times 10^{22}/\text{cm}^3$$

Student: Now I can find the Fermi energy.
Tutor: Be careful with nm and cm and m.

$$E_{\text{F}} = \left(\frac{4}{11}\ \text{nm}^2\cdot \text{eV}\right) n^{2/3} = \left(\frac{4}{11}\ \text{nm}^2\cdot \text{eV}\right)(8.5 \times 10^{22}/\text{cm}^3)^{2/3}\left(\frac{1\ \text{cm}}{10^7\ \text{nm}}\right)^2 = 7.03\ \text{eV}$$

Tutor: We can also find the density of states $N(E)$.

$$N(E) = (6.81/\text{eV}\cdot\text{nm}^3)\sqrt{E/(1\ \text{eV})} = (6.81\ /\text{eV nm}^3)\sqrt{(7.03\ \text{eV})/(1\ \text{eV})} = 18/\text{eV nm}^3$$

Tutor: We're more interested in the density per electron-volt.
Student: So I multiply by the volume?

$$N(E) = \left(18\ /\text{eV nm}^3\right) \times (1\ \text{cm})^3 \times \left(\frac{1\ \text{cm}}{10^7\ \text{nm}}\right)^3 = 1.8 \times 10^{22}\ /\text{eV}$$

Student: The energy levels are spaced 10^{22} eV apart?
Tutor: They are 10^{-22} eV apart, so that there are about 10^{22} eV of them over a space of 1 eV. Now, if the energy level spacing was such that there was 1000 state per eV, but the probability of a state being filled was 2/1000, about how many electrons would be in an electron-volt?
Student: $1000 \times 2/1000 = 2$, so there would be 2 electrons.
Tutor: About 2 electrons, because it's a probability. To find the highest energy electron, we could go until the density times the probability was about 1, or one electron over a whole electron-volt.
Student: So above that the probability is small enough that there are no electrons.
Tutor: Well, very few if any electrons. We could do a calculation to find the expected number of electrons with even higher energies. That would involve integrating the probability times the density of states.
Student: No thanks. I need to find where

$$P(E) = \frac{1}{e^{(E-E_{\text{F}})/kT} + 1} = \frac{1}{1.8 \times 10^{22}}$$

Tutor: I think you can safely neglect the +1 on the left.
Student: Oh, because the $e^{(E-E_{\text{F}})/kT}$ is *so* much bigger.

$$e^{(E-E_{\text{F}})/kT} = 1.8 \times 10^{22}$$

$$(E - E_{\text{F}})/kT = \ln\left(1.8 \times 10^{22}\right) = 51$$

$$E - E_{\text{F}} = 51kT = 51\left(\frac{1\ \text{eV}}{11,600\ \text{K}}\right)(300\ \text{K}) = 1.3\ \text{eV}$$

Tutor: At that energy, the density of states is a little higher, but only about 10%.
Student: Close enough for me.

In practice, semiconductors are "doped" with another material. Doping means to replace a small fraction of the atoms of the original substance with atoms of a difference substance. Each silicon atom has 4 valence electrons, and shares one of them with each of its four neighbors. Then each atom has eight electrons around it, and all of the electron shells are filled.

For each silicon atom that is replaced with something that has one more valence electron, the extra electron moves to the conduction band. One way to dope silicon is with phosphorus. Phosphorus has 5 valence electrons, so that there are enough to fill all of the electron shells, plus one extra electron. We saw earlier that only about a trillionth (1 in 10^{12}) of the atoms in a semiconductor contribute an electron to the conduction band. If only a millionth of the semiconductor atoms were replaced with something with an additional electron, these additional electrons would completely outnumber the conduction electrons from the semiconductor. Essentially all of the conduction electrons would be from the doping. This gives us good control of the density of conduction electrons when manufacturing doped semiconductors.

EXAMPLE

How much phosphorus must be added to 2 grams of silicon to create a doped semiconductor with a charge carrier density $n = 10^{21}\ /\text{m}^3$?

Student: The silicon itself contributes a density of $10^{16}\ /\text{m}^3$, so the doping must do all of the rest. But $10^{21} - 10^{16}$ is nearly the same as 10^{21}. All of the conduction electrons come from the phosphorus.
Tutor: Almost all. The error in saying "all" is in the sixth significant digit.
Student: I can live with that. So the density of phosphorus atoms needs to be $10^{21}/\text{m}^3$.
Tutor: Yes.
Student: If I knew the volume of the phosphorus, I could multiply by the density to get the number of phosphorus atoms.
Tutor: The phosphorus atoms are scattered among the silicon atoms. The vast majority of the silicon is still silicon, so the volume is determined by the silicon.
Student: So I find the volume of 2 grams of silicon and multiply by the density of phosphorus to find the number of phosphorus atoms.

$$V = \frac{m}{\rho} = \frac{2\ \text{g}}{2.33\ \text{g/cm}^3} = 0.86\ \text{cm}^3$$

$$N_\text{P} = (0.86\ \text{cm}^3)(10^{21}/\text{m}^3)\left(\frac{1\ \text{m}}{10^2\ \text{cm}}\right)^3 = 8.6 \times 10^{14}$$

$$m_\text{P} = (8.6 \times 10^{14})\left(\frac{1\ \text{mol}}{6.02 \times 10^{23}}\right)\left(\frac{31\ \text{g}}{1\ \text{mol}}\right) = 4.4 \times 10^{-8}\ \text{g}$$

Student: That's really small!
Tutor: Semiconductors are made in "clean rooms" to reduce the chances of contamination. It wouldn't take much to mess up the production of a semiconductor.
Student: Why are semiconductors so important?
Tutor: When we put semiconductors together we can form a transistor. Transistors allow us to control one current with a second current or voltage. We use transistors to make amplifiers and computer logic circuits.

Chapter 42

Nuclear Physics

We now examine the nucleus, located at the center of the atom. The nucleus is made up of protons and neutrons. Because there are only positive charges in the nucleus, the electric force tries to push the protons apart. The strong nuclear force holds the protons together, but only at short distances of about 10^{-15} m. The neutrons are necessary to create more space between the protons, so that the strong nuclear force will be greater than the electric repulsion.

Which element an atom is is determined by the charge of the nucleus or atomic number Z, equal to the number of protons. Atoms of the same element but with different numbers of neutrons are called isotopes. The atomic mass number A is the number of nucleons, which is the number of protons and neutrons. We write a nucleus or isotope ${}^{A}_{Z}\mathrm{El}$, Where El is the chemical symbol for the element. Because Z determines El, El and Z are redundant and Z is sometimes omitted. For example, ${}^{87}_{37}\mathrm{Rb}$ is rubidium with 37 protons and 40 neutrons, while ${}^{85}\mathrm{Rb}$ is rubidium with 37 protons and 38 neutrons.

Because everything wants to have less energy, remember that anything stable must have less energy together than apart, otherwise it would fall apart. Therefore, **the mass of the nucleus is less than the sum of the masses of the pieces**. The mass of the ${}^{87}_{37}\mathrm{Rb}$ nucleus is less than the sum of the masses of 37 protons and 40 neutrons. The difference in mass, times c^2, is how much energy we would have to put in to separate the nucleus into individual nucleons, and is called the binding energy.

To find the binding energy, we need to find this difference in mass. We add up the masses of the pieces of the nucleus. Then we find the mass of the nucleus itself. Nuclear physicists have measured these and we look up their results. We convert from atomic mass units (u) to kg, multiply by c^2, and convert to mega-electron volts (MeV). We can combine all of these factors into $c^2 = 931.5$ MeV/u.

Some Nuclear Masses			
${}^{0}_{-1}\mathrm{e}^-$	0.000549	${}^{235}_{92}\mathrm{U}$	235.043930
${}^{1}_{1}\mathrm{H}$	1.007825	${}^{238}_{92}\mathrm{U}$	238.050788
${}^{1}_{0}\mathrm{n}$	1.008665	${}^{239}_{94}\mathrm{Pu}$	239.052164
${}^{4}_{2}\mathrm{He}$	4.002603	${}^{131}_{53}\mathrm{I}$	130.906124
${}^{12}_{6}\mathrm{C}$	12.000000	${}^{102}_{39}\mathrm{Y}$	101.933558
${}^{56}_{26}\mathrm{Fe}$	55.934937	${}^{137}_{55}\mathrm{Cs}$	136.907089
		${}^{96}_{37}\mathrm{Rb}$	95.934272
(includes mass of electrons)			

The mass of a nucleus is not equal to the atomic mass number, but is approximately the same. Because a proton and a neutron each have mass of about 1 u, the mass of the nucleus (in atomic mass units u) is about equal to the number of nucleons. Note that some nuclei have more and some less mass than the atomic mass number A. The standard for an atomic mass unit is $\frac{1}{12}$ of the mass of a $^{12}_{6}$C nucleus, which includes the binding energy of that nucleus. Because the binding energies of the various isotopes is different, the deviation away from A is different for each isotope.

EXAMPLE

Find the binding energy of $^{238}_{92}$U.

Student: The binding energy is the difference in the mass.
Tutor: The binding energy is proportional to the difference, but what difference?
Student: I add up the mass of 92 protons and 238 neutrons. Then I subtract the mass of a $^{238}_{92}$U nucleus.
Tutor: A $^{238}_{92}$U nucleus has 92 protons and 238 nucleons, but not 238 neutrons.
Student: What's a nucleon?
Tutor: A nucleon is a proton or neutron in the nucleus.
Student: So there are only $238 - 92 = 146$ neutrons.
Tutor: Correct.
Student: And the mass of the pieces is

$$(92)(1.007825 \text{ u}) + (146)(1.008665 \text{ u}) = 239.984990 \text{ u}$$

Student: Is there a reason to keep so many significant figures?
Tutor: Yes. We're about to subtract a number that's pretty close to the number we have. When we do that, we'll get a much smaller number with two or three fewer significant figures. Try it.
Student: Okay. The sum of the masses of the pieces minus the mass of the nucleus is

$$239.984990 \text{ u} - 238.050788 \text{ u} = 1.934202 \text{ u}$$

Tutor: If we had only kept three or four significant figures, we would have one or two now.
Student: Okay. So the binding energy is 1.934202 u.
Tutor: The mass difference, sometimes called mass defect, is 1.934202 atomic mass units.
Student: Do we really need a new unit?
Tutor: It's handy to have a unit that's about the size of the thing that we're measuring. Just keep track of your units, as always.
Student: To get the binding energy, I multiply by 931.5.

$$(1.934202 \text{ u})(931.5 \text{ MeV/u}) = 1802 \text{ MeV}$$

Student: When we did atoms, the energy was negative. Is the binding energy negative?
Tutor: It's true that the binding energy is how much less energy the nucleus has than if the parts were separate. But we don't typically measure compared to all separate pieces, while we did measure the atom compared to all of the electrons removed. It's normal to specify binding energies as positive.
Student: The energy here is much greater than for an atom. Those were 13.6 eV.
Tutor: Hydrogen was 13.6 eV, and greater charge in the nucleus led to greater energies, but yes, this is much more. Nuclear reactions involve much more energy than chemical reactions. That's one reason to use nuclear energy. We'll work on nuclear energy in the next chapter.

Some nuclei are stable, and others are radioactive. Radioactive nuclei emit α, β, and γ particles, and occasionally neutrons. An alpha particle is the same as a helium-4 nucleus, a beta is the same as an electron, and gamma is a high-energy photon. Beta decay can also be the emission of a positron, an anti-electron with the same mass but a positive charge:

$$\alpha = {}^{4}_{2}\text{He}^{+2} \qquad \beta = {}^{\;\,0}_{-1}\text{e}^{-} \text{ or } {}^{\;\,0}_{+1}\text{e}^{+} \qquad \gamma = {}^{0}_{0}\gamma = \text{high-energy photon}$$

When a nucleus decays, the charge is conserved. The charge is expressed with the atomic number Z, so Z must be the same before and after. By adding the Z's on each side, we can determine one that is missing.

The mass number is also conserved in a nuclear reaction. This is not the same as the mass being conserved, since the mass is not exactly equal to the mass number. The binding energies before and after will be different, and this will cause the mass afterward to be different from the mass before. But the mass number A is conserved, so again we can determine any one missing mass number if we know the rest.

EXAMPLE

Astatine-219 ($^{219}_{85}$At) decays by α decay, followed by γ decay, followed by β decay. What isotope remains after the decays?

Student: I don't know where to start.
Tutor: So, write down the first reaction.
Student: But I don't know what comes out.
Tutor: Leave it blank, like a variable.

$$^{219}_{85}\text{At} \rightarrow {}^{4}_{2}\alpha + {}^{A}_{Z}?$$

Student: And the A's have to be the same, right? Does that mean that A is 219, like the At, or 215?
Tutor: The total A for each side is the same.
Student: So $219 = 4 + A$, and $A = 215$. Likewise $85 = 2 + Z$, so $Z = 83$.

$$^{219}_{85}\text{At} \rightarrow {}^{4}_{2}\alpha + {}^{215}_{83}?$$

Tutor: Good.
Student: What element is $Z = 83$?
Tutor: Look it up in a periodic table. There's no way to derive or deduce the symbol based on Z, you just learn them or look them up.
Student: Okay. $Z = 83$ is Bi for bismuth.

$$^{219}_{85}\text{At} \rightarrow {}^{4}_{2}\alpha + {}^{215}_{83}\text{Bi}$$

Student: Is alpha decay always a $^{4}_{2}\alpha$?
Tutor: Yes. All α particles are two protons and two neutrons, the same as a $^{4}_{2}$He nucleus.
Student: Now the bismuth-215 decays by gamma radiation.

$$^{215}_{83}\text{Bi} \rightarrow {}^{0}_{0}\gamma + {}^{A}_{Z}?$$

Student: How can the radiation particle have no mass and no charge?
Tutor: It's a photon, a high-energy photon. Photons don't have mass or charge.
Student: Since A and Z of the γ photon are zero, the isotope after should be the same as before.

$$^{215}_{83}\text{Bi} \rightarrow {}^{0}_{0}\gamma + {}^{215}_{83}\text{Bi}$$

Student: How can you get gamma decay? Nothing happens.
Tutor: Nuclei have excited states, just like the electrons in orbit around the nucleus do. An α or β decay can leave the nucleus in an excited state, and after a short time it decays, emitting a photon. The energies in the nucleus are much greater than for the electrons, so the photon is very high energy, a gamma ray. Sometimes we write a nucleus in an excited state with an asterisk.

$$^{215}_{83}\text{Bi}^{*} \rightarrow {}^{0}_{0}\gamma + {}^{215}_{83}\text{Bi}$$

Student: What's so interesting about gamma decay? Why do we care?
Tutor: γ rays will penetrate most things, so radioactive material can be detected from these gamma rays. So if you've recently had a nuclear medical procedure done, or are carrying kitty litter with you, you might

set off a security detector. Emitted gammas can be used to detect radioactive material.
Student: Kitty litter?
Tutor: Yes. It has small amounts of radioactive isotopes in it.
Student: Isn't that dangerous?
Tutor: If you swallow it, your chances of getting cancer will increase slightly.
Student: Well, I'm not going to do that. On to the beta decay. Is it β^+ or β^-?
Tutor: If it doesn't say, assume β^-.
Student: But how do you know?
Tutor: I know because I looked it up. Nuclear physicists are very good at compiling and disseminating this information.

$$^{215}_{83}\text{Bi} \rightarrow {}^{0}_{-1}\beta + {}^{A}_{Z}?$$
$$^{215}_{83}\text{Bi} \rightarrow {}^{0}_{-1}\beta + {}^{215}_{84}\text{Po}$$

Student: The atomic number goes up?
Tutor: Yes. In beta decay, the atomic number can go up.
Student: Is polonium-215 stable then?
Tutor: Most isotopes of polonium, including $^{215}_{84}$Po, are highly radioactive. A lethal dose of $^{210}_{84}$Po is only about 10^{-9} kg.
Student: What about $^{215}_{84}$Po?
Tutor: It has a half-life of 1.8 milliseconds.
Student: Where does it end?

$$^{219}_{85}\text{At} \xrightarrow{\alpha} {}^{215}_{83}\text{Bi}^* \xrightarrow{\gamma} {}^{215}_{83}\text{Bi} \xrightarrow{\beta} {}^{215}_{84}\text{Po} \xrightarrow{\alpha} {}^{211}_{82}\text{Pb} \xrightarrow{\beta} {}^{211}_{83}\text{Bi} \xrightarrow{\alpha} {}^{207}_{81}\text{Tl}^* \xrightarrow{\gamma} {}^{207}_{81}\text{Tl} \xrightarrow{\beta} {}^{207}_{82}\text{Pb}$$

Tutor: Within four hours, nearly all of the astatine-219 will have turned into lead-207.
Student: Do alpha and beta decay always alternate like that?
Tutor: No, just coincidence here.
Student: And where do you find all of this information.
Tutor: At the *nuclear wallet card*, found at *http://www.nndc.bnl.gov/wallet/* on the Web. You can also find it at Wikipedia by searching for *isotopes of polonium*, or whatever element you want.

When we said above that polonium-215 has a half-life of 1.8 milliseconds, it is not true that after precisely 1.8 milliseconds half of the polonium-215 turns into lead-211, or that after 3.6 milliseconds all of the polonium-215 will be gone. Nuclear decay is a random process, with each nucleus "deciding" independently when to decay. The result of this is that the rate at which the nuclei decrease in number R is proportional to the number of nuclei N.

$$R = \lambda N$$

λ is the disintegration constant or decay constant, and has nothing to do with wavelength (we just re-use the Greek letter). The units of λ are 1/time, typically 1/seconds but not always.

If we solve the ensuing differential equation, we get exponential decay. This may be expressed as

$$\frac{N}{N_0} = \frac{R}{R_0} = e^{-\lambda t} = e^{-t/\tau} = 2^{-t/T_{1/2}}$$

where N_0 is the size of the initial sample and R_0 is the initial decay rate in nuclei per time. As the number of nuclei decrease, so does the number that decay per time, but the disintegration constant does not change. τ is the mean life or natural lifetime. The meaning of τ is that, if the decay rate R remained the same (it won't, of course), then all of the nuclei would decay in one natural lifetime ($R = N/\tau$). We often use the half-life $T_{1/2}$ instead, which is the time needed for half of the sample to decay. Because these three forms of exponential decay are identical,

$$\tau = \frac{1}{\lambda} = \frac{T_{1/2}}{\ln 2}$$

EXAMPLE

Given a 100 μg sample of polonium-215, with a half-life of 1.781 ms, how long must you wait until the decay rate has dropped to 20 decays/second?

Student: There's N, N_0, R, R_0, $T_{1/2}$, τ, and λ. Where do we start?
Tutor: Are there any that you know?
Student: The half-life $T_{1/2}$ is 1.781 milliseconds.
Tutor: Okay. What's 20 decays/second?
Student: That's the final decay rate, so it's R.
Tutor: True. If you knew the initial decay rate R_0, or the final number N, then you could find the time.

$$R = \lambda N \qquad \frac{R}{R_0} = 2^{-t/T_{1/2}}$$

Student: I'll try to find the initial decay rate from the initial number N_0. 1 mole of polonium-215 would have a mass of 215 grams.

$$N_0 = (100 \times 10^{-6}\ \text{g}) \times \left(\frac{6.02 \times 10^{23}}{215\ \text{g}}\right) = 2.8 \times 10^{17}\ \text{atoms}$$

Tutor: Yes. 100 μg of polonium-215 has 2.8×10^{17} atoms.
Student: That seems like a lot, but 100 μg is really small.
Tutor: Consider that you have between 10^{27} and 10^{28} atoms in you, and 10^{17} isn't that much. A single cell has around 10^{12} to 10^{14} atoms.
Student: It just seemed like a big number. Now to find the initial rate R_0, I multiply the initial number N_0 by λ.

$$\frac{1}{\lambda} = \frac{T_{1/2}}{\ln 2} \qquad \longrightarrow \qquad \lambda = \frac{\ln 2}{T_{1/2}} = \frac{\ln 2}{1.781\ \text{ms}} = 0.389\ /\text{ms} = 389\ /\text{s}$$

Student: What does 0.389 /ms mean?
Tutor: It means that a fraction of 0.389, or 38.9%, decays in a time of 1 millisecond.
Student: That seems fast.
Tutor: It's a very short half-life. ${}^{238}_{92}$U has a half-life of about 5 billion years, so $\lambda \approx 5 \times 10^{-18}$ /s. Only a very small fraction of the uranium decays every second.
Student: But then 389 per second means that 389% decays in a second!
Tutor: It means that 38,900% decays in a second. Yes, the decay rate is such that, if it continued, they would all decay 389 times in a second. Of course, the decay rate does not stay constant. As the number of atoms declines, so does the decay rate, because there are fewer atoms to decay.

$$R_0 = \lambda N_0 = (389\ /\text{s})(2.8 \times 10^{17}\ \text{atoms}) = 1.09 \times 10^{20}\ \text{atoms/s}$$

Student: That's a lot, right? Because of the short half-life.
Tutor: Yes. But as the number of atoms decreases, this will eventually drop.
Student: And I need to find when it drops to only 20 per second.

$$\frac{R}{R_0} = 2^{-t/T_{1/2}} \qquad \longrightarrow \qquad \ln\left(\frac{R}{R_0}\right) = -\frac{t}{T_{1/2}} \ln 2 \qquad \longrightarrow \qquad t = -\frac{T_{1/2}}{\ln 2} \ln\left(\frac{R}{R_0}\right)$$

Tutor: Are you bothered by the minus sign?
Student: Is it because the number of atoms is decreasing?
Tutor: No. If the rate was negative, the sign would disappear when we divide the two rates.
Student: Ah! $R < R_0$, and the log of a number less than one is negative. We won't get a negative time.
Tutor: Very good.

$$t = -\frac{T_{1/2}}{\ln 2} \ln\left(\frac{R}{R_0}\right) = -\frac{1.781\ \text{ms}}{\ln 2} \ln\left(\frac{20/\text{s}}{1.09 \times 10^{20}/\text{s}}\right) = (1.781\ \text{ms})\,(62.2) = 111\ \text{ms}$$

Student: It only takes 111 milliseconds?
Tutor: It takes 62 half-lives, but the half-lives are very short. How many atoms are left when this happens?
Student: I can use $R = \lambda N$ or find the decay from the original number N_0.

$$N = N_0 2^{t/T_{1/2}} = (2.8 \times 10^{17}) 2^{-62.2} = 0.05$$

Student: How can you have less than one atom?
Tutor: You can't. Atoms, of course, are quantized.
Student: So what does this mean?
Tutor: Remember that even a single atom would have a decay rate of 389 decays per second.
Student: So you couldn't even get a decay rate of 20 per second.
Tutor: No. How long would it take until all of them have decayed?
Student: You can't say for sure when that happens, can you?
Tutor: No, but you can get a reasonably good estimate. Find the time that it takes to get to one atom, then add one natural lifetime.
Student: Why a natural lifetime instead of a half-life?
Tutor: That's the time it takes for all of them to decay once at the current rate, so it's the expected life of a single atom.

$$N = N_0 2^{-t/T_{1/2}} \quad \longrightarrow \quad t = -\frac{T_{1/2}}{\ln 2} \ln\left(\frac{N}{N_0}\right)$$

$$t = -\frac{1.781 \text{ ms}}{\ln 2} \ln\left(\frac{1}{2.8 \times 10^{17}}\right) = 103 \text{ ms}$$

$$\tau = \frac{T_{1/2}}{\ln 2} = \frac{1.781 \text{ ms}}{\ln 2} = 2.6 \text{ ms}$$

Student: $103 + 2.6 = 106$ ms for them to all decay, more or less.
Tutor: Correct, where more or less means it could be a few milliseconds more or less.
Student: But not everything decays that fast, does it?
Tutor: No.
Student: So how does this carbon dating work?
Tutor: People have carbon in their bodies, and some of it is carbon-14. When we find something organic that died, we can measure the ratio of carbon-14 to carbon-12, which is stable and doesn't decay. If the ratio of carbon-14 to carbon-12 is half what it would naturally be, then the thing died one half-life ago.
Student: But the half-life of carbon-14 is 5700 years.
Tutor: For things that died last week, so little of the carbon has decayed that carbon dating isn't useful. For things that died 50,000 years ago, there is so little carbon left that it is hard to measure.

How dangerous is radiation? We measure the energy that the radiation deposits in the body. Traditionally, this energy is measured in rads, where 1 rad = 0.01 J/kg of flesh. Of course, all units should be 1, so the new unit is 1 gray = 1 J/kg (1 Gy = 100 rad).

Some radiation does more damage than others. Alpha particles have more charge and mass and interact more strongly, while gammas have no charge and interact weakly. Because they interact strongly, alphas interact with the first thing they can, and dump all of their energy in a small volume. They also don't penetrate very far — you could defend yourself from a beam of alphas with one page of this book. To defend yourself from a beam of betas, you would want to block them with the whole book. Because gammas interact weakly, they penetrate much better. To block a beam of gammas, you would need a couple inches of lead.

Because the damage from alphas is localized and more extensive, we use rems rather than rads to describe the damage. To get the rems, multiply the rads by the quality factor. The quality factor for gammas is 1, but for alphas it is 10-20. The modern unit is the sievert, and grays times quality factor is sieverts. Anything old will be in rads and rems, and anything new and from the government will be in grays and sieverts.

How much radiation is too much? A person typically gets 300-400 millirem per year of radiation. 400-500 rem in a short period of time is fatal — it kills off the white blood cells and the victim dies of infection over the next couple of months, much like AIDS. Higher doses can cause other bodily malfunctions. The effect of small doses is incredibly hard to measure, because the effects are so small that they are indistinguishable from other effects.

Long half-lives are not the greatest danger. About half of the 300-400 millirems is from radon gas. (Other sources include medical x-rays, increased cosmic rays during air travel, even LCD wristwatches.) Uranium in the ground decays through a string of decays into radon gas. Radon has a half-life of 4 days — long enough to breathe it, but also short enough that it decays before you die of other causes. For something with a much longer half-life, like plutonium, you die of something else before most of the nuclei have decayed.

The EPA has a Web site where you can estimate your annual dose (*http://www.epa.gov/radiation/students/calculate.html*).

EXAMPLE

$^{238}_{92}$U has a half-life of 4.5 billion years and decays via α emission with an energy of 4.3 MeV. Estimate how much $^{238}_{92}$U you would have to swallow so that the added radiation dose would effectively double your annual dose.

Student: So the question is, how much $^{238}_{92}$U would cause a dose of 300 millirems?
Tutor: Yes.
Student: And 1 rem is 0.01 J/kg. I need to know the kilograms.
Tutor: One rad is 0.01 J/kg. What is the quality factor of the alphas?
Student: Oops. It's about 10, so it's only 30 millirads.
Tutor: Estimate. How much does a typical adult weigh?
Student: Oh, about 65 kg.

$$30 \text{ mrad} \times \frac{0.001 \text{ rad}}{1 \text{ mrad}} \times \frac{0.01 \text{ J/kg}}{1rad} \times (65 \text{ kg}) = 0.0195 \text{ J}$$

Student: That's not a lot of energy in a year.
Tutor: Radiation does damage at the molecular level, to proteins and DNA. It doesn't take many joules to break molecular bonds.
Student: Now I need to know how many decays it takes to get that much energy.
Tutor: Yes. Good.
Student: And each decay produces 4.3 MeV.

$$(0.0195 \text{ J}) \times \frac{1 \text{ eV}}{1.6 \times 10^{-19} \text{ J}} \times \frac{1 \text{ MeV}}{10^6 \text{ eV}} = 1.22 \times 10^{11} \text{ MeV}$$

Student: Is that a lot of MeV?
Tutor: That's why we do the math. It's impossible to avoid radiation altogether, and we want to know how much is dangerous.
Student: Is an extra 300 millirems dangerous?
Tutor: The 300 millirems we get can't be too dangerous, or we would all be dead. Doubling this dose shouldn't be too dangerous.
Student: If we divide by the energy in each decay, we can find the number of decays.

$$\frac{1.22 \times 10^{11} \text{ MeV}}{4.3 \text{ MeV}} = 2.83 \times 10^{10} \text{ decays}$$

Tutor: Remember that this is per year, because we want the 300 millirems to be in a year.
Student: So I need to convert it into seconds.

Tutor: You could. But we have the half-life in years, and years is a perfectly good unit of time.

$$\lambda = \frac{\ln 2}{T_{1/2}} = \frac{\ln 2}{4.5 \times 10^9 \text{ yr}} = 1.54 \times 10^{-10}/\text{yr}$$

Student: What about other decays? Doesn't uranium-238 turn into something radioactive, so that we need to include that as well?
Tutor: Excellent thinking. Yes, it is followed by two beta decays, but they have low energy, and it ends up as uranium-234. Uranium-234 has a half-life of a quarter of a million years, so very little of this will decay before we get hit by a truck or otherwise perish.
Student: Now that I have R and λ, I can find N.

$$R = \lambda N \quad \rightarrow \quad N = \frac{R}{\lambda} = \frac{2.83 \times 10^{10}/\text{yr}}{1.54 \times 10^{-10}/\text{yr}} = 1.8 \times 10^{20} \text{ atoms}$$

Tutor: What's the weight of a uranium-238 atom?
Student: 238 grams per mole, or Avogadro's number. Should we be worrying about other isotopes?
Tutor: It is true that about 1% of naturally occurring uranium is $^{235}_{92}\text{U}$, and that this isotope has a shorter half-life. But it's only seven times shorter, and 99 times less of it, so it doesn't change too much.

$$(1.8 \times 10^{20} \text{ atoms}) \times \frac{238 \text{ g}}{6.02 \times 10^{23} \text{ atoms}} = 0.073 \text{ g} = 73 \text{ mg}$$

Tutor: So don't go swallowing more than 70 milligrams of uranium.
Student: I'm not going to swallow anything radioactive!
Tutor: Many foods contain trace amounts of radioactive isotopes, like potassium-40. The average person gets 40 millirem from food each year.
Student: What about nuclear reactors and nuclear bombs?
Tutor: About 1 millirem/year from all past bomb tests, and about 0.01 millirem/year from nuclear power plants. Coal plants release more radiation than nuclear plants, because there are trace radioactive isotopes in the coal, and it goes into the air. In a nuclear power plant the radiation is contained.

Chapter 43

Energy from the Nucleus

How do we extract energy from the nucleus? The process works, from an energy point of view, much like extracting energy from fossil fuels.

Most of the energy that we use is chemical energy, and comes from the chemical reactions of fossil fuels. Consider the reaction when we burn methane:

$$CH_4 + 2O_2 \rightarrow CO_2 + 2H_2O$$

Each molecule has a binding energy. This is the energy that we would have to put in to separate the molecule into its pieces. (The energies in chemistry are not defined the same way as in physics, so positive energies are possible.) If we were to break the molecules into their pieces, it would take energy. They would then seek lower energies, form new molecules, and release energy. If the new molecules were more stable than the old ones, then more energy would be given off than we put in to start the reaction. Carbon dioxide (CO_2) and water (H_2O) are relatively stable molecules, so about 8 eV of energy is given off every time we do this reaction.

Now consider the process of nuclear fission, where a large nucleus splits into two pieces. The binding energy of the large nucleus is the energy that we have to put in to separate the nucleus into its pieces. The pieces then form into new nuclei. If the two pieces formed are more stable, and have lower energy or more binding energy, than the initial nucleus, then more energy comes off than we put in to start the process.

Because the number of nucleons does not change, we measure stability in binding energy per nucleon. The maximum for binding energy per nucleon occurs at $^{56}_{26}Fe$. If a large nucleus splits into two pieces that are closer to iron-56, we call it fission, and energy is given off. If two small nuclei merge into one larger piece that is closer to iron-56, we call it fusion, and energy is given off.

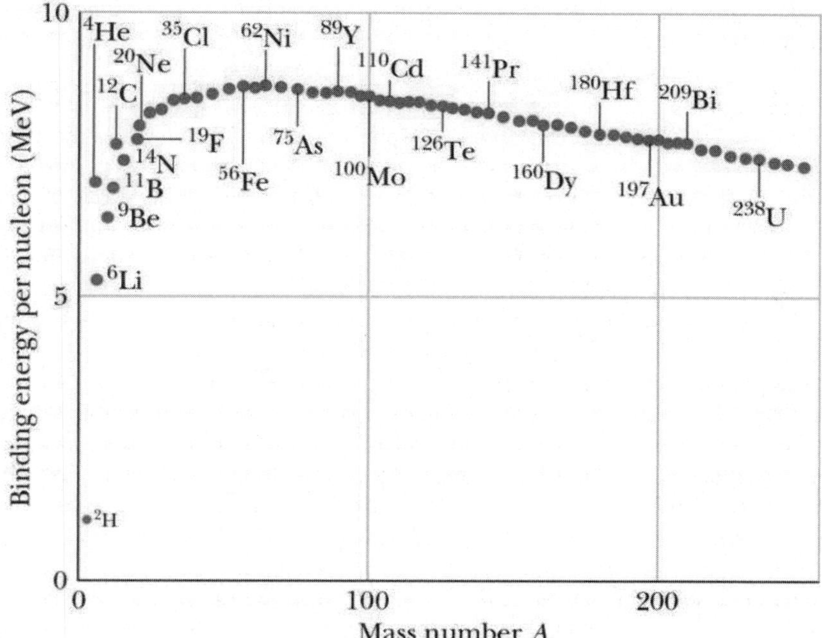

How do we split a large nucleus? We smack it with a neutron. This process results in two smaller pieces, plus two or three loose neutrons. These extra neutrons can smack other nuclei, causing more fissions, and a chain reaction that keeps the process going.

EXAMPLE

Determine the missing "fission fragment" and the energy released in the reaction

$$^1_0\text{n} + {}^{235}_{92}\text{U} \rightarrow {}^{137}_{55}\text{Cs} + {}^{A}_{Z}\,? \quad + 3\,{}^1_0\text{n}$$

Student: Figuring out the missing piece is just matching the A's and Z's, like previous chapter, right?
Tutor: Yes.
Student: A is 236 on the left, so $137 + A + 3 = 236$, and $A = 96$. Z is 92 on the left, so $55 + Z + 0 = 92$, and $Z = 37$. We look $Z = 37$ up in the periodic table, and it's rubidium.

$$^1_0\text{n} + {}^{235}_{92}\text{U} \rightarrow {}^{137}_{55}\text{Cs} + {}^{96}_{37}\text{Rb} \quad + 3\,{}^1_0\text{n}$$

Tutor: Very good.
Student: To find the energy, we find the difference in masses, just like previous chapter. Where do we find the masses?
Tutor: They are conveniently located in the previous chapter.

$$1.008665\text{ u} + 235.043930\text{ u} = 136.907089\text{ u} + 95.934272\text{ u} + 3(1.008665\text{ u}) + \Delta m$$

$$\Delta m = 0.185239\text{ u}$$

$$E = \Delta m \cdot c^2 = (0.185239\text{ u})(931.5\text{ MeV/u}) = 172\text{ MeV}$$

Student: So what's new?
Tutor: Nothing. Notice that the energy is 172 *mega* electron volts.
Student: So?
Tutor: The energy of burning methane was 8 electron volts. The energies involved in nuclear reactions

are 100 million times larger. So to get the same amount of electricity, we need to have one one-hundred millionth as many nuclear reactions.
Student: But what about the nuclear waste?
Tutor: The uranium and plutonium aren't the problem, because they have long half-lives so they decay slowly. The smaller pieces, called fission fragments, are the problem. They are highly radioactive, with much shorter half-lives.
Student: What about Chernobyl? Wasn't there a lot of radioactive material released?
Tutor: There was. But the most dangerous materials were the fission fragments with half-lives of days, months, or years. Radioactive isotopes of iodine, cesium, and strontium are the worst. Radioactive iodine lands on the grass, gets eaten by cows, goes into milk, is drunk by people, and then the iodine collects in the thyroid gland. There were about 8000 additional thyroid cancers in the area because of the accident. Fortunately thyroid cancer is treatable, and there were only 9 deaths due to thyroid cancer.
Student: Only 9 people died in Chernobyl?
Tutor: The United Nations did a study and concluded that the death toll was about 60. Also, there were perhaps 2000 more due to cancer, but about 40,000 of the surrounding population would be expected to die from cancer anyway, so 2000 more is hard to detect. Don't get me wrong — any death is a tragedy. But we have many more deaths from other causes, including coal-burning power plants, so this number is not substantially worse than what we already suffer. That many people die on American roads every month, for example.
Student: And by using nuclear power, we use 10^7 to 10^8 times less fuel.

Uranium-238 does not fission well. To achieve a substantial chain reaction, we need to use the isotope uranium-235. Uranium-235 comprises about 1% of naturally occurring uranium, but we need 4% ${}^{235}_{92}\text{U}$ for a reactor and 90% ${}^{235}_{92}\text{U}$ for a highly enriched uranium bomb. In an HEU bomb, once a neutron causes one fission, there are many neutrons around and the chain reaction happens very fast.

EXAMPLE

The bombs used in World War II had energy yields of about 13 kilotons. One kiloton is 2.6×10^{25} MeV. How much enriched uranium would you need to make such a bomb?

Student: Isn't this a little morbid?
Tutor: Again, any death is ...
Student: ... a tragedy.
Tutor: But reasonable estimates are that an invasion would have resulted in 50 times as many deaths as the explosions. More important, to prevent proliferation of bombs now, we need to know how much material is significant.
Student: So this is what the CIA looks for?
Tutor: Yes, for the capability to enrich uranium from 1% U-235 to 90% in substantial quantities.
Student: Okay. 13 kilotons is

$$(13)(2.6 \times 10^{25}\ \text{MeV}) = 3.38 \times 10^{26}\ \text{MeV}$$

Student: The fission above gave 172 MeV, but there could be others.
Tutor: Yes. There are hundreds of possible fission reactions. The average is about 200 MeV per fission.

$$(3.38 \times 10^{26}\ \text{MeV})/(200\ \text{MeV}) = 1.7 \times 10^{24}\ \text{fissions}$$

Tutor: How many kilograms is that?
Student: If all of the U-235 is fissioned, then we need 1.7×10^{24} atoms. Each mole is 235 grams.

$$(1.7 \times 10^{24}\ \text{atoms})\left(\frac{235\ \text{g}}{6.02 \times 10^{23}\ \text{atoms}}\right)\left(\frac{1\ \text{kg}}{1000\ \text{g}}\right) = 0.66\ \text{kg}$$

Student: That's not a lot!
Tutor: No it isn't. Fortunately, that isn't enough to make a working bomb. With a piece that size, too many neutrons leak out the side before causing more fissions, and the chain reaction doesn't turn into an explosion. Also, the enrichment process is very challenging, so building a bomb sounds easier than it is.
Student: That's a good thing. Can't we just ban them or something?
Tutor: The knowledge of how to build them is too widespread to ever do that — science works that way sometimes. We have to find another way to keep them from ever being used again.

Chapter 44

Quarks, Leptons, and the Big Bang

None of these things really exist, so let's save a little time and go straight to the appendix, which features some funny pictures and a mind-bending plot twist.

No, seriously.

Things are made out of molecules, and molecules are made out of atoms, and atoms are made out of protons, neutrons, and electrons, and protons are made out of . . . funny you should ask. But whatever the pieces of a proton are, they have less energy together than apart, because otherwise they wouldn't stay together. If we want to separate a proton into its constituent pieces, then we need to add energy. The typical method of doing this is to take a proton and something else, perhaps another proton, and slam them together at high speed. We do this hoping that the kinetic energy will change to internal energy and free the pieces of a proton so that we can examine them.

The most obvious result of slamming microscopic particles together is to create unusual particles like pions and kaons and exotic baryons. These particles have short lifetimes, so we don't encounter them very often and haven't needed them until now. The standard model of particle physics has been developed to explain these exotic particles, as well as protons and electrons. Over the last couple of decades it has been remarkably successful in interpreting the results of particle collisions.

Electrons are (currently believed to be) elementary particles — that is, they are not made of anything else. They are one of a group of particles called leptons. Other leptons include the muon and the tau.

There are also elusive leptons called neutrinos. Neutrinos interact very weakly, so that they are hard to detect. Most of the neutrinos from the Sun pass through the Earth without interacting. There are three neutrinos, and they are paired with the other leptons, so that we call them the electron neutrino, the muon neutrino, and the tau neutrino.

Student conception of a muon, before learning that muons are subatomic particles. (Ashley Hopkins)

The other six elementary particles are the quarks, named after an obscure literary reference. The six quarks are called up, down, strange, charm, bottom, and top. Particles made of quarks are called hadrons. Quarks are never seen alone, but only in groups. Groups of three quarks are called baryons, and include protons (uud) and neutrons (udd). Groups of one quark and one antiquark are called mesons.

Every particle also has an antiparticle. An anti-electron, for example, is a positron e^+. An antiproton is made of two anti-up quarks and an anti-down quark ($\bar{u}\bar{u}\bar{d}$).

Some Hadrons			
Baryons		Mesons	
proton p	uud	pion π^-	$\bar{u}d$
neutron	udd	pion π^0	$u\bar{u}$ or $d\bar{d}$
antiproton $\bar{p}$	$\bar{u}\bar{u}\bar{d}$	pion π^+	$u\bar{d}$
antineutron $\bar{n}$	$\bar{u}\bar{d}\bar{d}$	kaon K^+	$u\bar{s}$
Σ^+	uus	kaon K^0	$d\bar{s}$
Σ^0	uds	kaon $\bar{K}^0$	$s\bar{d}$
Σ^-	dds	kaon K^-	$s\bar{u}$
Ξ^0	uss		

There are many things that are conserved in all particle reactions.

Some Particle-Physics Conservation Laws
energy
momentum
charge
lepton# (in families)
baryon#
strangeness

EXAMPLE

Which, if any, conservation laws does the following reaction violate? Fix the reaction so that it doesn't violate any of the conservation laws.

$$n + p^+ \rightarrow e^+ + \nu_\mu + \Sigma^0 + \pi^0$$

Student: I go through the conservation laws one by one and check to see whether they're okay.
Tutor: Okay.
Student: First is energy. The energy shows up as mass, so if the mass on the right is more than the mass on the left, then the reaction can't happen.
Tutor: Good thinking, but there could be energy somewhere else.
Student: Ah, kinetic energy. Even if the rest mass afterward is greater, if the two initial particles are moving fast enough then there would be enough energy.
Tutor: Yes. What about momentum?
Student: How do I check momentum?
Tutor: While momentum is conserved, we don't have the momentums so we can't check that one.
Student: Okay. The charge beforehand is +1, and the charge afterward is +1 because of the positron e^+. So charge is okay.
Tutor: Good.
Student: Now we check lepton#. Neither of the things on the left is a lepton, but the positron and the neutrino ν_μ are leptons, so there are two on the right.
Tutor: Partially correct. The electron is a lepton, and the positron is an anti-electron.
Student: So the positron counts as −1 lepton?
Tutor: Yes.
Student: Then the number of leptons on the right is −1 plus +1 equals 0, so it's okay.
Tutor: Almost. The "in families" means that the number of electron-type leptons has to be the same, and the number of muon-type leptons has to be the same, and also the tau-type leptons.
Student: So there is −1 electron-type lepton and +1 muon-type lepton on the right, but no leptons on

the left, and it can't happen?
Tutor: Correct.
Student: Okay. Baryon# is next. There are two baryons on the left, and on the right only the Σ^0 is a baryon. There are fewer baryons afterward, so the reaction can't happen.
Tutor: Correct. We would need another baryon on the right.
Student: And last is strangeness.
Tutor: Just count the number of strange quarks on each side. A strange quark has strangeness of -1.
Student: The only strange quark is in the Σ^0 on the right, with strangeness -1.
Tutor: So to fix the reaction, you need to get the leptons conserved in families, and you need to add one baryon and $+1$ strange to the right, without adding charge.
Student: The leptons will be fixed if I change the ν_μ to a ν_e, won't they?
Tutor: Yes.
Student: And if I add another Σ^0 to the right, it will fix the baryon#.
Tutor: But then you'll have -2 strangeness on the right.
Student: What if I add an anti-Σ^0 on the right, a $\bar{\Sigma}^0$? That will have $+1$ strangeness.
Tutor: But then the baryon number on the right becomes zero.
Student: Oh, that's right.
Tutor: You need to add one baryon and one anti-strange quark to the right side.
Student: None of the baryons listed have strange quarks.
Tutor: A baryon has three quarks, so having an anti-strange quark makes it an anti-baryon.
Student: I'll add a Σ^0 to the right to fix the baryon number, but now I have -2 strangeness on the right. Now I'll add two kaon $\bar{\mathrm{K}}^0$ to the left so that the strangeness will be -2 on each side.
Tutor: You can't have more than two things on the left. It is very difficult to get more than two things to collide all at once.
Student: So I need to add two kaons K^0 to the right.

$$\mathrm{n} + \mathrm{p}^+ \rightarrow \mathrm{e}^+ + \nu_e + 2\ \Sigma^0 + \pi^0 + 2\ \mathrm{K}^0$$

Tutor: Yes. This is one of many reactions that are possible when a high-energy neutron and proton collide.
Student: If physicists understand all of this so well, why do they keep studying it?
Tutor: There are limits to what we know. For example, the Higgs boson is thought to exist, and to have a role in creating mass, but no one has ever observed one in an experiment. This is one of many areas of research that continue in physics today.
Student: And if I find the Higgs boson ...
Tutor: You'll receive the Nobel Prize in physics.

Mathematical Formulas

Right triangle:

$$a^2 + b^2 = c^2 \qquad \sin^2\theta + \cos^2\theta = 1 \qquad \tan\theta = \frac{\sin\theta}{\cos\theta}$$

$$\sin\theta = \frac{\text{opposite}}{\text{hypotenuse}} \qquad \cos\theta = \frac{\text{adjacent}}{\text{hypotenuse}} \qquad \tan\theta = \frac{\text{opposite}}{\text{adjacent}}$$

Quadratic equation:

$$\text{If} \quad ax^2 + bx + c = 0 \quad \text{then} \quad x = \frac{-b \pm \sqrt{b^2 - 4ac}}{2a}$$

Trigonometry identities:

$$\sin 2\theta = 2\sin\theta\cos\theta$$

$$\cos 2\theta = \cos^2\theta - \sin^2\theta = 2\cos^2 - 1 = 1 - 2\sin^2\theta$$

$$\sin(\alpha \pm \beta) = \sin\alpha\cos\beta \pm \cos\alpha\sin\beta$$

$$\cos(\alpha \pm \beta) = \cos\alpha\cos\beta \mp \sin\alpha\sin\beta$$

$$\sin\alpha \pm \sin\beta = 2\sin\left(\frac{\alpha \pm \beta}{2}\right)\cos\left(\frac{\alpha \mp \beta}{2}\right)$$

$$\cos\alpha + \cos\beta = 2\cos\left(\frac{\alpha + \beta}{2}\right)\cos\left(\frac{\alpha - \beta}{2}\right)$$

$$\cos\alpha - \cos\beta = -2\sin\left(\frac{\alpha + \beta}{2}\right)\sin\left(\frac{\alpha - \beta}{2}\right)$$

Some Common Taylor Series	
$(1+x)^n = 1 + nx + \frac{1}{2}n(n-1)x^2 + \cdots$	$(1+x)^2 = 1 + 2x + x^2$
$e^x = 1 + x + \frac{1}{2!}x^2 + \frac{1}{3!}x^3 + \cdots$	$(1+x)^3 = 1 + 3x + 3x^2 + \cdots$
$\ln(1+x) = x - \frac{1}{2}x^2 + \frac{1}{3}x^3 + \cdots$	$(1+x)^{-1} = 1 - x + x^2 - x^3 + \cdots$
$\sin x = x - \frac{1}{3!}x^3 + \frac{1}{5!}x^5 + \cdots$	$(1+x)^{-2} = 1 - 2x + 3x^2 - 4x^3 + \cdots$
$\cos x = 1 - \frac{1}{2!}x^2 + \frac{1}{4!}x^4 + \cdots$	$(1+x)^{1/2} = 1 + \frac{1}{2}x - \frac{1}{8}x^2 + \frac{1}{16}x^3 + \cdots$
$\tan x = x + \frac{1}{3}x^3 + \frac{2}{15}x^5 + \cdots$	$(1+x)^{-1/2} = 1 - \frac{1}{2}x + \frac{3}{8}x^2 - \frac{5}{16}x^3 + \cdots$

Some Useful Constants and Data		
acceleration of gravity	g	$9.80665\ \text{m/s}^2$
speed of light in vacuum	c	$2.998 \times 10^8\ \text{m/s}$
Newton's gravitation	G	$6.674 \times 10^{-11}\ \text{N·m}^2/\text{kg}^2$
Avogadro's constant	N_A	$6.022 \times 10^{23}/\text{mol}$
Boltzmann constant	k	$1.3807 \times 10^{-23}\ \text{J/K}$
elementary charge	e	$1.6022 \times 10^{-19}\ \text{C}$
universal gas constant	R	$8.3145\ \text{J/mol·K} = N_A k$
Planck constant	h	$6.626 \times 10^{-34}\ \text{J·s} = 4.136 \times 10^{-15}\ \text{eV·s}$
hc for photon energy	hc	$1239.84\ \text{nm·eV}$
permeability constant	μ_0	$4\pi \times 10^{-7}\ \text{T·m/A}$
permittivity constant	ϵ_0	$8.8542 \times 10^{-12}\ \text{C}^2/\text{N·m}^2$
Bohr radius	a_0	$0.529 \times 10^{-10}\ \text{m}$
Rydberg energy	E_R	$13.606\ \text{eV}$
Coulomb's law	k	$1/(4\pi\epsilon_0) = 8.99 \times 10^9\ \text{N·m}^2/\text{C}^2$
Stefan-Boltzmann constant	σ	$5.6704 \times 10^{-8}\ \text{W/m}^2\text{·K}^4$
electron mass	m_e	$9.109 \times 10^{-31}\ \text{kg} = 510,999\ \text{eV}/c^2$
proton mass	m_p	$1.6726 \times 10^{-27}\ \text{kg} = 938.27\ \text{MeV}/c^2$
neutron mass	m_n	$1.6749 \times 10^{-27}\ \text{kg} = 939.565\ \text{MeV}/c^2$
atomic mass unit	m_u	$1.6605 \times 10^{-27}\ \text{kg} = 931.494\ \text{MeV}/c^2$
mass of Earth	$M_\oplus$	$5.98 \times 10^{24}\ \text{kg}$
radius of Earth	$R_\oplus$	$6.37 \times 10^6\ \text{m} = (40.0 \times 10^6\ \text{m})/(2\pi)$
radius of Earth's orbit	AU	1.50×10^{11} m (AU = "astronomical unit")
speed of sound in 20°C air		343 m/s

SI Prefixes		
E	exa	10^{18}
P	peta	10^{15}
T	tera	10^{12}
G	giga	10^9
M	mega	10^6
k	kilo	10^3
c	centi	10^{-2}
m	milli	10^{-3}
μ	micro	10^{-6}
n	nano	10^{-9}
p	pico	10^{-12}
f	femto	10^{-15}

Conversion Factors
1 mile = 5280 ft (exact) = 1609 m
1 in = 2.54 cm (exact)
1 ft = 0.3048 m
1 y = 365.2425 d $\approx \pi \times 10^7$ s
1 m/s = 2.24 mph
360° = 2π radians
1 acre = 43,560 ft^2
1 hectare = $10^4\ \text{m}^2$
1 gallon (US) = 3.786 liter
1 pound = 4.45 N $\approx$ 0.4536 kg
1 eV = $1.602\ 10^{-19}$ J
1 atm = $1.013 \times 10^5\ \text{N/m}^2$ or pascals
1 cal = 4.1868 J
1 hp = 746 W
1 Tesla = 10^4 gauss

Units		
length	meter	m
time	second	s
mass	kilogram	kg
force	newton	$\text{N} = \text{kg·m/s}^2$
energy	joule	$\text{J} = \text{N m} = \text{kg·m}^2/\text{s}^2 = \text{W·s}$
power	watt	W = J/s = N·m/s = V·A
charge	coulomb	C = A·s
electric field		N/C = V/m
current	ampere	$\text{A} = \text{C/s} = \text{V}/\Omega$
potential	volt	$\text{V} = \text{J/C} = \text{A}\ \Omega = \text{T·m}^2/\text{s}$
resistance	ohm	$\Omega = \text{V/A}$
capacitance	farad	$\text{F} = \text{C/V} = \text{s}/\Omega$
inductance	henry	$\text{H} = \text{V·s/A} = \Omega\text{·s}$
magnetic field	tesla	T = N/A·m

circle: $C = 2\pi R \qquad A = \pi R^2$

sphere: $A = 4\pi R^2 \qquad V = \frac{4}{3}\pi R^3$

1 H																	2 He
3 Li	4 Be											5 B	6 C	7 N	8 O	9 F	10 Ne
11 Na	12 Mg											13 Al	14 Si	15 P	16 S	17 Cl	18 Ar
19 K	20 Ca	21 Sc	22 Ti	23 V	24 Cr	25 Mn	26 Fe	27 Co	28 Ni	29 Cu	30 Zn	31 Ga	32 Ge	33 As	34 Se	35 Br	36 Kr
37 Rb	38 Sr	39 Y	40 Zr	41 Nb	42 Mo	43 Tc	44 Ru	45 Rh	46 Pd	47 Ag	48 Cd	49 In	50 Sn	51 Sb	52 Te	53 I	54 Xe
55 Cs	56 Ba	*	72 Hf	73 Ta	74 W	75 Re	76 Os	77 Ir	78 Pt	79 Au	80 Hg	81 Tl	82 Pb	83 Bi	84 Po	85 At	86 Rn
87 Fr	88 Ra	**	104 Rf	105 Db	106 Sg	107 Bh	108 Hs	109 Mt	110 Ds	111 Rg							

	1	2	3	4	5	6	7	8	9	10	11	12	13	14	15
*	57 La	58 Ce	59 Pr	60 Nd	61 Pm	62 Sm	63 Eu	64 Gd	65 Tb	66 Dy	67 Ho	68 Er	69 Tm	70 Yb	71 Lu
**	89 Ac	90 Th	91 Pa	92 U	93 Np	94 Pu	95 Am	96 Cm	97 Bk	98 Cf	99 Es	100 Fm	101 Md	102 No	103 Lr